INSECT DIVERSITY AND ECOSYSTEM SERVICES

VOLUME 1

Importance, Threats, Conservation, and Economic Perspectives

Insect Diversity and Ecosystem Services: 2-volume set:
Volume 1: Importance, Threats, Conservation, and Economic Perspectives
Volume 2: Environmental Indicators, Molecular Approaches, and Management Strategies
ISBN: 978-1-77491-577-6 (hbk)
ISBN: 978-1-77491-579-0 (pbk)

Volume 1: Importance, Threats, Conservation, and Economic Perspectives
ISBN: 978-1-77491-580-6 (hbk)
ISBN: 978-1-77491-581-3 (pbk)
ISBN: 978-1-00346-720-5 (ebk)

Volume 2: Environmental Indicators, Molecular Approaches, and Management Strategies
ISBN: 978-1-77491-582-0 (hbk)
ISBN: 978-1-77491-583-7 (pbk)
ISBN: 978-1-00347-119-6 (ebk)

INSECT DIVERSITY AND ECOSYSTEM SERVICES

VOLUME 1

Importance, Threats, Conservation, and Economic Perspectives

Edited by

Younis Ahmad Hajam, PhD

Sajad Hussain Parey, PhD

Rouf Ahmad Bhat, PhD

AAP | APPLE ACADEMIC PRESS

First edition published 2025

Apple Academic Press Inc.
1265 Goldenrod Circle, NE,
Palm Bay, FL 32905 USA

760 Laurentian Drive, Unit 19,
Burlington, ON L7N 0A4, CANADA

CRC Press
2385 NW Executive Center Drive,
Suite 320, Boca Raton FL 33431

4 Park Square, Milton Park,
Abingdon, Oxon, OX14 4RN UK

Library and Archives Canada Cataloguing in Publication

Title: Insect diversity and ecosystem services / Younis Ahmad Hajam, Sajad Hussain Parey, Rouf Ahmad Bhat, editors.
Names: Hajam, Younis Ahmad, editor. | Parey, Sajad Hussain, editor. | Bhat, Rouf Ahmad, 1981- editor.
Description: First edition. | Includes bibliographical references and indexes. | Contents: Volume 1. Importance, threats, conservation, and economic perspectives.
Identifiers: Canadiana (print) 2024034314X | Canadiana (ebook) 20240343166 | ISBN 9781774915776 (hardcover ; set) | ISBN 9781774915790 (softcover ; set) | ISBN 9781774915806 (v. 1 ; hardcover) | ISBN 9781774915813 (v. 1 ; softcover) | ISBN 9781003467205 (v. 1 ; ebook)
Subjects: LCSH: Insects—Ecology. | LCSH: Insects—Conservation. | LCSH: Insects—Economic aspects. | LCSH: Insects—Molecular aspects. | LCSH: Biodiversity conservation.
Classification: LCC QL496.4 .I57 2024 | DDC 333.95/57—dc23

Library of Congress Cataloging-in-Publication Data

CIP data on file with US Library of Congress

ISBN: 978-1-77491-580-6 (hbk)
ISBN: 978-1-77491-581-3 (pbk)
ISBN: 978-1-00346-720-5 (ebk)

About the Editors

Younis Ahmad Hajam, PhD
Assistant Professor, Department of Life Sciences and Allied Health Sciences, Sant Baba Bhag Singh University, Padhiana, Jalandhar, Punjab, India

Younis Ahmad Hajam, PhD, is currently working as Assistant Professor in the Department of Life Sciences and Allied Health Sciences, Sant Baba Bhag Singh University, Padhiana, Jalandhar, Punjab, India. Formerly, Dr. Younis Ahmad had worked as an assistant professor and head, Department of Zoology, Career Point University, Hamirpur, Himachal Pradesh, India. He teaches undergraduate, postgraduate, and doctorate students. Dr. Hajam has published 35 research papers in leading national and international journals, 30 book chapters and four national proceedings. Dr. Hajam has published four books as editor in Taylor and Francis, etc. He has guided five MSc students under the thesis mode and 31 MSc students under the project mode. He has completed one major research project (Co-PI), three projects under HP Chief Minister start-up scheme. Dr. Hajam has also served as a Junior Research Fellow (JRF) of Zoology, Guru Ghasidas Central University, Bilaspur, Chhattisgarh, India. He has presented his research work in more than 35 national and international conferences and attended more than 30 national and international conferences, and five workshops. He has organized various conferences, seminars, workshops, and webinars at national and international levels. Dr. Hajam has also delivered invited lectures at international conferences. He has received national awards for his contributions to biological sciences. He is an editorial member of various international journals and is also a member of the Asian Council of Science editors. He holds a doctorate in Zoology from Guru Ghasidas Vishwavidayalaya (a Central University), Bilaspur, Chhattisgarh, India.

Sajad Hussain Parey, PhD
Assistant Professor, Department of Zoology, School of Biosciences and Biotechnology, Baba Ghulam Shah Badshah University Rajouri, Jammu and Kashmir, India

Sajad Hussain Parey, PhD, is currently working as an Assistant Professor in the Department of Zoology, School of Biosciences and Biotechnology, Baba Ghulam Shah Badshah University Rajouri, Jammu and Kashmir. Earlier, he worked as a Young Scientist at Sher-e-Kashmir University of Agricultural Sciences and Technology, Shalimar, Srinagar, under the fast-track scheme of the Science and Engineering Research Board from 2014 to 2017. He completed his PhD program at the Department of Zoology, Punjabi University, Patiala, Punjab, under the supervision of Prof. M. S. Saini. Dr. Parey has published more than 50 research articles in national and international journals. He has described more than 100 new species of caddisflies from the Indian Himalaya and set many new records, which were earlier reported in Nepal, Bhutan, China, Pakistan, and Sri Lanka. He has participated and presented his research work at more than 30 national and international conferences, seminars, and workshops.

He has supported MSc, MPhil, and PhD students in their dissertations and theses and successfully produced 26 MSc students, one M Phil, and currently four PhD students are pursuing their PhD program under his guidance. He has traveled to different countries, such as the United States of America, the Kingdom of Saudi Arabia, and Thailand. He has also visited and studied caddisflies at Smithsonian Institution, Washington, DC, and Rutgers University, New Jersey, USA. Dr. Parey has received extramural grants from different funding agencies, such as Science and Engineering Research Board (SERB-DST), National Mission on Himalayan Studies (NMHS), Ministry of Environment, Forest and Climate Change, New Delhi, and University Grants Commission (UGC), New Delhi.

Rouf Ahmad Bhat, PhD
Department of School Education, Government of Jammu and Kashmir, India

Rouf Ahmad Bhat, PhD, pursued his doctorate at Sher-e-Kashmir University of Agricultural Sciences and Technology Kashmir (Division of Environmental Sciences) and is presently working in the Department of School Education, Government of Jammu and Kashmir, India. Dr. Bhat has been teaching graduate and postgraduate students of environmental sciences for the past three years. He is an author of more than 55 research articles (h-index 30; i-index 44; total citations: 2800) and 40 book chapters, and has published more than 43 books with international publishers (Springer, Elsevier, CRC Press, Taylor and Francis, Apple Academic Press, John Wiley, IGI Global). He has his specialization in limnology, toxicology, phytochemistry, and phytoremediation. Dr. Bhat has presented and participated in numerous state, national, and international conferences, seminars, workshops, and symposia. In addition, he has worked as an associate environmental expert on a World Bank-funded Flood Recovery Project and also as an environmental support staff on Asian Development Bank (ADB) funded development projects. He has received many awards, appreciations, and recognition for his service to the science of water testing, air, and noise analysis. He has served as an editorial board member and a reviewer for reputed international journals. Dr. Bhat is still writing and experimenting with diverse capacities of plants for use in aquatic pollution remediation.

Contents

Contributors ... *xi*

Abbreviations ... *xv*

Foreword ... *xvii*

Preface ... *xix*

1. **Insect Biodiversity Conservation: Why It's Needed? ... 1**
Saniya Tyagi, Radheshyam Dhole, N. Srinivasa, and N. Vinay

2. **The Importance of Taxonomy in Biological Research ... 29**
Indu Sharma, Preeti Choudhary, and Rifat Hussain Raina

3. **Aquatic Insect Ecosystems: Diversity, Role, and Conservation ... 39**
Raman Jasrotia, Shivam Jasrotia, Menakshi Dhar, Preetika Gupta, Neha Jamwal, and Seema Langer

4. **Entomological Threats: Assessment, Approaches, and Challenges ... 59**
Neha Jamwal, Raman Jasrotia, Menakshi Dhar, and Sanjay Bhatia

5. **Insect-Based Value-Added Products: Current Scenario and Future Scope ... 79**
Aneela Ashiq, Muhammad Adnan Bodlah, Imran Bodlah, Ammara Gul-E-Fareen, and Muhammad Bilal Tahir

6. **Caddisfly (Trichoptera): Diversity and Ecosystem Services ... 147**
Manpreet S. Pandher, Deepti Garima, and Simarjit Kaur

7. **Trichoptera of Europe: Diversity, Importance, and Conservation ... 159**
Halil Ibrahimi

8. **Current Status and Future Prospects of the Aphidophagous Hoverflies (Diptera: Syrphidae) of Pakistan ... 173**
Muhammad Asghar Hassan, Imran Bodlah, Riaz Hussain, Zershina Maryam, Sajjad Hussain, Muhammad Qasim, and Tayyaba Naseem

9. Bumblebees: Diversity, Role in Pollination Ecology, Threats, and Conservation 189
Rifat H. Raina, Keshav Kumar, Purnima Pathak, Trilok Jangid, Preeti Choudhary, Aejaz Hussain Parrey, and Indu Sharma

10. Review on Antlions (Neuroptera: Myrmeleontidae) of India 201
T. B. Suryanarayanan and Bijoy C.

11. Industrial Entomology: Progress and Economic Perspectives 217
Raman Jasrotia, Menakshi Dhar, Seema Langer, and Neha Jamwal

12. Forensic Entomology: Progress and Implications 237
Rahul Das, Muskan Singhal, Prashant Singh, Rohit Verma, Mahipal Singh Sankhla, Mansi Saxena, Ekta B. Jadhav, Kapil Parihar, Kumud Kant Awasthi, and Chandra Shekhar Yadav

Index 265

Contributors

Aneela Ashiq
Department of Biochemistry, Khwaja Fareed University of Engineering and Information Technology, Rahim Yar Khan, Punjab, Pakistan

Kumud Kant Awasthi
Department of Life Science, Vivekananda Global University, Jaipur, Rajasthan, India

Sanjay Bhatia
Department of Zoology, University of Jammu, Jammu, Jammu and Kashmir, India

C. Bijoy
Shadpada Entomology Research Lab, Department of Zoology, Christ College (Autonomous), Irinjalakuda, Thrissur, Kerala, India

Muhammad Adnan Bodlah
Fareed Biodiversity Conservation Centre, Department of Agricultural Engineering, Khwaja Fareed University of Engineering and Information Technology, Rahim Yar Khan, Punjab, Pakistan

Imran Bodlah
Department of Entomology, Pir Mehr Ali ShahArid Agriculture University, Rawalpindi, Pakistan

Preeti Choudhary
Desert Regional Centre, Zoological Survey of India, Jodhpur, Rajasthan, India

Rahul Das
Department of Forensic Science, Vivekananda Global University, Jaipur, Rajasthan, India

Menakshi Dhar
Department of Zoology, University of Jammu, Jammu, Jammu and Kashmir, India

Radheshyam Dhole
Department of Entomology, Institute of Agricultural Sciences, Banaras Hindu University, Varanasi, Uttar Pradesh, India

Deepti Garima
Zoological Survey of India, New Alipore, Kolkata, West Bengal, India
Department of Zoology, University of Calcutta, Kolkata, West Bengal, India

Ammara Gul-E-Fareen
Department of Entomology, Pir Mehr Ali ShahArid Agriculture University, Rawalpindi, Pakistan

Preetika Gupta
Department of Zoology, University of Jammu, Jammu, Jammu and Kashmir, India

Muhammad Asghar Hassan
Department of Entomology, Pir Mehr Ali Shah Arid Agricultural University, Rawalpindi, Pakistan

Riaz Hussain
Department of Entomology, Pir Mehr Ali Shah Arid Agricultural University, Rawalpindi, Pakistan

Sajjad Hussain
Department of Biological Sciences, University of Baltistan, Skardu, Gilgit-Baltistan, Pakistan

Halil Ibrahimi
Department of Biology, Faculty of Mathematics and Natural Sciences, University of Prishtina "Hasan Prishtina," Prishtinë, Kosovo

Ekta B Jadhav
Department of Forensic Chemistry and Toxicology, Government Institute of Forensic Science, Aurangabad, Maharashtra, India

Neha Jamwal
Department of Zoology, University of Jammu, Jammu, Jammu and Kashmir, India

Trilok Jangid
Desert Regional Centre, Zoological Survey of India, Jodhpur, Rajasthan, India

Raman Jasrotia
Department of Zoology, University of Jammu, Jammu, Jammu and Kashmir, India

Shivam Jasrotia
Department of Zoology, Guru Nanak Dev University, Amritsar,

Simarjit Kaur
Zoological Survey of India, New Alipore, Kolkata, West Bengal, India
Post Graduate Government College for Girls, Chandigarh, India

Keshav Kumar
Desert Regional Centre, Zoological Survey of India, Jodhpur, Rajasthan, India

Seema Langer
Department Of Zoology, University Of Jammu, Jammu, Jammu and Kashmir, India

Zershina Maryam
Department of Biological Sciences, University of Baltistan, Skardu, Gilgit-Baltistan, Pakistan

Tayyaba Naseem
Department of Entomology, The University of Poonch, Rawalakot, Azad Kashmir, Pakistan.

Manpreet S. Pandher
Zoological Survey of India, New Alipore, Kolkata, West Bengal, India

Aejaz Hussain Parrey
Department of Zoology, School of Biosciences and Biotechnology, Baba Ghulam Shah Badshah University, Rajouri, Jammu and Kashmir

Kapil Parihar
State Forensic Science Laboratory, Jaipur, Rajasthan, India

Purnima Pathak
Desert Regional Centre, Zoological Survey of India, Jodhpur, Rajasthan, India

Muhammad Qasim
Department of Agriculture and Forestry, Kohsar University Murree, Punjab, Pakistan

Rifat H. Raina
Desert Regional Centre, Zoological Survey of India, Jodhpur, Rajasthan, India

Rifat Hussain Raina
Desert Regional Centre, Zoological Survey of India, Jodhpur, Rajasthan, India

Mahipal Singh Sankhla
Department of Forensic Science, Vivekananda Global University, Jaipur, Rajasthan, India

Mansi Saxena
Department of Forensic Science, Vivekananda Global University, Jaipur, Rajasthan, India

Indu Sharma
Desert Regional Centre, Zoological Survey of India, Jodhpur, Rajasthan, India

Prashant Singh
School of Forensic Sciences, National Forensic Sciences University, Gujarat, India

Muskan Singhal
Department of Forensic Science, Vivekananda Global University, Jaipur, Rajasthan, India

N Srinivasa
Department of Entomology, Institute of Agricultural Sciences, Banaras Hindu University, Varanasi, Uttar Pradesh, India

T B Suryanarayanan
Shadpada Entomology Research Lab, Department of Zoology, Christ College (Autonomous), Irinjalakuda, Thrissur, Kerala, India

Muhammad Bilal Tahir
Department of Physics, Khwaja Fareed University of Engineering and Information Technology, Rahim Yar Khan, Punjab, Pakistan

Saniya Tyagi
Department of Entomology, Institute of Agricultural Sciences, Banaras Hindu University, Varanasi, Uttar Pradesh, India

Rohit Verma
Dr. APJ Abdul Kalam Institute of Forensic Science & Criminology, Bundelkhand University, Jhansi, India

N. Vinay
Department of Entomology, Institute of Agricultural Sciences, Banaras Hindu University, Varanasi, Uttar Pradesh, India

Chandra Shekhar Yadav
Department of Life Science, Vivekananda Global University, Jaipur, Rajasthan, India

Abbreviations

ADD	accumulated degree days
ADV	aggregation dispersion variance
ANN	artificial neural network
CNN	convolutional neural network
DAET	Distributional Atlas of European Trichoptera
DAVIS	Digitize, Analyze, View, Insect Sounds
DRLSE	distance regularization level set evolution
EAFE	European Association for Forensic Entomology
EPT	Ephemeroptera, Plecoptera, and Trichoptera
ETS	Ephemeroptera, Trichoptera, and Simuliidae
FAO	Food and Agriculture Organization
FPOM	fine particulate organic matter
GHGs	greenhouse gases
GLCM	gray level co-occurrence matrix
GLRLM	gray-level run-length matrix
GMM	Gaussian mixture model
GNSS	global navigation satellite system
IHR	Indian Himalayan Region
IPM	integrated pest management
IUCN	International Union for Conservation of Nature
KNN	k-nearest neighbors
KVIC	Khadi and Village Industries Commission
MLP	multilayer perceptron
PMI	postmortem interval
RCC	river continuum concept
RIA	radioimmune analysis
SRT	speech recognition technique
SVM	support vector machine
TSD	time since death
WFD	water framework directive
WNF	West Nile fever

Foreword

This book titled, *Insect Diversity and Ecosystem Services (Vol-01): Importance, Threats, Conservation, and Economic Perspectives*, compiled and edited by Dr. Younis Ahmad Hajam, Dr. Sajad H. Parey, and Dr. Rouf Ahmad is a compendium of different insect species providing enormous free ecosystem services across the globe. Insects comprise the most diverse taxa of multicellular organisms on Earth, providing several ecosystem services in the form of pollination, decomposition/biodegradation, energy transfer through food chain, biological control of pests, providing food products in the form of honey, pollen, wax, etc.

Across the globe, about 900,000 different kinds of extant insects are known which represent approximately 80% of the world's species (which is around 1.5 million species). Insect biologists believe that this figure is around 2 million. New species in insects are being described every day, which reveals that a huge insect diversity is yet to be explored and named.

The first volume of this book titled *Insect Diversity and Sustainability of Ecosystem Services: Importance, Threats, Conservation and Economic Perspectives* contains 12 chapters contributed by different researchers from Asia and Europe and provides a balanced approach among different insect taxa by giving detailed information on insect taxonomy, diversity of aquatic and terrestrial insect fauna, insect-based value-added products, threats, and conservation of insects for biodiversity sustenance.

This is a wonderful book and a great companion for insect and nature lovers.

—**Prof. Akbar Masood**
Vice Chancellor
Baba Ghulam Shah Badshah University, Rajouri,
Jammu and Kashmir, India

Preface

Insects, the most abundant creatures on earth, both in terms of number and diversity are often taken for granted by the masses due to either negligence or the annoyance that a few species create. Insects are an imperative part of the majority of the food webs that exist across the globe and also play a significant role in the functioning of the ecosystem. Moreover, insects regulate environmental sustainability in many ways, such as by pollinating plants, cycling up nutrients, maintaining the structure and fertility of solum, decomposing various organic matter, etc. However, this harmony is now disturbed due to the worldwide increase in anthropogenic activities in the last few decades, leading to significant biodiversity losses in insects. Global pollution, climate change, habitat demolition are some of the major causes of insects decline. Thus, to maintain the ecological balance and also for our survival, new and efficient efforts on conservation of insects and their diversity are needed immediately.

Insect population, their relative incidence, and abundance are greatly influenced by various human activities either directly or indirectly, and the forecast is that in the forthcoming decades, it can lead to the extinction of 40% of insect species around the world. Insects have the largest population on earth and have evolved unique life characteristics, adaptations to the adverse environment, and evolutionary properties. Insects are considered the largest biological resources that are undiscovered by humans. Further exploration of insect's biological characteristics would help us manage the insect's resources at a large scale to produce insect-based value-added products in the agriculture sector and industrial field by advanced biological research. Moreover, insects and arthropods are found to be of utmost importance during the investigation of criminal cases. Some insect species are important for the medical field, as they produce some valuable compounds. Insects are promising assets for the production of novel antibiotics, as antibiotic-resistant human pathogens are currently on the rise. The mass rearing and the merchandising of useful insects is expanding gradually. It is a compendium of different insect species providing enormous free ecosystem services across the globe. Insects comprise the most diverse taxa of multicellular organisms on the planet Earth providing several ecosystem services in the form of pollination, decomposition/biodegradation, energy transfer through food chain, biological control of pests, providing food products in the form of honey, pollen, wax, etc.

This volume contains a total of 12 chapters from diverse fields, such as taxonomy, biogeography, terrestrial and aquatic insect diversity, applied and ecological studies. These chapters have been contributed by researchers across the globe from Europe to Asia, thus highlighting the importance of insects in sustaining the biodiversity.

—Dr. Younis Ahmad Hajam
Dr. Sajad Hussain Parey
Dr. Rouf Ahmad Bhat

CHAPTER 1

Insect Biodiversity Conservation: Why It's Needed?

SANIYA TYAGI, RADHESHYAM DHOLE, N. SRINIVASA, and N. VINAY

Department of Entomology, Institute of Agricultural Sciences, Banaras Hindu University, Varanasi, Uttar Pradesh, India

ABSTRACT

Insects, the most abundant creatures on earth, in terms of both number and diversity, are often taken for granted by the masses due to either negligence or the annoyance that a few species create. Insects are an imperative part of the majority of the food webs that exist across the globe and also play a significant role in the functioning of the ecosystem. Moreover, insects regulate environmental sustainability in many ways such as by pollinating plants, cycling up nutrients, maintaining the structure and fertility of solum, decomposing various organic matter, etc. However, this harmony is now disturbed due to the worldwide increase in anthropogenic activities in the last few decades, leading to significant biodiversity losses in insects. Global pollution, climate change, and habitat demolition are some of the major causes of insects decline. Thus, to maintain the ecological balance and for our survival also, new and efficient efforts on conservation of insects and their diversity are needed immediately.

1.1 INTRODUCTION

Insects are the most diverse and dominant organisms on the earth. So far nearly one million species have been described (Catalogue of Life, web).

Insect Diversity and Ecosystem Services, Volume 1: Importance, Threats, Conservation and Economic Perspectives. Younis Ahmad Hajam, Sajad Hussain Parey, and Rouf Ahmad Bhat (Eds)

They are the most dominant animal species on earth due to their small size, wide adaptability, metamorphosis, etc. In our rapidly changing world, we know the conservation of life forms is of utmost importance as the basic life support on which we depend has become a precarious ladder where insects are often forgotten (Dirzo et al., 2014). The insect extinction/decline is a hot topic among scientists around the globe. The decline of insect diversity is bad for the ecosystem and also for humans, as we depend on insects in various ways such as, for pest control, food, medicine, etc. (Ratcliffe et al., 2011). More than 75% loss in flying insect biomass over the last 20 years in 60 nature protected areas of Germany (Hallmann et al., 2017) is an alarming sign for insect conservation because if population declines in the natural areas there is a more chance of them being in the intensive agricultural areas. Similarly, long-term studies revealed a decline in the population of macro moths, beetles, and caddisflies in the Netherlands (Hallmann et al., 2020). There is a decline in the diversity of pollinators in Britain, especially in species associated with uplands (Powney et al., 2019). A study in the Puerto Rico rainforest revealed the decline of insect biomass (98% for ground foraging and 78% for canopy dwelling arthropods) over 36 years (Lister and Garcia, 2018). A nine-year survey of grasslands and 30 forest sites revealed a decline in the diversity of Arthropods over time. The annual survey of grasslands revealed a decline in arthropod biomass, abundance, and species diversity by 67%, 78%, and 34%, respectively, while in forest sites except for abundance, biomass and species diversity decreased by 41% and 36%, respectively. This indicates the decline of arthropod diversity is not restricted to agricultural areas but also extends to unattended areas (Seibold et al., 2019). According to an estimate, 40% of insect species are going to extinct in a few decades, among them the terrestrial insects like Lepidoptera, Hymenoptera, and dung beetles (Coleoptera) are the most affected taxa while aquatic insects, like Odonata, Ephemeroptera, Trichoptera, and Plecoptera already lost a considerable amount of species (Sánchez-Bayo and Wyckhuys, 2019). The reports of insect decline are mainly restricted to the temperate areas; however, the insect diversity in the tropical areas has not been studied well and more work in these areas will give a clear idea about their decline/extinction. The loss in insect biomass and diversity will have a negative impact on the complex food web and ecological services such as pollination, organic matter recycling, biological control, and seed dispersal provided by insects (Schowalter et al., 2018).

Insect populations, their relative incidence, and abundance are greatly influenced by various activities of humans either directly or indirectly and the forecast is that in the forthcoming decades it can lead to the extinction of 40% of insect species around the globe (Sánchez-Bayo and Wyckhuys, 2019). There are some biased opinions among different environmentalists and entomologists regarding the probable extinction of insect species from the different regions of the world. Few scholars estimated that around 500 thousand insect species might vanish up to the 25th century while others predicted that approximately 1/4th of the insect fauna would imperil within the century (Goulson, 2019). It was observed that in some parts of the European continent, local Lepidopteran insects like moths and butterflies extinct more rapidly than other floral and faunal species (Thomas et al., 2004). Moreover, there are symbiotic, mutual, and parasitic relationships of many insect species with those of vertebrates, which help in their coevolution and coexistence. These different forms of associations direct the development of an ecosystem (Dunn, 2005). Yet, only a few people have begun to realize just how vital insects are for the earth.

There is negligence toward insect conservation. First of all, knowledge about how many total insect species exist in an ecosystem and among them how many are facing the risk of extinction is very limited. For all we know is, diversity occurs more in class Insecta that comprise potentially over 90% of animal life forms (Devi and Indirakumar, 2020); they represent between six and ten million extant species. There is an enormous bias in taxonomic conservation research; much less emphasis has been given to the studies on invertebrates as compared to the studies of vertebrates and even plant studies (Clark and May, 2002). So, there are chances that much extinction had occurred or would occur, even before the species are described. Another limitation associated with insects' conservation efforts is the perception challenge. Insects are frequently symbolized as trivial and very little recognition is awarded to them as they are considered nasty and disgusting which create a nuisance (Hall and Martins, 2020). The general paucity of admiration for insects in human society goes on increasing from generation to generation; they do not want insects to be a part of their society and even hate their presence. With the rapid change in the ecosystem due to humans, the future of these hexapods lies on (1) acclimatization, (2) overcoming the human-induced barricade, or (3) die-off (Samways, 2018). Extinction of insect population from our ecosystem will vanish our presence too, and so it is indeed needed to develop the strategies that promote the sustainable growth and development of insect species, encourage their biodiversity, and indirectly ensure our livelihood in the future. In this chapter, we will first overview the role of insects in the ecosystem. Then, we highlight the various threats that are posed on insects due to anthropogenic activities and negligence and finally a look at conservation practices.

1.2 IMPORTANCE OF INSECTS IN THE ECOSYSTEM

In the last two decades, it became imperative to acknowledge, enumerate, and set out the various functions performed by the insects to invigorate the ecosystem (Díaz et al., 2013; Turner et al., 2007). The well-ordered life of every ecosystem has a remarkable association with the population of insects that exist in it as its key component (Schowalter, 2013; Weisser and Siemann, 2004). Insects perform several roles in an ecosystem that strengthen the basic framework of it and give essence to it. Services furnished by insects to the ecosystem of which they are part can be categorized into four major groups viz., equipping and stocking services (results in the development of vivacious habitat), aiding services (maintain the proper functioning of an ecosystem), synchronizing services (regulate the magnitude and flow of different growing parameters of an ecosystem), and spiritual services (provide aesthetic value) (Prather et al., 2013). The services rendered by insects help us to earn a quality life; they also provide an economic output of high monetary value, but still, the efforts of these small creatures are usually taken for granted (Beynon et al., 2015). Nonetheless, in the United States alone insects driven ecosystem services value around $57 billion annually (Losey and Vaughan, 2006), and the estimated value of the pollination services performed by them is approximately half of the trillion dollars per year (IPBES, 2016). The genuine valuation of total insect biomass present in a unit area for the delivery and maintenance of ecosystem services by their biodiversity lies on obtaining precise data with comprehensible knowledge regarding services-provision processes (Haines-Young and Potschin, 2010). Therefore, it is needed to have proper and correct data related to different types of functional roles performed by different groups of insects, so that the value of insects in ecosystem services can be analyzed. Despite the huge number and diversity of insects a very little research has been done that emphasized their role in the functioning and sustainability of the ecosystem in comparison to other flora and fauna (Schowalter, 2016).

Therefore, we lack an inclusive grasp of the performance of insects in various services and attributes of different ecosystems, like pollination by bees, scavenging by dung roller, decomposition by many minute insects, etc. (Boerema et al., 2017). Hence, evaluation of insect services to the ecosystem comprises substantial heterogeneous processes viz., on-ground observations, laboratory factsheets, remote sensing monitoring, speculations developed from substitutes of several outlooks of biodiversity (e.g. species abundance, morphological traits, total richness), expert opinions, and experimental findings cited from different labs and field studies (Noriega et al., 2018).

1.2.1 INPUTS TO UNDERGROUND SYSTEMS

Insects are the most crucial component of various communities found in an ecosystem like in some they form the structural unit while in others they act as a food source for the organisms of different strata. In some ecosystems, the cumulative biomass of insects is relatively small as compared to the mass of other organisms but whatever the total insect biomass present in that system, it exhibits major support in various vital cycles of that ecosystem. The forests of North America have seen noteworthy debris and detrital mass accumulation on the onset of the periodical outbreak in the population of cicadas (Yang, 2004). Not only live insects contribute a lot to the sustainable development of underground habitat and niche rather the remains of dead insects like their cadaver and carcass impart significant importance in below-ground units of the ecosystem by enhancing mineralization, aeration by creating favorable conditions for soil microbial fauna (Laigle et al., 2021). Collembola, termites, and ants are the major modifiers of soil processes by regulating dynamics and accessibility of different minerals, nutrients, and organic matter to the vegetation, thus increasing the availability of plant nutrients in the soil. Insects may affect ecosystem functions in various ways specifically by regulating the rate of putrefaction, recycling of minerals, and finally the per-unit production of crop yield (Neher and Barbercheck, 2019). These effects are comparatively inaccessible to study as they are underground, though, new techniques have allowed significant advances in both below- and aboveground processes of insects, such as controlled environment studies and improved stable isotope methods (Roy et al., 2021).

Insects physically break organic matter and make it available to the microbes for further decomposition within the soil with additional consequences for nutrient cycling. However, there are some indirect effects of insect biota in ecosystem regulation, like the effect of below-ground root-feeding herbivores on the rhizosphere, effects of above-ground folivores, and their relative feeding on nitrogen cycling (Hartley and Jones, 2008). Insects in addition to providing available nutrients to various flora, determine their accessibility within the area (Hunter, 2001). Moreover, the impact of insects on the overall growth and development of field crops or forest vegetation is indispensable.

1.2.2 POLLINATION

Insect pollinators and flora-bearing flowers show a symbiotic relationship. Pollinators help to transport pollen from anthers (male part of a flower) to stigmas (female part of a flower) for fertilization and in return get rewarded in the form of nectar and pollen (Dar et al., 2017). Crop pollination is essentially an ecosystem service provided by insects that assist in food production all over the world (Bommarco et al., 2013). About three-fourths of the total crop species used as food depend on pollination by insects to various extents (Klein et al., 2007). The phenomenon of pollination that regulates the ecosystem services, makes around 87% of plant species (and up to 94% in tropical regions) in the world depend upon other organisms (insects, birds, mammals) to some extent, particularly insects (Ollerton et al., 2011). Many groups of insects including species of Hymenoptera, Lepidoptera, Diptera, Coleoptera, Thysanoptera, etc. are admired as pollinating agents for bringing new generation plants. The bees which belong to the order Hymenoptera are the most recognized pollinators, comprising of 20,000 species worldwide, on the myriad of plant species, producing one-third of the world food crops worldwide (Stanley, 2020). However, other insect groups which are chiefly specialists for certain functions, at least for a part of their life cycles performed a vital function of pollinator in natural and agricultural environments (Rader et al., 2016).

Today, the existence of flora that bears flowers is in danger because of the declining species diversity of insects that act as pollinators for them.

There is a huge loss of insect diversity due to various anthropogenic factors like deforestation and forest degradation. Globally, we had signed several international conventions and treaties, conducted summits to increase the awareness about the sustainability of the ecosystem and factors affecting it, but despite all these, the diminishing rate of insect species is high and is especially more in tropics where several species of them had not recognized yet and are still unknown to the world (CBD, 2020).

1.2.3 ECOSYSTEM ENGINEERING

Ecosystem engineering by insects includes the creation and development of new habitat or modifying the earlier existing habitats in such a way that it will fulfil their needs and desires more effectively; mostly, it includes interlinking of unrelated strata with each other, alteration of boundaries, and some basic structures (Jones et al., 2010) relieving insects from various stresses like biotic and abiotic and facilitates the development of better niche for them (Romero et al., 2015). Insects play a crucial role in bringing diversity to a previously established community either by acting as tenants or by encroaching on the territory of the other organisms for shelter and food. These events are mostly observed with gall-formers, leaf-miners (Kozlov and Zvereva, 2017), and ants (Henriques et al., 2019) that forcefully occupy the shelters and mark their effect on community diversity. The use of shelters has strong effects on community associations. Data have shown that the shelters increase arthropod abundance approximately by 135% and richness by 128% (Cornelissen et al., 2016). There are many other insect species at their different growing stages that assist in the ecological engineering of the existing community such as the lepidopteran caterpillar construct different leaf structures while feeding on plants like leaf fold, leaf roll, leaf cylinder with patches and exit hole which later are used by other coleopteran or hymenopteran insect species as a shelter (Omer, 2016). Similarly, in the arid and semiarid regions or tropical forest and wetland areas, termites act as a chief agent that brings about putrefaction of different organic residues, leaf and wood litter which lead to the production of different chemical substances that invites the formation of various niches of micro-fauna for its further degradation (Jouquet et al., 2019). Termites also modify the structure, pH and pF values of soil where they are dwelling and create different habitat places like mud sheeting, sand rickles, and mounds with mini-micro galleries that are further used by other biotas in the environment. Some insects like Funnel ant *Aphaenogaster longiceps* have a large impact on soils; in arid areas of Australia it can move some 80% of the soil to the surface compared to all soil fauna combined (Shattuck, 2008).

1.3 BIOLOGICAL CONTROL

The role of insects in biological control is by acting as natural predators and parasitoids of insect pests, thereby controlling their population. This strategy has long been considered an important component of pest management. The chemical pesticides which are widely used to control insect pests also kill the nontargeted beneficial arthropods (Kim et al., 2021). Therefore, it is needed to adapt the biological method of pest control instead of the chemical one, which will help to check the population of targeted insect pests causing a menace to the farmer field by using and promoting the natural biological agents, i.e., parasites and predators, which is a more target-specific and eco-friendly approach and also enshrines the biodiversity in the environment.

Biological pest control by insects has been in action since the beginning of agriculture and researchers and farmers are seeking to increase its efficacy. In ancient China, populations of red ants were augmented as they were found to be effective predators of many citrus pests. There are several successful programs around the world using the importation and release of natural enemies which help in controlling the exotic insect pest (Nechols, 2020). During the initial periods of the 1960s, farmers of the United States faced the problem of severe infestation and damages caused by *Hypera postica* on alfalfa. As no insecticide gives better results in controlling its growing population, some researchers tried to discover its biological solution and found its 12 natural enemies at an out-station which was then systematically imported and released in the field at mass level, hence successfully eradicating the pest status of *Hypera postica* on alfalfa (Day, 1981). Suppression of pest populations by natural enemies instead of insecticides is based on the ecological principle that agricultural intensification should enhance the contribution of ecosystem services. Recently, the invasion of fall armyworm, *Spodoptera frugiperda* (Lepidoptera: Noctuidae) across the African continent and India, has attracted many natural enemies. In India, its management strategy is based on biological control principle via its parasitization by a newly reported dipteran parasitoid, humpbacked fly, *Megaselia scalaris* (Shylesha et al., 2018).

1.4 FOOD PROVISION

Insects are a major prey item for a variety of animals ranging from amphibians to mammals, as they are a great source of protein. Insects are the primary native consumers of plants acting as important links to food webs. Many birds, reptiles, chordates prefer insects in their diet as they fulfil their protein and other nutritional requirements (Tallamy and Shriver, 2021). Moreover, insects are an important source of diet for many other insects belonging to different orders; hence, there exists an intricate association between two different insect populations, one which feeds on plants and another which feeds on herbivore insects (Finke and Oonincx, 2014). Therefore, we can call insects a key component of any ecosystem as their presence determines the exact nature of various food chains and their functionality. Similarly, in the case of human beings also, insects are the source of nutrition, especially in tropical countries since time immemorial (DeFoliart, 2012). There might be around 2000 or more edible arthropod species that fall into several groups based on the family and order to which they belong, such as Coleoptera (31%) shares the highest percent of edible insect population, just next to Lepidoptera (17%) and Hymenoptera (15%) (Jongema, 2017).

Currently, the perception of humankind for insects is changing; they started to admire it as food, a direct source of various vitamins, minerals, and proteins. Insects are no longer food just for pets and for the fishery, poultry, or piggery industries but for humans also. The edible insects contain a higher amount of quality protein, vitamins, and amino acids in the same volume as compared to traditional food sources as they have a high food conversion rate. Additionally, they are more ecologically friendly by emitting far fewer greenhouse gases and ammonia. Due to deforestation, intensification of agriculture, and environmental pollution there is a consequent loss of insects and thereby in insect food which resulted in the decline of European birds, bats, and several other animals with their food sources getting lesser (Ramos-Elorduy, 2006).

TABLE 1.1 List of Major Insect Orders and Species with their Role in Ecosystem.

S. no.	Insect order	Common name	Species/genus/family	Role in ecosystem
1.	Hymenoptera	Honey bee	*Apis* sp.	Pollinators
		Bumblebee	*Bombus sp.*	
		Smallbee	*Mallotamunda*	
			Tachinidaesp.	
			Takanoahakusana	
		Ants	Formicidae sp.	Predators, scavengers
		wasps	Vespidae sp.	
			Ichnemonidae sp.	Parasitoids
			Braconidae sp.	
			Chalcidoidea sp.	
2.	Diptera	Flies	*Mallotamunda*	Pollinators
			Tachinidaesp.	
			Takanoahakusana	
			Muscaconducens	
			Many species (Calliphoridae, Sarcophagidae, Muscidae, etc.)	Scavengers
3.	Coleoptera	ground beetles	Carabidaesp.	Predators
		ladybugs	Coccinellidaesp.	
		Smallbees	*Hopliamoerens*	Pollinators
			Arthromacraviridissima	
4.	Lepidoptera	Moths, butterflies	Many species	Pollinators
5.	Blattodea	Termites, cockroaches, weevils	Many species	Decomposers

1.5 THREATS TO INSECTS

1.5.1 ENVIRONMENTAL POLLUTION

The ever-increasing human population demands more energy and other resources posing detrimental effects on ecosystems and organisms. Pollution is lethal to plant species diversity ultimately affecting dependent insect species also. There are direct and sublethal effects of environmental contamination (mainly through pesticides) on insects. For example, larva and adult butterflies when exposed to 1/640 of the field recommended dose of the insecticide Deltamethrin showed reduced longevity (Samways, 2007). The pesticidal negative effects are mainly implicated in flying insects (van Lexmond et al., 2015; Goulson et al., 2015). Insecticide Fipronil was found to be responsible for the mass mortality of honey bees in France during the 1990s (Holder et al., 2018). The neonicotinoid insecticides such as imidacloprid, thiacloprid, and clothianidin interfere with the navigation of bees (Fischer et al., 2014). Also, bees prefer to feed on neonicotinoid treated sugars and they cannot differentiate between the two, leading to more undesirable effects (Kessler et al., 2015). There is also an indirect impact through habitat alteration (mainly herbicides). Bioaccumulation of contaminants, in addition, is more harmful to insects, by posing unknown changes in their physiology and behavior (Hayes and Hansen, 2017). Despite the magnitude of environmental pollution, little is studied about its effect on insect survival as the responses to a specific pollutant and concentration vary among species. In addition, impacts of contamination form an adverse interaction with other forms of stress, such as habitat fragmentation. The various types of environmental pollution faced by insects are in the form of chemicals used in pesticides, fertilizers, certain industries etc.; or/and in the form of noise and light, especially impacting nocturnal insects. The long-term exposure to pollutants also results in threatening of many beneficial insect species like pollinators, decomposers, or natural enemies. The water pollution affects the fresh-water population of invertebrates severely, causing more than 41% of insects' species to show on the IUCN Red List (Darwall et al., 2012).

1.5.2 LOSS OF NATURAL HABITAT

Fragmentation, degradation, and loss in habitation are possibly the major reasons for biodiversity losses (Dirzo et al., 2014; Habel et al., 2019a). The land transformation due to deforestation, urbanization, and agricultural expansion are the prime reasons for the loss of insect habitats across the world (Habel et al., 2019b). Mining activities particularly restrict subterranean insect species movements and affect their population diversity (Mammola et al., 2019). The conversion of natural land to agriculture and increase in eutrophication is synergistic with habitat fragmentation, pollution, global warming, and invasive alien organisms' impact. Due to these situations, specialists are likely to decline the most (Koh et al., 2004), inferring that not all species are affected equally. For example, extinction of highly abundant Rocky Mountain locust, *Melanoplus spretus* in the North American Midwest during the last century (Lockwood, 2009).

The degree of habitat fragmentation affects different species based on their mobility. Less mobile and highly mobile species may tolerate such isolation (Samways, 2006), whereas human-made landscape medley possibly cause difficulties to the species with intermediate mobilities (Thomas, 2000). The reduction in tropical forests cover harboring more than half of the total insects is of the greatest concern. At present, we are losing around 130,000 km^2 of forests annually, with researches showing forest-to-farmland conversion causing a huge impact, predominantly on primary specialists (Shumi et al., 2018). Indeed, the function of the ecosystem alters along with the tree canopy (Lewinsohn et al., 2005). In other ecosystems too, insect species are at risk, such as freshwater habitats suffering from increased siltation from agricultural runoff (Goulson, 2019), in addition to flow regulation, cause habitat loss of stream-dwelling species (Jones et al., 2012). Pond ecosystems are also threatened mainly due to urbanization and agricultural land drainage (Wood et al., 2003).

1.5.3 PERVASION OF INVASIVE ALIEN ORGANISMS

Sometimes due to many anthropogenic reasons, a certain species is introduced outside its native location, disturbing the prevailing environmental or socioeconomic situation of the new area (Turbelin et al., 2017). These introduced species referred to as invasive alien species can pose a serious threat to the endemic species (Clavero and Garc'ıa-Berthou, 2005). Impacts may be direct and/or indirect. Direct impacts by non-native species can be through a series of interactions; they can predate and/or compete with the local species or they can act as a vector of certain disease-causing extinction in the local populations (Havel et al., 2015), whereas indirect impacts on local populations are through cascading effect at various trophic levels and coextinction of insect pest or natural enemy hosts etc. In 1773, local insect populations of Britain were diminished due to the invasion of the ladybird, *Harmonia axyridis* Pallas, which outcompeted with the local coccinellids (Comont et al., 2014). The introduced species can flood the ecosystems by replacing the local fauna, therefore inevitably reducing indigenous insect diversity (Samways and Taylor, 2004), which is reversible when the alien species are removed (Samways et al., 2005). The common wasp, *Vespula vulgaris,* have invaded the ecosystems in New Zealand and had a disastrous effect on local arthropods (Brown and Groenteman, 2017). Invasive insects may eventually direct to the loss of local insects, of which specialists or those species exhibiting narrow geographic distributions are highly vulnerable (Wagner and Van Driesche, 2010). Invasive species including plants and pathogens are also responsible for local insect extinction. The invasive plants often have a major impact on insect communities by reducing the quantity or quality of food, therefore lowering resources vital for them (Havel et al., 2015). Invasive fungal pathogen of the bees, *Nosema bombi,* (European strains) is the impending cause of extensive fall down of bumblebees in North America (Cameron and Sadd, 2020).

1.5.4 WORLDWIDE CLIMATE CHANGE

Climate change is not just simply a universal temperature rise rather it is a complex phenomenon posing threats to all living beings including insects and their respective ecosystems, whether terrestrial, aquatic, or subterranean (Burrows et al., 2011; Woodward et al., 2010). It directs to a series of ecological processes, involving changes in species distributions (Chen et al., 2011), unforeseen relations among remote species (Krosby et al., 2015), phenological shifts (Forrest, 2016), extinctions (Dirzo et al., 2014), and other erratic cascading effects in the ecosystem processes (Peñuelas et al., 2013). For example, between the 1980s and 1990s, the phenological shifts of many British butterflies were recorded as the butterflies were emerging earlier from their cocoon (Roy and Sparks, 2000). Climate change is responsible for altered trophic interactions affecting functional feeding group diversity, which can be associated with changes in food webs (Jourdan et al., 2018). These changes, in addition, affect all elements of food chains from host, pathogens, predators, to parasitoids, directly or indirectly (Kehoe et al., 2020). In a study, insect pests that consumed less nutritious plant material grown in an increased carbon dioxide environment showed slower growth and increased mortality even by feeding on higher amounts (Jactel, 2019). Climate change lead to a reshuffling of communities as insects can shift their location quickly from tropical to temperate areas through migration (Demain, 2021). Plant species of the temperate regions are expected to have new encounters with particular herbivores. The biggest threat is the synergistic effect of climate change with adversities like habitat loss, pollution, causing many fold irreversible impact on insects. This can be well understood by an example of Britain where only 30% of the larger moths had survived since mid-twentieth century possibly due to the interactive and synergistic effect of climate change with intensive agriculture practices especially excessive chemicals use (Conrad et al., 2004).

1.5.5 OVEREXPLOITATION OF INSECTS

Overexploitation of insects through their unsustainable harvesting, such as petting them and using them as ornaments (as souvenirs and jewels), or by consuming them as food or in traditional medicine may lead to insect biodiversity loss. Various insect groups, such as crickets, ants, antlions, caterpillars, praying mantids, mealworms, etc. are kept as pets (Smithsonian, 2019), and some of them are traded illegally. Ornamental insects as preserved decorations, jewellery, embroidery, and pottery are in demand in many places. In such places, this demand has become the reason for the decline of much-demanded species like Lepidoptera and Coleoptera (Huang, 2014). The overexploitation of insects as food and alternative medicine is another driver of their decline, as the number of individuals assembled, frequently exceeds the rate at which they regenerate (Cerritos, 2009).

1.5.6 CO-EXTINCTION

In Co-extinction, extinction of a single-insect species is rarely an isolated event. Instead, numerous insect lineages which have become co-dependent by diversifying with vertebrates and invertebrates, as parasites, commensal coprophages, or epizoic mutualists, face the risk of co-extinction as their hosts decline and fail (Dunn et al., 2009). Primary extinctions can cause secondary and further extinction cascades due to numerous trophic levels interactions in ecological networks. For example, mammals decline resulting from synergistic causes such as habitat destruction, climate change also caused dung beetles' co-decline (Bogoni et al., 2019). Similarly, trading and exploitation of pet birds caused a threat to their related insect fauna (Eaton et al., 2015).

1.5.7 BIOLOGICAL CONTROL

Biological control for insect pest control involving the introduction of exotic natural enemies to a foreign land has inevitably carried some risks for nontarget organisms despite its economic and environmental benefits (Kenis, 2017). Even when bio-control agents are species or genus-specific, they can still have substantial nontarget effects on indigenous biotas through food web interactions (Pearson and Callaway, 2003). This problem can be solved by more effective bio-control agents which improve both control and safety. After the introduction and establishment of natural enemies in classical bio-control programmes, the biggest threat is that the new species can alter the host landscape permanently as these practices are irreversible (Lockwood et al., 2001). For instance, the introduction of the cactus moth, *Cactoblastis cactorum*, in the United States for the control of local prickly pear cacti (*Opuntia* spp.) is now becoming the cause of its decline in some areas of North America (Espinosa-Zaragoza, 2020).

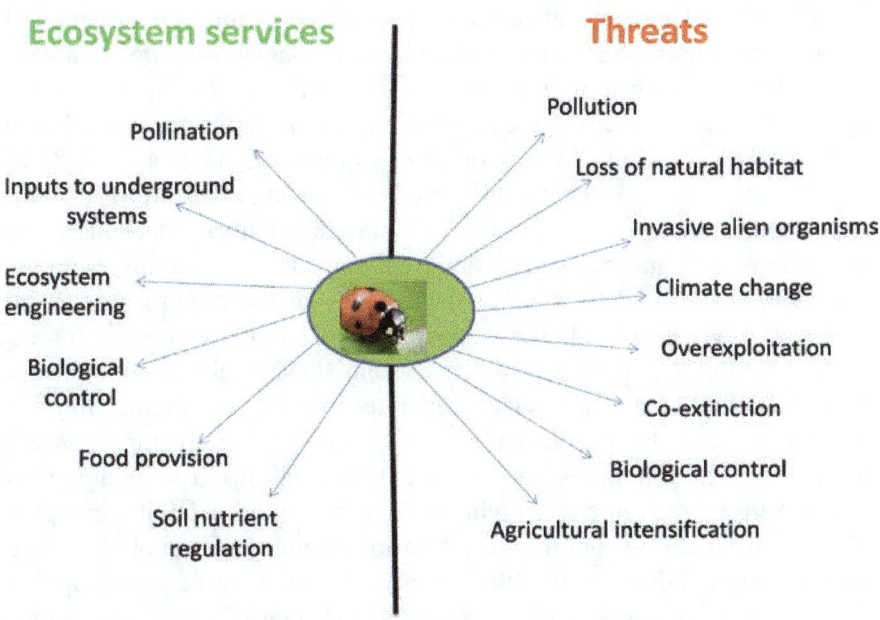

FIGURE 1.1 Depiction of various ecosystem services provided by insects (left) against numerous threats that are imposed on them (right).

1.6 INSECT CONSERVATION

The philosophy of insect conservation is based on two aspects. First, insects must be conserved for themselves, as they have an intrinsic value since they play an important role in biodiversity. We must value insects by acting as a wise, responsible, and caring species. Second, insects have instrumental value, which is solely for us, by rendering irreplaceable services and resources. Instead of this, very few people realize this value. With declining global ecosystems, we must not continue the same trend and require an immediate change in mentality and action in favor of their conservation.

1.6.1 INSECTS MONITORING: NEED AND APPROACHES

Except for the deep benthic zone of oceans, insects construct the main biotic component for all terrestrial ecosystems and niches (Grimaldi and Engel, 2005). Therefore, it becomes imperative to monitor insects as it forms a vital part of conservation and management programs (Lamarre et al., 2020). The awareness regarding the occurrence, establishment, and spread of insects (precisely the population dynamics and factors affecting it) is elementary while formulating the strategy (Jiang et al., 2008) for biodiversity conservation and integrated pest management (IPM).

Various techniques already exist for insect monitoring like light trap, pheromone trap, automatic detection traps, sensors, etc., but more innovations are still needed to enhance the efficacy of the monitoring system. Recently, modern technologies came up in field surveys of various insects, such as Global Navigation Satellite System (GNSS) for fauna mensuration and assessment (telemetry), radar techniques, for monitoring pest transhumance, video accoutrements, to observe flying creatures (usually hexapods), chemiluminescent tags, and thermal infrared imaging for tracking insect movement at night hours or in darkness, habitat mapping, and use of an echo-sounding detection instrument to monitor the movement of larvae (Jiang et al., 2008). For real-time identification, applying the Aggregation Dispersion Variance (ADV), Gaussian Mixture Model (GMM), and Distance Regularization Level Set Evolution (DRLSE) to the pictorial data obtained from a specially designed movable robot car with the camera, it is possible to monitor several Lepidopteran insects sat field level with an accuracy of up to 95% (Zhao et al., 2019).

Different classifiers like binary decision tree classifier, Artificial Neural Networks (ANNs), Support Vector Machine (SVM), Bayesian classifier, and k-nearest neighbor, added with different statistical extraction methods such as Gray-Level Co-Occurrence Matrix (GLCM) and Gray-Level Run Length Matrix (GLRLM) were observed to have practical utility with a high percent of accuracy in the monitoring of various sucking insect pests in the various terrestrial ecosystems (Dey et al., 2016). Acoustic sensors along with speech recognition technique (SRT) presume to have a wide utility in the identification and detection of various insects belonging to order Coleoptera, as every other species possesses a unique sound that can be differentiated and isolated from the sounds of other insects and the environment (Martin et al., 2015). All these novel insect automatic detection and monitoring systems perform real-time surveillance functions on the field level viz., in the forest, agriculture, biospheres, national parks, grasslands, etc. as they can be easily connected to the internet or wireless networking sensors (Internet of Things) or cloud computing services which help in further decision-making while planning different biodiversity conservation programmes at large scale.

1.6.2 HABITAT AND LANDSCAPE CONSERVATION

For the conservation of insects, a wide range of habitats and landscapes have to be conserved. Today, most landscapes are devoid of insect conservation. Their appropriate management across the landscape can be achieved with awareness followed by necessary actions, for example, conservation headlands. Strategies required to restore ecosystem services particularly for insects require biodiversity action plans, enhancing urban environments, integrated pest management, organic farming, conservation agriculture, judicious use of pesticides, insect parks in the protected area, captive breeding programmes, etc (Leather and Helden, 2005). Scientists keep searching for new and better ways for conserving insect diversity, by researching the most favorable environmental conditions for insect survival. Due to anthropogenic modifications, the landscape matrix within and around a habitat may not have essential resources for insects, so affecting their generational survival adversely. Research drive in the

TABLE 1.2 Application of Various Sensors with Automatic Monitoring and Detection Systems in Insect Biodiversity Conservation.

Order	Insect	Landscape	Sensor	Automatic monitoring and detection systems	Potency	References
Coleoptera	*Anoplophora glabripennis*	hardwood forests	Rivet with piezoelectric detector [model 723 Oyster], amplifier [model 1485 ENC], and an audio recorder (model TCD-D10 Pro II)	Specifically designed signal probing software DAVIS [Digitize, Analyze, View, Insect Sounds]	81.5%	Mankin et al., 2008
	Rhynchophorus ferrugineus	Plantation	Microphone with piezoelectric diode and amplifier [12v]	Mixture simulation with Vector mensuration, Gaussian and nearest-neighborhood classifier	95%	Pinhas et al., 2008
Lepidoptera	19 species of the family Pieridae	Agriculture and Forest	Image database	Gray level co-occurrence matrix (GLCM) analysis added with texture and hue filter	92.8%	Kaya and Kayci, 2014
	5 species of the family Papilionidae	Grassland	Image database	ANNs with di-patterns	84%	Kaya et al., 2015
Hemiptera	*Bemisia tabaci*	Pastureland Agricultural region	Smart camera and blue and yellow sticky traps	2-D Fourier transfiguration spectrum	96%	Sun et al., 2017
	Trialeurodes vaporariorum	Protected cultivation Greenhouse Glasshouse	Auto-microcontrollers, aided with cameras, tripods, USB drives to generate data	Multilayer perceptron (MLP) and k-nearest neighbor (KNN)	73.8%	Gutierrez et al., 2019
	Aphis glycines	Agriculture and Grassland	Digital camera	Hue-chroma intensity algorithm	92%	Maharlooei et al., 2017
Diptera	*Bactrocera oleae*	Orchard	Auto detector webcams provided with internet	Blur filter [Gaussian], algorithm threshold [OTSU] and counting	78%	Doitsidis et al., 2017

TABLE 1.2 *(Continued)*

Order	Insect	Landscape	Sensor	Automatic monitoring and detection systems	Potency	References
	Ceratitis capitata	Orchard	Specially designed traps with optoelectronic sensor	Adaboost meta classifier, support vector classifier	86%	Potamitis et al., 2017
Thysanoptera	*Frankliniella occidentalis*	Forest Grasslands Agro-field habitat	Digitalized sticky traps added with microsensors camera	System of prewittedge detection with AANs and 2-D transformation	96%	Sun et al., 2017
Non-insect (Acarina)	*Petrobia latens*	Pasture	Image-processing database	Convolutional neural network (CNN)	90%	Li et al., 2020

twenty-first century should be on making a cordial environment for insects so that the reconciliation of both human and insects needs can occur.

KEYWORDS

- **insect conservation**
- **insects decline**
- **pollination**
- **threats on insects**
- **biological control**

REFERENCES

Alexander, K. Wood Decay, Insects, Palaeoecology, and Woodland Conservation Policy and Practice: Breaking the Halter. *Antenna* **2005,** *29,* 171–1778.

Alroy, J. How Many Named Species are Valid? *Proc. Natl. Acad. Sci. USA.* **2002,** *99,* 3706–3711.

Beynon, S. A.; Wainwright, W. A.; Christie, M. The Application of an Ecosystem Services Framework to Estimate the Economic Value of Dung Beetles to the UK Cattle Industry. *Ecol. Entomol.* **2015,** *40,* 124–135.

Bickford, D.; Lohman, D. J.; Sodhi, N. S.; Ng, P. K.; Meier, R.; Winker, K.; Ingram, K. K.; Das, I. Cryptic Species as a Window on Diversity and Conservation. *Tr. Ecol. Evol.* **2007,** *22,* 148–155.

Boerema, A.; Rebelo, A. J.; Bodi, M. B.; Esler, K. J.; Meire, P. Are Ecosystem Services Adequately Quantified? *J. Appl. Ecol.* **2017,** *54,* 358–370.

Bogoni, J. A.; da Silva, P. G.; Peres, C. A. Co-Declining Mammal–Dung Beetle Faunas Throughout the Atlantic Forest Biome of South America. *Ecography* **2019,** *42,* 1803–1818.

Bommarco, R.; Kleijn, D; Potts, S. G. Ecological Intensification: Harnessing Ecosystem Services for Food Security. *Trends Ecol. Evol.* **2013,** *28* (4), 230–238.

Brandenburg, A.; Dell'Olivo, A.; Bshary, R.; Kuhlemeier, C. The Sweetest Thing: Advances in Nectar Research. *Curr. Opinion Plant Bio.* **2009,** *12* (4), 486–490.

Brown, B.; Groenteman, R. September. Poster 31: Vespula Biocontrol in New Zealand Revisited. In *Symposium on Biological Control of Arthropods*, 2017; p 306.

Burrows, M. T.; Schoeman, D. S.; Buckley, L. B.; Moore, P.; Poloczanska, E. S.; Brander, K. M. The Pace of Shifting Climate in Marine and Terrestrial Ecosystems. *Sci.* **2011,** *334,* 652–655.https://doi.org/10.1126/science.1210288

Cameron, S. A.; Sadd, B. M. Global Trends in Bumble Bee Health. *Ann. Rev. Entomol.* **2020,** 65. https://doi.org/10.1146/annurev-ento-011118–111847

Cerritos, R. Insects as Food: An Ecological, Social and Economical Approach. CAB Rev.: *Perspect. Agric., Veterin. Sci. Nutri. Nat. Res.* **2009,** *4,* 1–10. https://doi.org/10.1079/PAVSNNR20094027

Chen, I.-C.; Hill, J. K.; Ohlemüller, R.; Roy, D. B.; Thomas, C. D. Rapid Range Shifts of Species Associated with High Levels of Climate Warming. *Science* **2011,** *333,* 1024–1026. https://doi.org/10.1126/science.1206432

Choosai, C.; Mathieu, J.; Hanboonsong, Y.; Jouquet, P. Termite Mounds and Dykes are Biodiversity Refuges in Paddy Fields in North-eastern Thailand. *Environ. Conserv.* **2009,** 71–79.

Clark, J. A.; May, R. M. Taxonomic Bias in Conservation Research. *Science* **2002,** *297,* 191–192.

Clavero, M.; García-Berthou, E. Invasive Species are a Leading Cause of Animal Extinctions. *Trends Ecol. Evol.* **2005,** *20,* 110.

Comont, R. F.; Roy, H. E.; Harrington, R.; Shortall, C. R.; Purse, B. V. Ecological Correlates of Local Extinction and Colonisation in the British Ladybird Beetles (Coleoptera: Coccinellidae). *Biol. Invasions.* **2014,** *16,* 1805–1817. https://doi.org/10.1007/s10530-013-0628-3

Conrad, K. F.; Woiwod, I. P.; Parsons, M.; Fox, R.; Warren, M. S. Long-Term Population Trends in Widespread British Moths. *J. Insect Conserv.* **2004,** *8,* 119–136.

Cornelissen, T.; Cintra, F.; Santos, J. C. Shelter-building Insects and Their Role as Ecosystem Engineers. *Neotrop. Entomol.* **2016,** *45,* 1–12.

Costello, M. J.; Houlding, B.; Wilson, S. As in Other Taxa, Relatively Fewer Beetles are Being Described by an Increasing Number of Authors: Response to Lobl and Leschen. *Syst. Entomol.* **2014,** *39,* 395–399.

Costello, M. J.; Wilson, S.; Houlding, B. More Taxonomists Describing Significantly Fewer Species per Unit Effort may Indicate that Most Species have been Discovered. *Syst. Biol.* **2013,** *62* (4), 616–624.

Dar, S. A.; Hassan, G. I.; Padder, B. A.; Wani, A. R.; Parey, S. H. Pollination and Evolution of Plant and Insect Interaction. *J. Pharmacogn. Phytochem.* **2017,** *6* (3), 304–311.

Darwall, W.; Seddon, M.; Clausnitzer, V.; Cumberlidge, N. Freshwater Invertebrate Life. In *Spineless: Status and Trends of the World's Invertebrates*; Collen, B., Böhm, M., Kemp, R., Baillie, J. E. M., Eds.; Zoological Society of London: UK, 2012; pp 26–33.

DeFoliart, G. The Human Use of Insects as a Food Resource: A Bibliographic Account in Progress [Online]. 2012. http://www.food-insects.com/

Demain, J. G. Insect Migration and Changes in Venom Allergy due to Climate Change. *Immunol. Allergy Clin.* **2021,** *41* (1), 85–95.

Devi, M.; Indirakumar, K. Why Insects are Dominance in the Biosphere? *Biotica Res. Today* **2020,** *2* (11), 1174–1178.

Dey, A.; Bhoumik, D.; Dey, K. N. Automatic Detection of Whitefly Pestusing Statistical Feature Extraction and Image Classification Methods. *Int. Res. J. Eng. Technol.* **2016,** *3,* 950–959.

Díaz, S.; Purvis, A.; Cornelissen, J. H. C.; Mace, G. M.; Donoghue, M. J.; Ewers, R. M.; Jordano, P.; Pearse, W. D. Functional Traits, the Phylogeny of Function, and Ecosystem Service Vulnerability. *Ecol. Evol.* **2013,** *3,* 2958–2975.

Dirzo, R.; Young, H. S.; Galetti, M.; Ceballos, G.; Isaac, N. J. B.; Collen, B. Defaunation in the Anthropocene. *Science* **2014,** *345,* 401–406. https://doi.org/10.1126/science.1251817

Doitsidis, L.; Fouskitakis, G. N.; Varikou, K. N.; Rigakis, I. I.; Chatzichristofis, S. A.; Papafilippaki, A. K.; Birouraki, A. E. Remote Monitoring of the *Bactrocera oleae* (Gmelin) (Diptera: Tephritidae) Population Using an Automated McPhail Trap. *Comput. Electron. Agric.* **2017,** *137,* 69–78.

Dunn, R. R. Modern Insect Extinctions, the Neglected Majority. *Conserv. Biol.* **2005,** *19,* 1030–1036.

Dunn, R. R.; Harris, N. C.; Colwell, R. K.; Koh, L. P.; Sodhi, N. S. The Sixth Mass Coextinction: Are Most Endangered Species Parasites and Mutualists? *Proc. R. Soc. B.* **2009,** *276,* 3037–3045. https://doi.org/10.1098/rspb.2009.0413

Eaton, J. A.; Shepherd, C. R.; Rheindt, F. E.; Harris, J. B. C.; Van Balen, S.; Wilcove, D. S.; Collar, N. J. Trade-driven Extinctions and Near-extinctions of Avian Taxa in Sundaic Indonesia. *Forktail* **2015,** *31,* 1–12.

Espinosa-Zaragoza, S.; López-Martínez, V.; D Hight, S.; Varone, L. Ecological Niche Modeling to Calculate Ideal Sites to Introduce a Natural Enemy: The Case of *Apanteles opuntiarum* (Hymenoptera: Braconidae) to Control *Cactoblastis cactorum* (Lepidoptera: Pyralidae) in North America. *Insects* **2020,** *11* (7), 454.

Finke, M. D.; Oonincx, D. Insects as Food for Insectivores. In *Mass Production of Beneficial Organisms*; Academic Press, 2014; pp 583–616.

Fischer, J.; Mueller, T.; Spatz, A. K.; Greggers, U.; Gruenewald, B.; Menzel, R. Neonicotinoids Interfere with Specific Components of Navigation in honeybees. *PLoS One* **2014,** *9* (3), e91364.

Forrest, J. R. Complex Responses of Insect Phenology to Climate Change. *Curr. Opin. Insect Sci.* **2016,** *17,* 49–54. https://doi.org/10.1016/j.cois.2016.07.002

Goulson, D. The Insect Apocalypse, and Why It Matters. *Curr. Biol.* **2019,** *29* (19), 67–71.

Goulson, D.; Nicholls, E.; Botías, C.; Rotheray, E. L. Bee Declines Driven by Combined Stress from Parasites, Pesticides, and Lack of Flowers. *Science* **2015,** *347* (6229).

Grimaldi, D.; Engel M. S. *Evolution of the Insects*; Cambridge University Press: New York, 2005; p 755.

Gutierrez, A.; Ansuategi, A.; Susperregi, L.; Tubío, C.; Rankić, I.; Lenža, L. A Benchmarking of Learning Strategies for Pest Detection and Identification on Tomato Plants for Autonomous Scouting Robots Using Internal Databases. *J. Sens.***2019,** *2019,* 5219471.

Habel, J. C.; Samways, M. J.; Schmitt, T. Mitigating the Precipitous Decline of Terrestrial European Insects: Requirements for a New Strategy. *Biodivers. Conserv.* **2019a,** *28,* 1343–1360. https://doi.org/10.1007/s10531-019-01741-8

Habel, J. C.; Trusch, R.; Schmitt, T.; Ochse, M.; Ulrich, W. Long-term Large-scale Decline in Relative Abundances of Butterfly and Burnet Moth Species Across Southwestern Germany. *Sci. Rep.* **2019b,** *9,* 14921. https://doi.org/10.1038/s41598-019-51424-1

Haines-Young, R. H.; Potschin, M. The Links Between Biodiversity, Ecosystem Services and Human Well-being. In *Ecosystem Ecology: A New Synthesis, BES Ecol. Rev. Series*; Raffaelli, D., Frid, C., Eds.; Cambridge University Press: Cambridge, 2010; pp 110–139.

Hall, D. M.; Martins, D. J. Human Dimensions of Insect Pollinator Conservation. *Curr. Opin. Insect Sci.* **2020,** *38,* 107–114.

Hallmann, C. A.; Zeegers, T.; van Klink, R.; Vermeulen, R.; van Wielink, P.; Spijkers, H. E. JONGEJANS: Declining Abundance of Beetles, Moths and Caddisflies in the Netherlands. In *Insect Conservation and Diversity*; 2019.

Hartley, S. E.; Jones, T. H. Insect Herbivores, Nutrient cycling and Plant Productivity. In *Insects and Ecosystem Function;* Springer: Berlin, Heidelberg, 2008; pp 27–52.

Havel, J. E.; Kovalenko, K. E.; Thomaz, S. M.; Amalfitano, S.; Kats, L. B. Aquatic Invasive Species: Challenges for the Future. *Hydrobiologia* **2015,** *750,* 147–170. https://doi.org/10.1007/s10750-014-2166-0

Hayes, T. B.; Hansen, M. From Silent Spring to Silent Night: Agrochemicals and the Anthropocene. *Elem. Sci.* Anth. **2017,** *5,* 57. https://doi.org/10.1525/elementa.246

Henneman, M. L.; Memmott, J. Infiltration of a Hawaiian Community by Introduced Biological Control Agents. *Science* **2001,** *293,* 1314–1316.

Henriques, N. R.; Cintra, F.; Pereira, C. C.; Cornelissen, T. Indirect Effects of Ecosystem Engineering by Insects in a Tropical Liana. *Arthropod Plant Interact.* **2019,** *13* (3), 499–504.

Hewitt, G. The Genetic Legacy of the Quaternary Ice Ages. *Nature* **2004,** *405,* 907–913.

Hight, S. D.; Carpenter, J. E.; Bloem, S.; Pemberton, R. W.; Stiling, P. Expanding Geographical Range of *Cactoblastis cactorum* (Lepidoptera: Pyralidae) in North America. *Fla. Entomol.* **2002,** *85,* 527–529.

Holder, P. J.; Jones, A.; Tyler, C. R; Cresswell, J. E. Fipronil Pesticide as a Suspect in Historical Mass Mortalities of Honey Bees. *Proc. Natl. Acad. Sci.* **2018,** *115* (51), 13033–13038.

Huang, J. P. Modeling the Effects of Anthropogenic Exploitation and Climate Change on an Endemic Stag Beetle, *Lucanus miwai* (Lucanidae), of Taiwan. *J. Asia Pac. Entomol.* **2014,** *17,* 423–429. https://doi.org/10.1016/j.aspen.2014.03.009

Hunter, M. D. Insect Population Dynamics Meets Ecosystem Ecology: Effects of Herbivory on Soil Nutrient Dynamics. *Agric. For. Entomol.* **2001,** *3* (2), 77–84.

Jactel, H.; Koricheva, J.; Castagneyrol, B. Responses of Forest Insect Pests to Climate Change: Not so Simple. *Curr. Opin. Insect Sci.* **2019,** *35,* 103–108.

Jiang, J. A.; Tseng, C. L.; Lu, F. M.; Yang, E. C.; Wu, Z. S.; Chen, C. P.; Lin, S. H.; Lin, K. C.; Liao, C. S. A GSM-based Remote Wireless Automatic Monitoring System for Field Information: A Case Study for Ecological Monitoring of the Oriental Fruit Fly, *Bactrocera dorsalis* (Hendel). *Comput. Electron. Agric.* **2008,** *62,* 243–259.

Jones, C. G.; Gutiérrez, J. L.; Byers, J. E.; Crooks, J. A.; Lambrinos, J. G.; Talley, T. S. A Framework for Understanding Physical Ecosystem Engineering by Organisms. *Oikos* **2010,** *119,* 1862–1869.

Jones, D. T.; Susilo, F. X.; Bignell, D. E.; Hardiwinoto, S.; Gillison, A. N.; Eggleton, P. Termite Assemblage Collapse Along a Land-use Intensification Gradient in Lowland Central Sumatra, Indonesia. *J. Appl. Ecol.* **2003,** *40,* 380–391.

Jones, J. I.; Murphy, J. F.; Collins, A. L.; Sear, D. A.; Naden, P. S.; Armitage, P. D. The Impact of Fine Sediment on Macro-invertebrates. *River Res. Appl.* **2012,** *28,* 1055–1071.

Jongema, Y. *List of Edible Insect Species of the World.* Laboratory of Entomology, Wageningen University: The Netherlands [Online], 2017. http://www.ent.wur.nl/UK/Edible+insects/ Worldwide+species+list/

Jouquet, P.; Harit, A.; Cheik, S.; Traoré, S.; Bottinelli, N. Termites: Soil Engineers for Ecological Engineering. *C. R. Biol.* **2019,** *342* (7–8), 258–259.

Joyce, T. W.; Durban, J. W.; Claridge, D. E.; Dunn, C. A.; Fearnbach, H.; Parsons, K. M.; Andrews, R. D.; Ballance, L. T. Physiological, Morphological, and Ecological Tradeoffs Influence Vertical Habitat Use of Deep-diving Toothed-whales in the Bahamas. *PLoS One* **2017,** *12* (10), e0185113.

Kaya, Y.; Kayci, L. Application of Artificial Neural Network for Automatic Detection of Butterfly Species Using Color and Texture Features. *Vis. Comput.* **2014,** *30,* 71–79.

Kaya, Y.; Kayci, L.; Uyar, M. Automatic Identification of Butterfly Species based on Local Binary Patterns and Artificial Neural Network. *Appl. Soft Comput.* **2015,** *28,* 132–137.

Kehoe, R.; Frago, E.; Sanders, D. Cascading Extinctions as a Hidden Driver of Insect Decline. *Ecol. Entomol.* **2020.**

Kenis, M.; Hurley, B. P.; Hajek, A. E. Cock, M. J. Classical Biological Control of Insect Pests of Trees: Facts and Figures. *Biol. Invasions*. **2017,** *19* (11), 3401–3417.

Kessler, S. C.; Tiedeken, E. J.; Simcock, K. L.; Derveau, S.; Mitchell, J; et al. Bees Prefer Foods Containing Neonicotinoid Pesticides. *Nature* **2015,** *521,* 74–76.

Kim, J, Huebner, C. D.; Reardon, R.; Park, Y. L. Spatially Targeted Biological Control of Mile-a-Minute Weed Using *Rhinoncomimus latipes* (Coleoptera: Curculionidae) and an Unmanned Aircraft System. *J. Econ. Entomol.* **2021.**

Klein, A. M.; Vaissiere, B. E.; Cane, J. H.; Steffan-Dewenter, I.; Cunningham, S. A.; Kremen, C.; Tscharntke, T. D. Importance of Pollinators in Changing Landscapes for World Crops. *Proc. Biol. Sci.* **2007,** *274,* 303–313. https://doi.org/10.1098/rspb.2006.3721

Koh, L. P.; Sodhi, N. S.; Brook, B. W. Ecological Correlates of Extinction Proneness in Tropical Butterflies. *Conserv. Biol.* **2004,** *18* (6), 1571–1578.

Kozlov, M. V.; Zvereva, E. L. Background Insect Herbivory: Impacts, Patterns and Methodology. *Prog. Bot.* **2017,** *79,* 313–355.

Krosby, M.; Wilsey C. B.; McGuire, J. L.; Duggan, J. M.; Nogeire, T. M.; Heinrichs, J. A.; Tewksbury, J. J.; Lawler, J. J. Climate-induced Range Overlap Among Closely Related Species. *Nat. Clim. Chang.* **2015,** *5,* 883–886. https://doi.org/10.1038/nclimate2699

Labandeira, C. C.; Johnson, K. R.; Wilf, P. Impact of the Terminal Cretaceous Event on Plant-insect Associations. *Proc. Natl. Acad. Sci. USA.* **2002,** *99,* 2061–2066.

Labandeira, C. C.; Sepkoski, J. J., Jr. Insect Diversity and the Fossil Record. *Science 261,* 310–15.

Laigle, I.; Moretti, M.; Rousseau, L.; Gravel, D.; Venier, L.; Handa, I. T.; Messier, C.; Morris, D.; Hazlett, P.; Fleming, R.; Webster, K. Direct and Indirect Effects of Forest Anthropogenic Disturbance on Above and Below Ground Communities and Litter Decomposition. *Ecosystems* **2021,** 1–22.

Lamarre, G. P.; Fayle, T. M.; Segar, S. T.; Laird-Hopkins, B. C.; Nakamura, A.; Souto-Vilarós, D.; Watanabe, S.; Basset, Y. Monitoring Tropical Insects in the 21st Century. *Adv. Ecol. Res.* **2020,** *62,* 295–330.

Leather, S. R.; Basset, Y.; Hawkins, B. A. Insect Conservation: Finding the Way Forward. *Insect Conserv. Diversity.* **2008.**

Leather, S. R.; Helden, A. J. Roundabouts-Our Neglected Nature Reserves? Roundabouts are Common, but Often Unnoticed Features of our Towns and Cities, but are These Islands of Biodiversity Actually Our Future Nature Reserves? *Biologist (London)* **2005,** *52* (2), 102–106.

Lewinsohn, T. M.; Freitas, A. V. L.; Prado, P. I. Conservation of Terrestrial Invertebrates and their Habitats in Brazil. *Conserv. Biol.* **2005,** *19,* 640–645.

Li, W.; Chen, P.; Wang, B.; Xie, C. Automatic Localization and Count of Agricultural Crop Pests Based on an Improved Deep Learning Pipeline. *Sci. Rep.* **2019,** *9,* 1–11.

Lister, B. C.; Garcia, A. Climate-driven Declines in Arthropod Abundance Restructure a Rainforest Food Web. *Proc. Natl. Acad. Sci.* **2018**. https://doi.org/10.1073/pnas.1722477115 (in press).

Lockwood, J. A. *Locust: the Devastating Rise and Mysterious Disappearance of the Insect that Shaped the American frontier*; Basic Books, 2009.

Losey, J. E.; Vaughan, M. The Economic Value of Ecological Services Provided by Insects. *Bioscience* **2006,** *56,* 331–323.

Maharlooei, M.; Sivarajan, S.; Bajwa, S. G.; Harmon, J. P.; Nowatzki, J. Detection of Soybean Aphids in a Greenhouse Using an Image Processing Technique. *Comput. Electron. Agric.* **2017,** *132,* 63–70.

Mammola, S.; Cardoso, P.; Culver, D. C.; Deharveng, L.; Ferreira, R. L.; Fišer, C.; Galassi, D. M.; Griebler, C.; Halse, S.; Humphreys, W. F.; Isaia, M. Scientists' Warning on the Conservation of Subterranean Ecosystems. *Bio. Sci.* **2019,** *69* (8), 641–650.

Mankin, R. W.; Smith, M. T.; Tropp, J. M.; Atkinson, E. B.; Jong, D. Y. Detection of *Anoplophora glabripennis* (Coleoptera: Cerambycidae) Larvae in Different Host Trees and Tissues by Automated Analyses of Sound-impulse Frequency and Temporal Patterns. *J. Econ. Entomol.* **2008,** *101,* 838–849.

Martin, B.; Shaby, S. M.; Premi, M. G. Studies on Acoustic Activity of Redpalm Weevil the Deadly Pest on Coconut Crops. *Procedia Mater. Sci.* **2015**, *10,* 455–466.

May, R. M. Tomorrow's Taxonomy: Collecting New Species in the Field Will Remain the Rate Limiting Step. *Philos. Trans. R Soc. Lond. B Biol. Sci.* **2004,** *359,* 733–734.

Millennium Ecosystem Assessment. *Ecosystems and Human Well-being; A Framework for Assessment*; Island Press: Washington, D.C., US, 2003; p 266.

Nechols, J. R. The Potential Impact of Climate Change on Non-target Risks from Imported Generalist Natural Enemies and on Biological Control. *BioControl.* **2020,** *25,* 1–8.

Neher, D. A.; Barbercheck, M. E. Soil microarthropods and soil health: Intersection of decomposition and pest suppression in agroecosystems. *Insects*. **2019,** *10* (12), 414.

Noriega, J. A.; Hortal, J.; Azcárate, F. M.; Berg, M. P.; Bonada, N.; Briones, M. J.; Del Toro, I.; Goulson, D.; Ibanez, S.; Landis, D. A.; Moretti, M. Research Trends in Ecosystem Services Provided by Insects. *Basic Appl. Ecol.* **2018,** *26,* 8–23.

Ollerton, J.; Winfree, R.; Tarrant, S. How Many Flowering Plants are Pollinated by Animals? *Oikos* **2011,** *120* (3), 321–326. https://doi.org/10.1111/j.1600-0706.2010.18644.x

Omer, A. M. Biodiversity, the Green Ecology, Economics and Ecosystem Engineering. *Agric. Adv.* **2016,** *5* (2), 227–250.

Pearson, D. E.; Callaway, R. M. Indirect Effects of Host-specific Biological Control Agents. *Trends Eco. Evol.* **2003**, *18* (9), 456–461.

Peñuelas, J.; Sardans, J.; Estiarte, M.; Ogaya, R.; Carnicer, J.; Coll, M.; Barbeta, A.; Rivas-Ubach, A.; Llusià, J.; Garbulsky, M.; Filella, I. Evidence of Current Impact of Climate Change on Life: A Walk from Genes to the Biosphere. *Glob. Change Biol.* **2013**, *19,* 2303–2338. https://doi.org/10.1111/gcb.12143

Pinhas, J.; Soroker, V.; Hetzroni, A.; Mizrach, A.; Teicher, M.; Goldberger, J. Automatic Acoustic Detection of the Red Palm Weevil. *Comput. Electron. Agric.* **2008,** *63,* 131–139.

Potamitis, I.; Rigakis, I.; Tatlas, N. A. Automated Surveillance of Fruit Flies. *Sensors* **2017,** *17,* 110.

Powney, G. D.; Carvell, C.; Edwards, M.; Morris, R. K. A.; Roy, H. E.; Woodcock, B. A.; Isaac, N. J. B. Widespread Losses of Pollinating Insects in Britain. *Nat. Commun.* **2019,** *10,* 1018. https://doi.org/10.1038/s41467-019-08974-9

Prather, C. M.; Pelini, S. L.; Laws, A.; Rivest, E.; Woltz, M.; Bloch, C. P.; Del Toro, I.; Ho, C. K.; Kominoski, J.; Newbold, T. A. S.; Parsons, S.; Joern, A. Invertebrates, Ecosystem Services and Climate Change. *Biol. Rev.* **2013,** *88,* 327–348.

Rader, R.; Bartomeus, I.; Garibaldi, L. A.; Garratt, M. P.; Howlett, B. G.; Winfree, R.; Cunningham, S. A.; Mayfield, M. M.; Arthur, A. D.; Andersson, G. K.; Bommarco, R. Non-bee Insects are Important Contributors to Global Crop Pollination. *Proc. Natl. Acad. Sci. USA.* **2016,** *113* (1), 146–151.

Ramos-Elorduy, J. Energy Supplied by Edible Insects from Mexico and Their Nutritional and Ecological Importance. *Ecol. Food Nutr.* **2008,** *47,* 280–297.

Ratcliffe, N. A.; Mello, C. B.; Garcia, E. S.; Butt, T. M.; Azambuja, P. Insect Natural Products and Processes: New Treatments for Human Disease. *Insect Biochem. Mol. Biol.* **2011,** *41,* 747–769. https://doi.org/10.1016/j.ibmb.2011.05.007

Romero, G. Q,; Gonçalves-Souza, T.; Vieira, C.; Koricheva, J. Ecosystem Engineering Effects on Species Diversity Across Ecosystems: A Meta-analysis. *Biol. Rev.* **2015,** *90,* 877–890.

Roy, D. B.; Sparks, T. H. Phenology of British Butterflies and Climate Change. *Glob. Change Biol.* **2000,** *6,* 407–416. https://doi.org/10.1046/j.1365-2486.2000.00322.x

Roy, J.; Rineau, F.; De Boeck, H. J.; Nijs, I.; Pütz, T.; Abiven, S.; Arnone, J. A.; Barton, C. V.; Beenaerts, N.; Brüggemann, N.; Dainese, M. Ecotrons: Powerful and Versatile Ecosystem Analysers for Ecology, Agronomy and Environmental Science. *Glob. Chang. Biol.* **2021,** *27* (7), 1387–407.

Samways, M. J. Insect Conservation for the Twenty-First Century. In *Insect Science-Diversity, Conservation and Nutrition*; IntechOpen, 2018.

Samways, M. J. Insect Conservation: A Synthetic Management Approach. *Annu. Rev. Entomol.* **2007,** *52,* 465–487.

Samways, M. J. Insect Extinctions and Insect Survival. *Conserv. Biol.* **2006,** *20,* 245–246.

Samways, M. J.; Taylor, S. Impacts of Invasive Alien Plants on Red-listed South African dragonflies (Odonata). *S. Afr. J. Sci.* **2004,** *100,* 78–80.

Samways, M. J.; Taylor, S.; Tarboton, W. Extinction Reprieve Following Alien Removal. *Conserv. Biol.* **2005,** *19,* 1329–1330.

Sánchez-Bayo, F.; Wyckhuys, K. A. Worldwide Decline of the Entomofauna: A Review of Its Drivers. *Biol. Conserv.* **2019,** *232,* 8–27.

Schowalter, T. D. *Insect Ecology: An Ecosystem Approach*, 4th ed.; Academic Press, 2016; p 774.

Schowalter, T. D. Insects and Sustainability of Ecosystem Services. In *Social Environmental Sustainability Series*; CRC Press/Taylor & Francis Group: Boca Raton, FL, USA, 2013; p 362.

Schowalter, T. D.; Noriega, J. A.; Tscharntke, T. Insect Effects on Ecosystem Services—Introduction. *Basic Appl. Ecol.* **2018,** *26,* 1–7. https://doi.org/10.1016/j.baae.2017.09

Seibold, S.; Gossner, M. M.; Simons, N. K.; Blüthgen, N.; Müller, J.; Ambarlı, D.; Ammer, C.; Bauhus, J.; Fischer, M.; Habel, J. C.; Linsenmair, K. E.; Nauss, T.; Penone, C.; Prati, D.; Schall, P.; Schulze, E. D.; Vogt, J.; Wöllauer, S.; Weisser, W. W. Arthropod Decline in Grasslands and Forests is Associated with Landscape-level Drivers. *Nature* **2019,** *574* (7780), 671–674. https://doi.org/10.1038/s41586-019-1684-3

Shattuck, S. O. Australian Ants of the Genus Aphaenogaster (Hymenoptera: Formicidae). *Zootaxa* **2008,** *1677* (1), 25–45.

Shumi, G.; Schultner, J.; Dorresteijn, I.; Rodrigues, P.; Hanspach, J.; Hylander, K.; Senbeta, F.; Fischer, J. Land Use Legacy Effects on Woody Vegetation in Agricultural Landscapes of South-western Ethiopia. *Divers. Distrib.* **2018,** *24* (8), 1136–1148.

Shylesha, A. N.; Jalali, S. K.; Gupta, A.; Varshney, R.; Venkatesan, T.; Shetty, P.; Ojha, R.; Ganiger, P. C.; Navik, O.; Subaharan, K.; Bakthavatsalam, N. Studies on New Invasive Pest *Spodoptera frugiperda* (JE Smith) (Lepidoptera: Noctuidae) and Its Natural Enemies. *J. Biol. Control.* **2018,** *32* (3), 145–151.

Smith, M. A.; Rodriguez, J. J.; Whitfield, J. B.; Deans, A. R.; Janzen, D. H.; Hallwachs, W.; Hebert, P. D. Extreme Diversity of Tropical Parasitoid Wasps Exposed by Iterative Integration of Natural History, DNA Barcoding, Morphology, and Collections. *Proc. Natl. Acad. Sci. USA.* **2008,** *105* (34), 12359–12364.

Smith, M. A.; Wood, D. M.; Janzen, D. H.; Hallwachs, W.; Hebert, P. D. DNA Barcodes Affirm that 16 Species of Apparently Generalist Tropical Parasitoid Flies (Diptera, Tachinidae) are Not All Generalists. *Proc. Natl. Acad. Sci. USA.* **2007,** *104* (12), 4967–4972.

Smithsonian. *Bug Info: Insects as Pets. Department of Systematic Biology, Entomology Section, National Museum of Natural History, in Cooperation with Public Inquiry Services.* Smithsonian Institution [Online], 2019. https://www.si.edu/spotlight/ buginfo/pets

Stanley, D. A.; Msweli, S. M.; Johnson, S. D. Native Honeybees as Flower Visitors and Pollinators in Wild Plant Communities in a Biodiversity Hotspot. *Ecosphere* **2020,** *11* (2), e02957.

Sun, Y.; Cheng, H.; Cheng, Q.; Zhou, H.; Li, M.; Fan, Y.; Shan, G.; Damerow, L.; Lammers, P. S.; Jones, S. B. A Smart-vision Algorithm for Counting Whiteflies and Thrips on Sticky Traps Using a Two-dimensional Fourier Transform Spectrum. *Biosyst. Eng.* **2017,** *153,* 82–88.

Tallamy, D. W.; Shriver, W. G. Are Declines in Insects and Insectivorous Birds Related? *The Condor* **2021,** *23* (1), duaa059.

Thomas, C. D. Dispersal and Extinction in Fragmented Landscapes. *Proc. Biol. Sci.* **2000,** *267,* 139–145.

Thomas, J. A.; Telfer, M. G.; Roy, D. B.; Preston, C. D.; Greenwood, J. J. D. Comparative Losses of British Butterflies, Birds and Plants and the Global Extinction Crisis. *Science* **2004,** *303,* 1879–1881.

Turbelin, A. J.; Malamud, B. D.; Francis, R. A. Mapping the Global State of Invasive Alien Species: Patterns of invasion and Policy Responses. *Glob. Ecol. Biogeogr.* **2017,** *26,* 78–92. https://doi.org/10.1111/geb.12517

Turner, W. R.; Brandon, K.; Brooks, T. M.; Costanza, R.; da Fonseca, G. A. B.; Portela, R. Global Conservation of Biodiversity and Ecosystem Services. *Bio. Sci.* **2007,** *57,* 868–873.

van Lexmond, M. B.; Bonmatin, J.-M.; Goulson, D.; Noome, D. A. Worldwide Integrated Assessment Onsystemic Pesticides. *Environ. Sci. Pollut. Res.* **2015,** *22,* 1–4.

Wagner, D. L.; Van Driesche, R. G. Threats Posed to Rare or Endangered Insects by Invasions of Non-native Species. *Ann. Rev. Entomol.* **2010,** *55,* 547–568. https://doi.org/10.1146/annurev-ento-112408–085516

Weisser, W. W.; Siemann, E.; Ed. *Insects and Ecosystem Function*; Springer-Verlag: Berlin, Germany, 2004; p 415.

Wood, P. J.; Greenwood, M. T.; Agnew, M. D. Pond Biodiversity and Habitat Loss in the UK. *Area* **2003,** *35* (2), 206–216.

Woodward, G.; Perkins, D. M.; Brown, L. E. Climate Change and Freshwater Ecosystems-Impacts Across Multiple Levels of Organization. *Philos. Trans. R. Soc. Lond. B Biol. Sci.* **2010,** *365,* 2093–2106. https://doi.org/10.1098/rstb.2010.0055Yang, L. H. Periodical Cicadas as Resource Pulses in North American Forests. *Science* **2004,** *306,* 1565–1567.

Zhao, Y.; Wang, Y.; Wang, J.; Hu, Z.; Lin, F.; Xu, M. In *GMM and DRLSE Based Detection and Segmentation of Pests: A Case Study, Proceedings of the 2019 4th International Conference on Multimedia Systems and Signal Processing*, Guangzhou, China, 10–12 May 2019; pp 62–66.

CHAPTER 2

The Importance of Taxonomy in Biological Research

INDU SHARMA, PREETI CHOUDHARY, and RIFAT HUSSAIN RAINA

Desert Regional Centre, Zoological Survey of India, Jodhpur, Rajasthan, India

ABSTRACT

The initial aim in the field of biological research is to identify an area of species origin, pests and their control are Taxonomy. The presence of indigenous fauna and the growth of biological control programs in the targeted area is working out with the help of taxonomy all through pre- and post-release surveys. Proper guidance and instructions to researchers should be highlighted for proper collection, preservation, and identification of exotic species. This chapter highlights the importance of taxonomy in education, research, biodiversity conservation, agriculture, pest management, natural enemies, fisheries, parasitology, veterinary science, medicine, conservation of flora, etc. Therefore, taxonomy is an important tool to know the biodiversity and subsequently to its conservation of human existence on this earth.

2.1 INTRODUCTION

Botanist A.P. de Candolle (1813) for floral classification first proposed the term Taxonomy. Classification of existing and vanished organisms is known as Taxonomy. Term taxonomy is derived from the Greek words, "taxis," and "nomos," which means "arrangement" and "law"

Insect Diversity and Ecosystem Services, Volume 1: Importance, Threats, Conservation and Economic Perspectives. Younis Ahmad Hajam, Sajad Hussain Parey, and Rouf Ahmad Bhat (Eds)

respectively. Therefore, the principles of systematics and its methodology in the hierarchy of superior and subordinate groups of flora and fauna are taxonomy.

Simpson (1961) defines taxonomy as the theoretic study of classification including its principles, methods, and procedure, while Mayr (1969) describes it as a principle theory to classify flora and fauna. Thus, classification is termed taxonomy. According to Christoffersen (1995), taxonomy is the practice of identifying, naming, and ordering taxa with its relationship with the environment.

In general, the classification of living things develops due to their necessity and is frequently superficial. For example, Saxon termed the worm as a creeping thing like snake or earthworm and fish as any aquatic object like crayfish, shellfish, but they have anatomical differences.

The process of taxonomy involves two steps:

- Accurate definition, recognition of organisms and their relationships;
- Use of a suitable title for organisms and different groups to include.

The earlier classification includes the study of characters and grouping of individuals while the latter is termed nomenclature.

2.2 BACKGROUND

An ecologist is very familiar with the various components of the native flora and fauna, their importance in the environment, and knows higher vertebrate groups like Pisces, Aves, and Mammalia.

The catalog of 365 species of medicinal plants is now used in hydrological studies in China. In the previous era, generally, these medicinal plants are used to treat illness and injuries.

According to Aristotle, western classification is in the field of logical science of about 2000 years in regular contact with aquatic life. Aristotle explained a large number of ecological groups and placed them from simple to complex, but it was not in an evolutionary order. Due to the lack of a microscope, he did not describe miniatures but differentiated invertebrate groups.

Classification following the Aristotelian method was followed up in the 19th century. According to which several specimens have shallow resemblance with adaptable and persistent characters. It develops the core of living things that is irreversible. In the 12th century, a similar arrangement contains true plates of medicinal plants to create standard and modern observations. The first flowering of the Renaissance and human anatomy of Vesalius treatises were introduced in 1543 and 1545 respectively to produce flourishing flora and fauna. While in the 17th century, systematics with effective classification was introduced by John Ray. The study of monocot and dicot plants examined the species concept. Biological classification with its basic unit was explained by true affinities of whales and modified experimental observation according to Aristotle.

2.2.1 LEVELS OF TAXONOMY

Three levels of taxonomy are:

2.2.1.1 ALPHA TAXONOMY: Naming and characterization of species.

2.2.1.2 BETA TAXONOMY: An arrangement of species of a particular group or order of organisms.

2.2.1.3 GAMMA TAXONOMY: Evolutionary sequencing, a variation, and the analysis of organic diversity.

The taxonomy has two levels according to Mayr and Ashlock (1991):

2.2.1.4 MICRO TAXONOMY: It deals only with the problems, related to species.

2.2.1.5 MACRO TAXONOMY: It deals with the problems and principles of higher taxa (from subgenus and above) only.

Furthermore, macro-taxonomy includes:

i. Phonetics or numerical taxonomy
ii. Phylogenetic systematics or Cladistics and
iii. Evolutionary systematics.

2.2.1.6 NUMERICAL TAXONOMY: Overall characters based on their similarities and dissimilarities characterize organisms.

2.2.1.6.1 Phylogenetic Systematics: Organisms are ranked and classified according to a phylogenetic tree. Taxa based on synapomorphic characters originated from a common ancestor. It is a process to identify unique phyletic connections or clades of phylogenetic analysis.

2.2.1.6.2 Evolutionary Systematics: On the basis of evolution, a valid taxon is described but the characteristic feature does not depend on evolution. It is a branch of biological classification in which the organisms are classified using a combination of phylogenetic relationships and overall similarity. Thus, it considers taxa more significant rather than solitary species.

Usually, taxonomy is of two types:

2.2.1.7 CLASSICAL TAXONOMY: Naming, classifying flora and fauna based on morphological features description including genitalia, anatomy, embryology, etc.

2.2.1.8 EXPERIMENTAL TAXONOMY OR NEO-TAXONOMY: Genetic studies take place to describe taxon based on the gene pool and differentiate two taxa. It includes modern steps to collect functional data morphologically.

2.2.1.9 BIOCHEMICAL TAXONOMY: It deals with taxonomic characters obtained from chemical analysis of enzymes, hormones, and proteins with peptides, nucleic acids, amino acids, and sugars. Amino acid sequences of proteins are diverse in different organisms and are helpful to differentiate the diverse species. Numerous techniques are applied to the study of constituent bio-molecules such as enzymes, hormones, nucleic acids, amino acids, etc., and help in the systematics. To study chemical substances, various procedures such as immunological, chromatography, electrophoretic methods are used in systematics. Chromatography is a different technique for the separation of a complex liquid mixture such as biological fluids (e.g., amino acids, steroids, carbohydrates, etc). This moves in separate columns for absorption of different components and mixtures.

2.2.1.10 CYTOTAXONOMY: It deals with taxonomic characters obtained from cytological studies based on the structure, the number of chromosomes. The position of the centromere is an important feature of chromosome structure that helps in taxonomic studies.

2.3 SIGNIFICANCE

Taxonomy is the basic tool used to identify and group organisms based on their similar morphological characters. A significant barrier and effecting resolutions of the convention at the national level are due to the lack of taxonomists identified by CBD. The annual rate of extinction of the species of flora and fauna is estimated to be 1000–10000 (Wilson, 2003). So, biodiversity is precious and should be protected with the world's biodiversity difference of 5-100 million and in the middle era 14 million (Cherian, 2004). Out of which 2 million species are identified. Biodiversity conservation will result in species richness. Its innovation and affiliation with the basic unit of classification are used to identify (Narendran, 2006, 2008).

The utility of imported flora and fauna is determined by quarantine agencies and they restrict the entry of harmful organisms. It includes field-oriented work with the need for observation and analysis to determine taxon up to species level.

Molecular taxonomy deals with DNA barcoding, isozymes, molecularity, genetic, etc. Presently DNA barcoding is used for taxonomic identification. A genetic marker in mtDNA is used to identify species.

Variability in the group due to a long period of independence is a serious barrier for molecular taxonomy (Narendran, 2006). DNA barcoding provides reliable information up to species level. DNA barcoding is used in molecular taxonomy while classical uses, holistic approaches, including comparison with other species in the environment (Grimaldi and Engel, 2007).

In molecular techniques, DNA sequencing helps to establish a holistic approach and its relationship with barcodes (Katz, 2005) while genomic barcoding is used to differentiate sibling species. According to Grimaldi and Engel, 2007 molecular description is important to start innovation while Ogura (1964) supported classical taxonomy to be continued.

2.3.1 ROLE OF TAXONOMY IN INVESTIGATION AND EDUCATION

To study and investigate any species' correct scientific name helps in gathering quantities and quality information for easy reference (Narendran, 2000).

2.3.2 ROLE OF TAXONOMY IN BIODIVERSITY CONSERVATION

Less than 10% of flora and fauna present in the ecosystem are identified (Winston, 2000). It will take many years to identify the whole as a result, many species may get extinct before their identification. Therefore, biodiversity conservation is necessary from a taxonomic point of view.

2.3.3 ROLE OF TAXONOMY IN AGRICULTURE AND PEST MANAGEMENT

For the management of crop pests by biological means, taxonomy plays a significant role. True identification of pests and their enemies has great importance to control pests at the time of their transport from one region to another. Workers must know:

The correct identification of pests and their probability of spread

- Natural enemies of the imported pest and inventory on them to substitute their hosts.
- Identification key with host-parasite lists.

- Relevant hidden necessary information under obsolescent species names.
- Difference between natural enemies, pests, and their effect to control biodiversity.

Schauff and LaSalle (1998) described several errors in biological control, in which workers utilize lots of currency and energy in breeding, transport, curation of species for hyperactive parasites that do not harm targeted hosts. Thus, true identification of target pests is more important before instigating any biological control program.

2.3.4 ROLE OF TAXONOMY IN FISHERIES, PARASITOLOGY, AND VETERINARY SCIENCE

Taxonomic information helps to locate food availability for aquatic organisms with their richness. It is due to allied taxonomic publications and explores hidden data. In veterinary science as well as in medical sciences taxonomy plays a dynamic role in the identification of man's parasites.

2.3.5 ROLE OF TAXONOMY IN THE CONSERVATION OF FLORA AND FAUNA

Quantity destruction of habitats causes the damage and extinction of a large number of species. The rate of loss of species each year is about 27000, 74 per day and 3 per hour (Gadagkar, 1998). Documentation of faunal diversity should be prepared before its decay (Gupta, 1987). Getting familiar with the ecosystem results from less destruction of it (Narendran and Cherian, 2004).

2.3.6 ROLE OF TAXONOMY IN MEDICINE

For effective treatment of transmitted disease, it is important to identify the true organism. Many floral species are used in Ayurvedic treatment. In the past natural methods are used to identify the cause of disease, and identification of flora and their variety to utilize them as traditional medicine.

Major steps to remove taxonomic barriers are:

1. Taxonomy must be involved as an essential topic in the course and prospectus in a higher level of education.
2. Preparing monographs and scientific literature should be funded.
3. Providing enough work with great opportunities to sign the biodiversity document.

The development of taxonomy in all aspects should be in priority.

"Taxonomy is no less attractive, challengingly difficult, satisfying and productive than most sophisticated, spectacularly dramatic biological experiments currently in fashion here is an unknown and a new world, literally at our doorstep, for discovery, exploration and conquest" (Mani, 1989).

It is summarized that the main limitation in the field of the taxonomy of manpower, adequate funding, training in taxonomy, library facilities, and sufficient taxonomic centers for identification as well as proper training of taxonomy. Elimination of these barriers is very important for the invention and conservation of our biodiversity.

ACKNOWLEDGMENT

We are grateful to Dr. Dhriti Banerjee, Director, Zoological Survey of India, Kolkata for providing the necessary facilities.

KEYWORDS

- **taxonomy**
- **biological**
- **research**
- **importance**
- **species**

REFERENCES

Candolle, A. P. *Theorieelementaire de la Botanique*; Chez Deterville: Paris, 1813; p 500 + 27.

Cherian, P. T. Getting the Measure of India's Insect Diversity: 41–54. In *Perspectives on Biosystematics and Biodiversity*; Rajmohana, K., Sudheer, K., Girish, P., Santosh, S, Eds.; TCN Com. vol. March 2004, pp 1–666.

Christoffersen, M. L. Cladistic Taxonomy, Phylogenetic Systematics and Evolutionary Ranking. *Syst. Biol.* **1995,** *44* (3), 440–454.

Mayr, E.; Ashlock, P. D. *Principles of Systematic Zoology*, 2nd ed.; McGraw-Hill: New York, 1991; pp xx + 475.

Gadagkar, R. Biology in the 21 51 Century—Back to Stamp Collection? *Scampus* **1998,** *98,* 5.

Grimaldi, D. A.; Engel, M. S. Why Descriptive Science Still Matters. *BioScience* **2007,** *57* (8): 645–646.

Gupta, V. K. Development of Taxonomic Resources in the Tropics. *Insect Sci. Applic.* **1987,** *8,* 407–412.

Katz, R. The Problem with DNA Barcoding: The Whole is More Than Its Parts. *Flower Essence Society* [Online], 2005. http://flowersociety.org/DNA-barcoding.htm.

Mani, M. S.; From the Editor; Hexapoda 1 (1&2): Narendran, T. C. 2000. The Importance of Systematics. *Resonance* **1989,** *5* (6), 60–68.

Mayr, E. *Principles of Systematic Zoology*; McGraw-Hill: New York, 1969.

Narendran, T. C. *Parasitic Hymenoptera and Biological Control*; Palani Paramount Publications: Palani, 2001; pp 1–201.

Narendran, T. C. Importance of Systematics. *Resonance*, **2000,** *5,* 60–68.

Narendran, T. C. Taxonomic Entomology: Research and Education in India. *Curr. Sci.* **2001,** *8* (5), 445–447.

Narendran, T. C. Role of Taxonomy in Biological Control and Insect Pest Management in India. In *Biological Control of Insect Pests*; Ignacimuthu, Jayaraj, S., Eds.; 2003, pp 250–225.

Narendran, T, C. *An Introduction to Taxonomy*; Director Zoological Survey of India, Ed.; Zoological Survey of India publ: Kolkatta, 2006; pp 1–80.

Narendran, T. C.; Importance of Taxonomy in Conservation of Biodiversity In *Wild Life Biodiversity Conservation*; Reddy, M. V, Ed.; Daya Publishing House: Delhi 1-390, 2008; pp 14–19.

Narendran, T. C.; Cherian, P. T. On Some Misconceptions on Conservation of Insects. *Zoos' Print* **2002,** *17* (1), 687–688.

Ogura, Y.; Comparative Morphology and Classification of Plants. *Phytomorph* **1964,** *14,* 240–247.

Schauff, M. E.; La Salle, J. In *The Relevance of Systematics to biocontrol: Protecting the Investment in Research; Pest Management-Future Challenges*, Proceedings of the Sixth Austrasian Applied Entomological Research Conference; Zalucki, M. P., Drew, R. W. I., White, G. G., Eds.; Brisbane, 1998, pp 425–436.

Simpson, G. G. *Principles of Animal Taxonomy*; Columbia University Press: New York. Sneath, P. H. A., Sokal, R. R. 1973, *Numerical Taxonomy—the Principles and Practice of Numerical Classification*, W. H. Freeman and Co.: San Francisco, Calif, 1961.

Wilson, E. O. *What is Nature Worth For?* Part 1: Span ZLIV-4: 54-57 Part 2: XLIV: 2003, pp 23–29.

Winston, J. E. *Describing Species: Practical Taxonomic Procedure for Biologists*; Columbia University Press: Columbia, 2000.

CHAPTER 3

Aquatic Insect Ecosystems: Diversity, Role, and Conservation

RAMAN JASROTIA[1], SHIVAM JASROTIA[2], MENAKSHI DHAR[1], PREETIKA GUPTA[1], NEHA JAMWAL[1], and SEEMA LANGER[1]

[1]*Department of Zoology, University of Jammu, Jammu, India*

[2]*Department of Zoology, Guru Nanak Dev University, Amritsar, India*

ABSTRACT

Freshwaters are tremendously diversified and embrace pools, ponds, lakes, streams, wetlands, springs, rivers, bogs, swamps, and reservoirs, whereas the marine habitats encompass estuaries, oceans, coral reefs, intertidal zones, and seabeds. Aquatic insects inhabit both the lotic and lentic water bodies and thus act as the primary biotic element of an aquatic ecosystem. Across the globe, nearly 45,000 insect species are known to inhabit freshwater ecosystems. Of all the known species of insects, approximately 3–5% lives in intimate association with aquatic habitats. Aquatic insects play a significant role in nutrient cycling and serve as an important food item for the vertebrate and invertebrate organisms dwelling in that ecosystem. The diversity of these insects depends upon their ecological role, the physical conditions of habitat, and the availability of food. Several insects inhabiting the aquatic ecosystem are known to scale down or banish the noxious, invasive aquatic weeds. The distribution of the aquatic fauna is reflective of the water quality and the overall health of the environment. Rapid development in cities, industrial areas, and the increasing levels of sewage and industrial effluents set out the threat to aquatic insect diversity. Any change in the aquatic ecosystem upshots a transition in relative

Insect Diversity and Ecosystem Services, Volume 1: Importance, Threats, Conservation and Economic Perspectives. Younis Ahmad Hajam, Sajad Hussain Parey, and Rouf Ahmad Bhat (Eds)

abundance and diverseness of the insects in that ecosystem. Humans and insects are entwined with each other, so as to avert the decline in the insect population; effective solutions of conservation are to be explored.

3.1 INTRODUCTION

Insects account for the highly diversified group in the Kingdom-Animalia. There are approximately 7,51,000 species of insects all over the world, found in almost every type of habitat. Some of these insect species inhabit the water bodies and are regarded as model organisms as they aid in scrutinizing the role and structure of an aquatic ecosystem. These ecosystems hold up a considerable part of the earth's biodiversity. Insects are generally found in the littoral zone of a water body; very few are reported from the profundal and limnetic zones (McCafferty, 1981). The aquatic insects are found in both fast-flowing and stagnant water bodies and according to Daly (1998), with the increase in nutrients in stagnant water bodies, the diversity of insects increases. Cheng (2003) reported that several insect species are found in marine habitats, such as biting lice and sucking lice that are present on the seabirds.

Insects living in water bodies are the key components of any ecosystem as they play a significant role in the environment. They help in the recycling of nutrients and they are the prime source of food for the fishes, thus playing an important role in the transfer of energy between various trophic levels (Uniyal, 1988). Across the globe, about 45,000 insect species are known to live in freshwater bodies (Balaram, 2005). According to Cheng (1976), approximately 3% of the insects, found all over the world, are known to spend some part of their life stages in water. Insects living in waterbodies are crucial because of various reasons. Various artificial baits resembling the aquatic insects are used in the seafood industry, as they help in catching the fish in water systems. These insects play a significant role in food webs by predating upon various invertebrates, small and large fishes, unwanted plants, algal matter, and dead organic substances. Besides acting as a predator, these insects are preyed upon by aves, reptiles, amphibians, and some fishes. The number and the type of aquatic insects habiting the water body indicate the status of that aquatic ecosystem. Various dipterans such as midges feed on the algal matter. Both midges and moth flies remain associated with the sewage treatment plants and this way these insects prevent the clogging of drains. According to McCafferty (1981) aquatic insects that are herbivorous feed on unwanted aquatic plants. Insects belonging to the order-Odonata and some diving beetles are known to control the population of mosquitoes in aquatic systems.

Almost all important insect orders are represented in a wet environment. Only a few species spend their lives uninterruptedly in water. Some live out of the water only as pupae. But most pass through all their developmental stages in water and take to the air as adults. Various insect orders that are known to occupy the aquatic ecosystems exclusively in their larval stages include Odonata, Plecoptera, Trichoptera, and Ephemeroptera. Some other insect orders such as Hemiptera, Coleoptera, and Diptera spend some part of their lives in aquatic habitats.

3.2 MAJOR INSECT ORDERS ASSOCIATED WITH AQUATIC ECOSYSTEMS

3.2.1 ORDER DIPTERA—TRUE FLIES

The insects of this order show holometabolous development and are known to inhabit all types of aquatic ecosystems. 90% of the species of this order inhabit freshwater ecosystems and only a few species are present in the saline water bodies. Larvae of the midges are used to evaluate the water quality, and few species are used in toxicological studies (Solanki and Shukla, 2017).

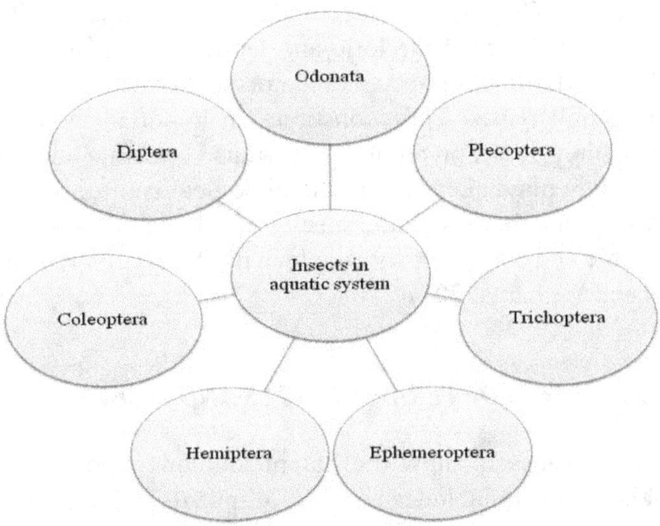

FIGURE 3.1 Principal insect orders related to aquatic environment.

An increase in the abundance of *Chironomus sps.* in any aquatic habitat reveals the decreased concentration of the oxygen levels in any water body. Larval forms of Order-Diptera inhabit aquatic and semi-aquatic habitats. Adult and larval stages of *Hemerodromia* feed on mosquitoes and buffalo gnats. Aquatic Diptera constitutes the important source of food for various vertebrate and invertebrate organisms. Cibrowski et al. (2003) reported that the larva of *Chironomus sps* is a prime source of food for fishes and water birds. Some species of *Ephydra* are found in highly saline lakes and the larvae of moth flies are found in sewage treatment plants. Flies associated with the dead bodies of animals in aquatic ecosystems play an important role in forensic entomology (Adler and Courtney, 2019). They are responsible for the transmission of various communicable diseases as they act as vectors for disease-causing organisms. Herbivores dipterans are used for the biological control of aquatic weeds.

3.2.2 ORDER HEMIPTERA—WATER BUGS

The order Hemiptera comprises semi-aquatic bugs that live on the surface of the water and the aquatic bugs that remain inundated in water. Across the globe, there are about 3800 species of water bugs. They show incomplete metamorphosis and include three stages viz., egg, nymph, and adult in their life cycle. They possess sucking type of mouthparts. These insects are known to inhabit both the lotic and lentic ecosystems. They form a part of fish food and also act as predators for other species of aquatic insects and small fishes. They constitute an important part of the food web because they act as predators for various vertebrate and invertebrate organisms. They play a key role in the biological control of the larvae of mosquitoes in water bodies. Some species of water striders and backswimmers are edible and consumed by people in different countries of the world (Williams and Williams, 2017).

3.2.3 ORDER COLEOPTERA—BEETLES AND WEEVILS

The coleopteran insects show the holometabolous type of development and so their life cycle includes egg, larval, pupal, and adult stages. They inhabit both the terrestrial and freshwater ecosystems. Mouthparts of these insects are well developed and thus show a wide variety of feeding habits.

Diving beetle or true water beetles, approximately 2.5 cm in length, are highly predatory and commonly found in water filled with aquatic weeds. The whirligig beetles form large groups and swim on the water surface. These aquatic insects play a key role in the functioning of an ecosystem. They form a source of food for the fishes that feed near the bottom of a water body. They are responsible for the decomposition of organic matter and the recycling of nutrients in an aquatic ecosystem. Water beetles act as biological indicators and so are useful in monitoring the health of an ecosystem (Sharma et al., 2013).

3.2.4 ORDER ODONATA—DRAGONFLIES AND DAMSELFLIES

The order-Odonata is small and includes only about 6300 species across the globe. They are closely related to the order-Ephemeroptera. Insects of this order show incomplete metamorphosis and the pupal stage is absent. Larval forms of Odonates are aquatic and are found inhabit ponds, marshes, banks of rivers, streams, and shallow regions of water bodies. They possess the chewing type of mouthparts and are highly predaceous. They are known to control the populations of mosquitoes and other pests in the environment. The larval forms show the phenomenon of cannibalism and thus help in controlling the growth of their species. Various studies done by Jacob et al. (2017) have depicted that dragonflies and damselflies can be used to control the population of insects that have aquatic larval forms. The larvae of a species of dragonfly, *Hemianaxephippiger* are known to control the snail population in an ecosystem as observed by Younes et al. (2016). Odonates are known to feed on the vectors of various diseases and crop pests. They act as indicators of the quality of aquatic habitat. The gut content analysis had revealed the consumption of blood worms by the insects of this order (Tennessen, 2009).

3.2.5 ORDER EPHEMEROPTERA—MAYFLIES

The insects belonging to this order show hemimetabolous development and include only 3300 species. They are found all across the world inhabiting both the fresh and marine water ecosystems and most commonly these are found in tropical and warm temperate regions. Nymphal forms of mayfly are found under the stones whereas some of them actively swim in stagnant and fast-flowing waters. The growth rate of nymphs of mayflies is controlled by the temperature of the aquatic ecosystem. They act as an important food item for the fish. Grant (2001) reported that various predators such as leeches, caddisflies, and crayfishes feed on the nymphs of mayflies due to which the population of mayflies decreases. According to Novikova and Kluge (1997), spillage of the oil and petroleum in water bodies poses a significant threat to the survival of these insects. Chemical substances used to kill fishes in water bodies lead to the death of mayflies also (Arnekleiv et al., 2001). Alteration in temperature leads to the disruption in the life cycle of these flies. Any change in water temperature lessens the egg-laying capacity of mayflies. They render various indispensable services to the biota such as decaying of dead plant and animal matter, changing the texture of sediments, cycling of nutrients in a freshwater ecosystem, or between terrestrial and aquatic ecosystem and thus act as the indicators of the health of an aquatic ecosystem.

3.2.6 ORDER PLECOPTERA—STONEFLIES

The insects belonging to this order have originated about 300 million years ago and show incomplete metamorphosis. They are generally found under the stones in fast-moving well-oxygenated water streams. They are known to inhabit lotic water systems. They feed on the algae and lichens. Mouthparts are generally reduced but the antennae are long. Adult forms of these insects prefer to inhabit the banks of water bodies and are not able to tolerate pollution, so they can be used as indicators of the aquatic ecosystems. They act as an important source of food for various fish species. These stoneflies play a key role as primary or secondary consumers and as prey in a food web in the aquatic ecosystem (Figueroa and López-Rodríguez, 2019).

3.2.7 ORDER TRICHOPTERA—CADDISFLIES

The caddisflies make up the order Trichoptera. They undergo complete metamorphosis and the wings in such insects develop inside the body. During the process of metamorphosis, their larvae and pupae are exclusively aquatic forms. According to Lehmkuhl (1979) these flies inhabit all types of aquatic ecosystems and are unable to tolerate water pollution.

Adult forms of these insects have membranous hairy wings. The mouth-parts and antennae of these insects are reduced. Larval and adult forms of caddisflies act as the source of food for many vertebrate organisms. In the food web, they play an important role by consuming organic matter. Few larval forms pierce and suck the living aquatic plants. *Hydropsyche-contubernali, inhabiting the rivers and streams, are capable of removing suspended organic matter from the eutrophic river systems.* Larvae of the caddisfly possess labial silk glands that are involved in the production of silk which sticks to underwater objects (Ashton et al., 2012). *Some species of this order have been reported as the pests of rice plants and watercress* (Wiggins et al., 2004).

3.3 ECONOMIC IMPORTANCE OF AQUATIC INSECTS

The importance which aquatic insects enjoy in fisheries, limnology, and aquaculture is well known. They serve as food for fishes, frogs, birds, and men. Some insects have proven to be useful in the field of medicine, while some are used in the biological control of aquatic plants and pests. Various insects are used to attract and catch the fish, some are used as pollution indicators of the water body. On the other hand, some aquatic insects are harmful to man as they not only cause painful bites on occasions but also are sometimes responsible for killing a large number of small fishes, their fries, and fingerlings.

3.3.1 USE OF INSECTS AS FOOD FOR FISH

As investigated by Wilson (1923), insects constitute an important source of food for the small- and large-sized fishes, amphibians, reptiles, aves, and small mammals and therefore play a major role in the energy flow of a community. Insects also play fundamental roles in aquatic communities as do in terrestrial ones, but it is in a variety of freshwater ecosystems that insects stand out as an important source of food. Insect grazers and filter feeders such as mayfly nymphs, caddis fly larvae, stonefly larvae, and various midge larvae serve as the main converters of plant material into animal protoplasm. Along with a variety of scavenging and predacious species, they comprise a major part of the staple diet of fishes, particularly in shallow lakes, ponds, and streams. Mayflies are said to be the single most important dietary component of trout, which, like many fishes, feed not only on the aquatic larvae of insects but also on the adult as they emerge. Among all aquatic insects, Ephemeroptera, Coleoptera, Odonata, Hemiptera, and Diptera constitute the major component of the fish species. Various fishes such as *Catlacatla* (adult), *Labeokontius* (adult), *Barbuscarnaticus* (fingerling and adult), *Anabustestudincus* (fingerling and adult), *Channa striatus* (fry and fingerling), *Channamarulius* (fry and fingerling), *Heteropneustesfossilis* (fingerling and adult) have been reported by Jhingran (1975) from the Indian plains that feed on aquatic insects in one or more stages of their life cycle.

3.3.2 USE OF INSECTS AS FOOD FOR OTHER ANIMALS

Besides serving as food for many fishes, many aquatic insects can be used to supplement the diet of many animals including man also. For example, eggs of two North American species of Hemiptera, *Arctocorixamerconana* and *Arctocorixaabdominalis*, serve as food for the natives, and their adults are dried and exported in tones for the consumption of the poultry and fishes. In certain South American and Asian countries, the giant water bugs are eaten by the natives besides manufacturing medicines out of them. Even their eggs are collected from the water bodies and are used as food. Ozdal et al. (2012) reported the consumption of various kinds of beetles by humans in different countries across the world. These insect species are prepared in different ways such as some sps are roasted, some are consumed as salad, some are only boiled, and few of them are grinded and mixed with eggs in different countries all over the world. They are highly nutritious and freely available to people living in rural areas.

3.3.3 USE TO CONTROL PESTS AND WEEDS

The order-Odonata and Hemiptera are known to play a vital role to limit the growth of insects that prove harmful for the ecosystem and are known to feed on the larvae of mosquitoes, tiny flies, and nymphal stages of the mayfly. Yuval and Bouskila (1993) reported that as the sun begins to set, western malaria mosquito undergoes swarming which is then attacked by the aquatic insects viz., spot winged glider and the western pondhawk. Insect species of the family-Nepidae, Notonectidae, and Belostomidae are regarded as the predaceous bugs. Insects that swim upside down, i.e., backswimmer insects largely prey upon the larval forms of mosquitoes and thus aid in the control of mosquitoes. According to Murdoch et al. (1984), an aquatic bug, *Notonectasps.* and *Ranatrasps.* feed on the mosquitoes and thus act as a restraint for their growth. Chandra et al. (2008) observed that aquatic insects of the order-Coleoptera are interlinked with the larvae of mosquitoes in a variety of habitats. Most commonly, insects of the family-Hydrophilidae and Dytiscidae (order-Coleoptera) act as predators for the larvae of mosquitoes. According to the studies done by Aditya et al. (2006), water beetles are the most potent agents for mosquito control. These insects have a great utility in the control of the insect population by natural methods. Instead of using herbicides, water beetles can be utilized for restricting the growth of unwanted aquatic plants such as water hyacinth, alligator weeds, and kariba weeds. For example, a leaf beetle (*Agasicleshygrophila*) is used to limit the growth of alligator weed. Various studies have demonstrated that the milfoil weevil (*Eurychiopsislecontei)* is an important biological control agent of the watermilfoil (*Myriophyllum*spp. Haloragaceae) in the U.S. and Canada.

3.3.4 MUSE FOR THE CONTROL OF POLLUTION

This aspect is dependent on the detritus feeding of the insects due to their suspension-feeding habit. Some of them have filtering devices like nets and mucus and thereby passively collect the particles which are swayed by running water. The hemipterans and other insects inhabiting the bottom of streams, lakes, and ponds feed on the detritus and other organic matter, debris, and humus and thus constitute an important pathway of energy flow and are thus very beneficial for the productivity of the aquatic environment. By ingesting a large mass of detritus and other organic matter, bottom-dwelling insects decrease the rapidity with which the freshwater system turned shallow.

3.3.5 USE OF INSECTS FOR BIOLOGICAL ENVIRONMENTAL MONITORING

Freshwater environments are more fragile than most terrestrial ones. When man disturbs a terrestrial ecosystem, the impact of the disturbance is frequently rather visible. But we often disturb aquatic ecosystems more subtly and may not notice the damage done until it is too late. A wide range of chemical wastes have been entering freshwater systems for many years, but still, we do not have full information on the effects that varying concentrations of thousand of different compounds have on even test specimens. It is fully within our capability to monitor changes in water chemistry, but since these changes occur as a result of irrigation discharges of waste products, the data may not mean much in terms of estimating long-term impacts. However, any good aquatic entomologist knows which insect requires what conditions. Certain insects are found in clear, well-oxygenated water, whereas others do best when the organic content is high and dissolved oxygen is low. As a result, aquatic biologists are studying change in the insect fauna that occurs as bodies of water become polluted. Insects that are adapted to average conditions rather than to momentary fluctuations serve as useful biological indicators of the condition that prevail over a large period.

Insects inhabiting the water conjoin to form groups that differ from one geographical region to the other region depending upon the ecological functions. The communities of insects living in lakes or rivers are repressed by the physicochemical parameters of that ecosystem. Oxygen plays a very important role and any alteration in the availability of oxygen results in the formation of different types of insect communities. For example, different insect communities are known to inhabit well-oxygenated areas of water bodies. In the oxygen-demanding region of the water bodies, the communities show less number of species. Identical variations are seen in the insect communities due to alterations in other physicochemical parameters such as temperature, substrate, and the deposition of heavy metals in the water bodies. Insects are regarded as advantageous for the biological monitoring of an aquatic ecosystem (Kour, 2013). Aquatic insects depict various responses when their communities are disrupted which include:

i. When the level of dissolved oxygen decreases, there is an increase in the abundance of blood worms (*Chironomus*).
ii. Rise in the temperature of water leads to a decrease in the nymphs of stoneflies.
iii. Run-off of the pesticides from the agricultural fields into the water bodies causes a significant reduction in diversity and abundance of aquatic insects.

iv. Eutrophication of the water bodies decreases the insect diversity of any water body.
v. Due to an increase in the pollutant levels in an aquatic system, caddisflies and mayflies become more abundant.

Aquatic beetles act as the indicators of the quality of the water and the other habitats that are under the threat of extinction. These beetles are efficacious for various conservation assessments (Moog and Jach, 1995).

3.3.6 USE OF INSECTS TO FEED ON DECAYING MATTER

When plants and animals die they become parts of the ecosystem as they accumulate in the form of organic matter called detritus. These contain a large quantity of energy in the form of chemical bonds. Many aquatic insects feed on dead plants and animals and thereby contribute to its decomposition and the recycling of its nutrient constituents.

3.3.7 USE OF INSECTS AS PREDATORS FOR OTHER ORGANISMS

The insects living in the water bodies constitute only about 4% of the entire number of living insectspecies present across the globe. Myriad numbers of aquatic insects inhabit the ponds that are used for the rearing of the eggs of carps, particularly in the rainy season. But they are known to destruct the eggs of carps in these ponds as described by Ganguly and Mitra (1961). Various larval and adult forms of the insects either feed directly on the spawn and fry of carp or some insect species may face off spawns and fry to meet their food requirement. The eleven different orders of the class-Insecta are known to inhabit aquatic ecosystems. However, the members of order-Hemiptera, Coleoptera, and Odonata are relatively more dominant as far as this aspect of fisheries damage is concerned. Spawn of the carps in nursey ponds suffers a lot due to the predatory habit of backswimmers, giantwater bugs, and water scorpions that belong to the families Notonectidae, Belostomatidae, and Nepidae, respectively. Species of the family-Notonectidae are known to kill the spawn of carps, small fishes, amphibians, and other vertebrate species. Gorai and Chaudhuri (1962) studied the feeding behavior of the notonectid bug in the ponds and found that this bug feed on the phytoplankton, zooplankton, *Chironomus*, nymphs of various insect species, and the larval forms of fish. These bugs can kill the fry of a carp which is about 10–13 mm in length (Julka, 1965). In order to kill the prey, these backswimmers use their sharp beak to pierce the body of the fry and suck out the fluids completely which may result in the death of the host. Freshwater hemipteran insects are extremely predatory in habit and so they prey upon the fishes and frogs by the release of poisonous substances in their saliva. Aquatic Heteropteran insects belonging to the family Nepidae prey and feed upon the fry of the carp. Some species of this family feed on the fry and spawn of the carp. *Anisops* sp. (14–15 mm) and *Ranatra*sp (45–48 mm) consume 162–182 carp eggs within 24 hours as reported by Tripathi and Sharaf (1974).

3.4 ROLE OF AQUATIC INSECTS IN AQUACULTURE

Insects living in the water habitats are known to influence the health of any aquatic ecosystems and they form an important source of food in fisheries and aquaculture. Insects carry out various functions in an ecosystem, i.e., some recycle nutrients by feeding on detritus, and few species act as a source of food for various invertebrates and vertebrates. Several insect species are voracious feeders and feed on fishes, amphibians, and other animals inhabiting the water bodies. Insects living in freshwater habitats are known to play a key role in aquatic ecosystems. Some of the aquatic hemipterans, such as *Anisopsbouvieri* and *Notonecta glauca*, are detrimental to fish culture (Julka, 1965; Daham, 1972). While certain groups namely Belostomatids, Notonectids, and Nepids may be used as bio-agents in the control of mosquito larvae (Hoffmann, 1927; Menke, 1979) and yet other groups of aquatic hemipterans such as Gerridae, Corixidae, Pleidae, and Nepidae nourish fishes. Insects constitute an important part of the fish diet because of this, they are of considerable importance for the aquaculturists for increasing the population of fishes in the pond.

Insects living in the aquatic ecosystem symbolize prime linkers in the food chain as they act both as the predator and prey. Many aquatic insects viz., water beetles (Coleoptera), larvae of chironomids, mosquitoes and Tabanus (Diptera), nymphs of dragonflies and damselflies (Odonata), and adults of water bugs *Laccotrephes* and *Ranatra*(Hemiptera) serve as a majorfood source for fish *Channa punctatus* (Reddy, 1980) and larvae of Ephemeroptera were found to be the most abundant food item of *Tor tor* (Dasgupta, 1990). Fish that move from sea to freshwater for spawning is entirely dependent for food on various insect species. Some estuarine fish also devour insects. However, it is the freshwater fish that make the maximum utilization of insects in their diets. The relationship between a population of insects and a population of fish may be direct or indirect. Insects living in the freshwater ecosystem are used as the source of food by the fishes in their early stages of development; as they get older their dietary requirements may change. For example, the large-mouth black bass eats insects until it is about 4 inches in length. After that, small fish are sought as a food source (Uniyal, 1988). Also, the feeding habit of *Channa punctatus* revealed a distinct variation in food intake at different life stages. In the fry stages, the most dominant food item was zooplankton followed by insects, annelids, crustaceans, and plant matter. Similarly, in the juvenile stage, the insect was the most preferred food item and in adults, fish was found to be the most dominant item. Again, the main food items of *Channa punctatus* were insect eggs (Saikia et al., 2012).

Wide varieties of aquatic insects are known to be present in abundance in the various aquatic ecosystems. Trout subsist on these; however, they also depend on terrestrial insects that fall into the water. Young salmon was found to feed mainly on mayfly nymphs, caddis larvae, and chironomids as reported by Frost (1950). The stomach contents of such stream-inhabiting fish usually contain stoneflies, bugs, mayflies, caddisflies, and nymphal forms of dragonflies. The stomach contents of forage fish (like minnows, dace, roaches, chubs, blackfish, squawfish, and shiners) include insect matter as the primary component. Insects constitute the basic food of the fingerlings and sexually immature forms of Garhwal mahseer, as studied by Nautiyal and Lal (1984). A few other non-game fish (such as suckers, sticklebacks, and sculpins) also feed upon insects to varying degrees. Bluegills, at times, have had 90% of their stomach contents composed of insect matter (Uniyal, 1988). Other spiny-rayed fish like the black crappie, long-eared sunfish, rockbass, and large-mouthed bass feed voraciously on insects. Other stream-dwelling fish such as sturgeon, suckers, and white-fish use insects as a dietary staple. These feeding habits differ from region to region; however, nearly any freshwater fish population would starve or be forced to adapt to other types of food if they were deprived of insects as a food source.

Some insects, especially members of Hemiptera, Coleoptera, and Odonata, have habits of viciously attacking and feeding upon fish. Various aquatic beetles and giant water bugs are well known for their fish feeding habits. Some carnivorous beetles in the families Dytiscidae, Hydrophilidae, and Gyrinidae are destructive by feeding on fish spawn. *Cybister confuses*, *C. rufipes* and *Eretesstictus* pose a special threat to fish spawn and fry. Larvae of *Sternolophusrufipes*, a scavenger beetle of the family Hydrophilidae, habitually feed on fish spawn. Members of the genus *Dinuetes* (Gyrinidae) are known to surround and subdue fish, even when there is an abundance of other more easily obtained food.

The true water bugs also rank high in their predation of fish. These insects attack their prey and use their piercing-sucking mouthparts to remove the body fluids. Fifth instar nymphs of the backswimmer *Anisops-bouveri* are known to attack the larval stages of fish. Giant water bugs belonging to the family-Belostomatidae are also well known for their ability to overpower young fish. 4-inch giant water bugs have been known to feed upon 12-inch trout. Various species of fish have been reported to die because of the feeding habits of giant water. The water scorpion *Laccotrephes griseus*, eats fish fry. Similarly, *Ranatraelongata* feeds on the spawn and fry of various species of carps. The water bugs secrete toxic saliva that kills the prey.

Dragonflies and damselflies belonging to the order-Odonata have also been reported to consume the various species of fish. Nymphal stages of large *Aeschna* sp. and *Anax* sp. are notorious. Some species of the fish are pounced and eliminated by the immature forms of the caddis fly. The net spinning larval stages of these predacious insects snare and entangle the fry of trout and salmon.

The inception of various fish species in the aquatic bodies where aquatic insects undergo breeding can obstruct the growth of these obnoxious insects. The growth of the larval and pupal stages of the mosquitoes has been restrained by this method. Gnats have been controlled by the introduction of fish into the infested waters. This procedure is an indispensable part of mosquito abatement programs in many tropical areas of the world. Two families of fish viz., Poeciliidae and Cyprinodontidae, are used in these programs and they are widely utilized for suppressing the growth of the larvae of mosquitoes. *Gambusia* rates foremost among these and this fish is used across the globe for diminishing the growth of larval forms of insect. *Gambusia affinis* can feed on various types of food such as small organisms drifting with water currents, insects inhabiting freshwater ecosystems, various species of fish, molluscs, and some other organisms that dwell on the land but are occasionally present on the surface of the water (Hayes and Rutledge, 1991). Most commonly, *Gambusia affinis* is used for the control of mosquitoes as a part of integrated pest management strategy (Linden and Cech, 1990) and the early instars of aquatic insects belonging to the family Corixidae and Notonectidae are preferably consumed by this fish (Mansfield and Mcardle, 1998). Some other fish of importance in this connection are *Lebistes*, *Molliensia*, *Poecilia,* and *Heterandria*. *Fundulus*, *Rivulus,* and *Cyprinodon*, members of the family Cyprinodontidae have also been used extensively for killing mosquito larvae. Other fish of comparatively less importance in this respect are *Oryzias*, *Aplocheilus*, *Aphanius*, *Horaichthys*, *Pygidium,* and *Carassius*. All these fish have short life cycles, small adult size, large fecundity, and surface feeding habits. Also, the ease in their distribution makes them readily useable. Some other fish species like *Danio*, *Rasbora,* and *Esomus* can also be utilized for mosquito control because of their larvivorous habits.

3.5 CONSERVATION OF AQUATIC INSECTS

Several steps are required to be taken for the conservation of aquatic insects, such as: Invasive species should not be introduced in the water bodies as they impart negative influence on the behavior of native species already inhabiting that ecosystem. Laws and regulations should be strictly implemented for the conservation of aquatic ecosystems. Anthropogenic activities lead to the pollution of water sources, so human activities should be minimized near the water bodies. Weedicides that are used to remove unwanted aquatic plants should be used with great care so that they do not have a negative effect on aquatic insects. The introduction of organic pollutants, directly into the water bodies, should be restricted as they lead to the eutrophication of the aquatic ecosystem and ultimately affect the aquatic insects. Insecticides and pesticides should be used in minimal amounts in the agricultural fields present near the water bodies as run-off pesticides and insecticides will harm the insects inhabiting that waterbody.

3.6 CONCLUSION

Considering immense services provided by the biodiversity to mankind, it becomes pertinent to conserve the dwindling biodiversity on the planet. A common approach with which diversity of lifeforms can be conserved is by the establishment of protected areas. It has been estimated that species loss in freshwater is distressfully high and a very little evaluation has been done to decipher the capacity of protected areas to conserve aquatic insects (Abellan et al., 2007). Use of species distribution models is very effective in fulfilling the conservation strategies as they provide significant approach to identify the under-tested areas. With their abundant number and diversity, aquatic insects dominate aquatic ecosystems. Their economic importance and role in ecology persuade conservation biologists to frame strategies for their preservation and conservation.

KEYWORDS

- **aquatic insects**
- **conservation**
- **ecological role**
- **diversity**
- **lentic**
- **lotic**
- **nutrient cycling**

REFERENCES

Abellan, P.; Sanchez-Fernandez, D.; Velasco, J.; Millan, A. Effectiveness of Protected Area Networks in Representing Freshwater Biodiversity: The Case of a Mediterranean River Basin (South-Eastern Spain). *Aquat. Conserv.: Mar. Freshw. Ecosyst.* **2007,** *17,* 361–374.

Aditya, G.; Ash, A.; Saha, G. K. Predatory Activity of *Rhanyussikkimensis* and Larvae of *Toxorhynchitessplendens*on Mosquito Larvae in Darjeeling, India. *J Vector Borne Dis.* **2006,** *43,* 66–72.

Adler, P. H.; Courtney, G. W. Ecological and Societal Services of Aquatic Diptera. *Insects* **2019,** *10,* 1–23.

Arnekleiv, J. V.; Dolmen, D.; Ronning, L. Effects of Rotenone Treatment on Mayfly Drift and Standing Stocks in Two Norwegian Rivers. In*Trends in Research in Ephemeroptera and Plecoptera*; Dominguez, E., Ed.; Kluwer Academic/Plenum Publishers: New York, 2001; pp 77–88.

Ashton, N. N.; Taggart, D. S.; Stewart, R. J. Silk Tape Nanostructure and Silk Gland Anatomy of Trichoptera. *Biopolymers* **2012,** *97,* 432–445.

Balaram, P. Insects of Tropical Streams. *Cur. Sci.* **2005,** *89,* 914–922.

Chandra, G.; Mandal, S. K.; Ghosh, A. K.; Das, D.; Banerjee, S. S.; Chakraborty, S. Biocontrol of Larval Mosquitoes by *Aciliussulcatus*(Coleoptera: Dytiscidae). *BMC Infect. Dis.* **2008,** *8,* 1–8.

Cheng, L. Marine Insects. In *Encyclopedia of Insects*; Resh, V. H., Cardé, R. T., Eds.; Academic: St. Louis, MO, **2003**; pp 679–682.

Cheng, L. *Marine Insects*; North-Holland Publishing Company: Amsterdam, 1976; p 581.

Cibrowski, J. J. H.; Corkum, L. D. *United Earth Fund. Appendix 9: Sediment-Zoobenthos Interactions*; 2003; pp 78–82.

Daham, E. On the Biology of *Notonecta glauca* (Insecta : Hemiptera) with Special Reference to Its Damage to Fisheries. *Int. Rev. Gesamten. Hydrobiol.* **1972,** *57,* 429–461.

Daly, H.; Doyen, J.; Purcell, A. *Introduction to Insect Biology and Diversity*; Oxford University Press: New York, 1998; p 680.

Dasgupta, M. A Study on the Food and Feeding Habits of the Mahseer *Tor tor* (Hamilton). *Indian J. Fish.* **1990**, *37,* 297–304.

Figueroa, J. M. T.; Lopez-Rodríguez, M. J. Trophic Ecology of Plecoptera (Insecta): A Review. *Eur. Zool. J.* **2019,** *86,* 79–102.

Frost, W. E. The Growth and Food of Young Salmon *Salmo Salar* and Trout *S. trutta* L. *J. Fish Priol.* **1950,** *15,* 43–53.

Ganguly, D. N.; Mitra, B. Observations on the Fish-Fry Destroying Capacity of Certain Aquatic Insects and Suggestions for their Eradication. *Indian Agric.* **1961,** *5,* 184–188.

Gorai, A. K.; Chaudhuri, D. N. Food and Feeding Habits of Anisopsbouvieri Kirk. (Heteroptera: Notonectidae). *J. Asiatic Soc.* **1962,** *4,* 135–139.

Grant, P. M. Mayflies as Food. In *Trends in Research in Ephemeroptera and Plecoptera*; Dominguez, E., Ed.; Kluwer Academic/Plenum Publishers: New York, 2001; pp 107–124.

Hayes, J. W.; Rutledge, M. J. Relationship Between Turbidity and Fish Diets in Lakes Waahi and Whangape, New Zealand. *N. Z. J. Mar. Freshwater Res.* **1991,** *25,* 297–301.

Hoffman, W. E. Biological Notes on *Laccotrephes. Ling. Agri. Rev.* **1927,** *4,* 77–94.

Jacob, S.; Thomas, A. P.; Manju, E. K. Bio Control Efficiency of Odonata Nymphs on *Aedes aegypti* Larvae. *IOSR J. Environ. Sci. Toxicol. Food Technol.* **2017,** *11,* 1–4.

Jhingran, V. G. *Fish and Fisheries in India*; Hindustan Publishing Corporation: (Indian), Delhi, 1975; p 954.

Julka, J. M. Observation on the Biology of Aquatic Bugs (Hemiptera) Injurious to the Pond Fishes, *Anisopsbouvieri* Kirk. *Proc. Indian Acad. Sci. Bangalore.* **1965**, *61,* 49–61.

Kour, R. Diversity and Bionomics of Aquatic Insects of Jammu. Ph.D. Thesis, University of Jammu, 2013.

Lehmkuhl, D. M. *How to Know the Aquatic Insects*. Wm. C. Brown Co.: Dubuque, IA, 1979; p 168.

Linden, A. L.; Cech, J. J. Prey Selection by Mosquitofish (*Gambusia affinis*) in California Rice Fields: Effect of Vegetation and Prey Species. *J. Am. Mosq. Control Assoc.* **1990,** *6,* 115–120.

Mansfield, S.; Mcardle, B. H. Dietary Composition of *Gambusia affinis* (Family Poeciliidae) Populations in the Northern Waikato Region of New Zealand. *N. Z. J. Mar. Freshwater Res.* **1998,** *32,* 375–383.

McCafferty, W. P. *Aquatic Entomology: The Fishermen's and Ecologists' Illustrated Guide to Insects and Their Relatives*. Science Books International: Boston, MA, 1981; p 448.

Menke, A. S. The Semi Aquatic and Aquatic Hemiptera of California (Heteroptera: Hemiptera). *Bull. Calif. Insect Survey* **1979,** *21,* 1–166.

Moog, O.; Jach, M. A. Elmidae. In *Fauna Aquatica Austriaca,* May 1995 ed.; Moog, O., Ed.; Wasserwirtschaftskataster, Bundesministeriumfür Land- und Forstwirtschaft: Wien, 1995.

Murdoch, W. W.; Scott, M. A.; Ebsworth, P. Effects of the General Predator, *Notonecta* (Hemiptera) Upon a Fresh Water Community. *J. Anim. Ecol.* **1984,** *53,* 791–808.

Nautiyal, P.; Lal, M. S. Food and Feeding Habits of Fingerlings and Juveniles of Mahseer (*Tor putitora* Ham.) in Nayar River. *Bombay Nat. Hist. Soc.* **1984,** *81,* 642–646.

Novikova, E. A.; Kluge, N. J. Mayflies (Ephemeroptera) of the Western Siberian lowland and oil pollution. In *Ephemeroptera & Plecoptera: Biology–Ecology–Systematics*; Landolt, P., Sartori, M., Eds.; Mauron+Tinguely & Lachat SA: Fribourg, 1997; pp 269–274.

Ozdal, M.; Incekara, U.; Polat, A.; Gur, O.; Kurbanoglu, E. B.; Tasar, G. E. Isolation of Filamentous Fungi Associated with Two Common Edible Aquatic Insects, *Hydrophiluspiceus* and *Dytiscusmarginalis*. *J. Microbiol. Biotechnol. Food Sci.* **2012,** *2,* 95–105.

Reddy, P. B. Food and Feeding Habits of *Channa punctata* (Bloch) from Guntur. *Indian J. Fish.* **1980,** *27,* 123–129.

Saikia, A. K.; Abujam, S. K.; Biswas, S. P. Food and Feeding Habit of *Channa punctatus* (Bloch) from the Paddy Fields of Sivsagar District, Assam. *Bull. Environ. Pharmacol. Life Sci.* **2012,** *1,* 10–15.

Sharma, S.; Pandey, P.; Dave, V.; Role of Aquatic Beetles for Water Quality Assessment. *Int. J. Recent Sci. Res.* **2013,** *4,* 1673–1676.

Solanki, R.; Shukla, A. Aquatic Insects for Biomonitoring Freshwater Ecosystems: A Report. *Int. J. Sci. Res.* **2017,** *6,* 2056–2058.

Tennessen, K. J.; Odonata: Dragonflies, Damselflies. In *Encyclopedia of Insects*, 2nd ed.; 2009, pp 721–729.

Tripathi, S. D.; Sharaf, R. K. Predatory Role of Weed Fishes vis-a-vis Aquatic Insects, Tadpole and Froglets. *Jawaharlal Nehru Krishi Vishwa Vidyalaya Res. J.* **1974,** *8,* 123–127.

Uniyal, V. P. *Insects and Their Relationship with Fish*, Vol 5; Y.E.S. Quarterly; Michigan State University: USA, 1988, pp 22–23.

Wiggins, G. B. *Caddisflies—The Underwater Architects*; University of Toronto Press; Toronto, ON, Canada: Buffalo, NY, USA: London, UK, 2004; p 292.

Williams, D. D.; Williams, S. S. Aquatic Insects and their Potential to Contribute to the Diet of the Globally Expanding Human Population. *Insects* **2017,** *8,* 1–20.

Wilson, C. B. *Life History of the Scavenger Water Beetle, Hydrous (Hydrophilus) triangularis, and its Economic Relation to Fish Breeding*; Bulletin of the Bureau of Fisheries: Fairport, IA, 1923; p 38.

Younes, A.; El-Sherief, H.; Gawish, F.; Mahmoud, M. Experimental Evaluation of Odonata Nymph in the Biocontrol of Schistosomiasis Intermediate Hosts. *Asian Pac. J. Trop. Biomed.* **2016,** *6,* 995–1000.

Yuval, B.; Bouskila, A. Temporal Dynamics of Mating and Predation in Mosquito Swarms. *Oecologia* **1993,** *95,* 65–69.

CHAPTER 4

Entomological Threats: Assessment, Approaches, and Challenges

NEHA JAMWAL, RAMAN JASROTIA, MENAKSHI DHAR, and SANJAY BHATIA

Department of Zoology, University of Jammu, Jammu, Jammu and Kashmir, India

ABSTRACT

Insects represent one of the principal biotic communities in an ecosystem. They have rendered themselves adaptive to almost all kinds of environments, that is, both aquatic as well as terrestrial. This ability has led them to be pests of various crops, vectors of many diseases, and as predators and parasitoids. The contribution of beneficial insects to mankind is also significant. Considering their economic and ecological importance, the protection of insect species is pivotal. There is scaling up concern over the decline in the insect population due to the conversion of their habitat to intensive agriculture. Agrochemical pollutants, non-native species, and climate change are the other causes of the population decline. Predation of the endemic insect species by the invasive introduced species of insects is a major challenge for the native population as it leads to niche competition. The rate of extinction among insects is several times higher than the mammals, birds, and reptiles. Approximately, more than 40% of the insect species have declined, which can even lead to the collapse of an ecosystem. The decline in insect diversity leads to the deprivation of key services to mankind. Efforts to save insect species are crucial, both for flora and fauna and for the survival of humans. Plantation of native tree species and restoration of the diminishing landscapes need to be done

Insect Diversity and Ecosystem Services, Volume 1: Importance, Threats, Conservation and Economic Perspectives. Younis Ahmad Hajam, Sajad Hussain Parey, and Rouf Ahmad Bhat (Eds)

to improve the insect population. The diversity of pollinator species can be increased by the revival of insect farms, conservatories of insect, and other natural areas. Extinction of species is a never-ending process and a decrease in the abundance of insects is the pioneer step in this process. Humans and insects are interconnected to each other, so it is the need of an hour to halt the further decline in the insect population. Conservation of insects will prove beneficial for the overall maintenance of the ecosystem. Peril to the diverseness of insects is rising swiftly and several species will not be able to get through the ongoing anthropogenic pressure, strict decisions are needed to be taken for insect conservation.

4.1 INTRODUCTION

Insects inhabit almost every habitat on the planet. Abundance of insects surpass all forms of visible life and contributes to about 75% of known and documented fauna on this planet. Nearly, one million insect species have been named worldwide, yet, the entomologists propose the total living insects to be 5.5 million (Stork, 2018), which indicates that as many as 80% of the world's insect species remains undiscovered and awaiting nomenclature (Gaston, 1994). Insects are known to be the first animals to evolve flight mechanism and have diversified and dispersed as plant eaters and parasites to mammals, almost across the globe (Condamine et al., 2016; Eggleton, 2020). The body of the insect has a general plan of body segmentation having head, thorax and abdomen. Head consists of pair of antennae, compound eyes, ocelli and feeding units (the mouthparts) depending upon the feeding ground. Thorax bear two pairs of wings (except Diptera and Strepsiptera having one pair of wings and few wingless insect species) and three pairs of legs. The abdomen comprises the last body segments in insects, having reproductive organs internally and spiracles and tracheoles for passive breathing externally (Gullan and Cranston, 2020). Besides having the general body plan, many insects have developed variations in their structures due to allied habitats and different feeding guilds, such as predators, parasites, herbivores, and decomposers,. This diversification has led to insects being split into 29 present orders, and further on the basis of developmental pattern, Insects have been divided into three basic groups, that is, Holometabolous (complete metamorphosis; eggs, larvae, pupae, and adults), Hemimetabolous (incomplete metamorphosis; nymphs appearing more likely like adults but without wings) and Ametabolous (no nymphs, no distinction between adults and larvae). Most of these insects are holometabolous; Diptera, Lepidoptera, Coleoptera, and Hymenoptera. Coleoptera contains the largest number of described species so far (Eggleton, 2020).

4.2 INSECTS AND HUMANS

Insects display many different lifestyles, based on their diversity. Some spend their early life inside the body of another insect, a few attached to rocks while many spend their entire life hidden in the front yard grass. It is very much obligate that a few of those one million species can and will probably eat our clothes, food (Bebber et al., 2014; Stejskal et al., 2015), damage household items, induce allergies and transmit various diseases (Lounibos, 2002; Grabowskwi and Klein, 2017). Many other insects are simply a nuisance because they are visible around our living space (Ali, 1980).

As long as we have lived in confined habitats, there has been an increase in organisms that is being housed close by us. A diverse group of uninvited and invited guests ranging from large vertebrates, like pets and livestock to microorganisms (Martin et al., 2015), shares our dwelling spaces. Insects and other arthropods represent the most abundant group of multicellular life encountered inside home.

Association of insects and other arthropods with humans has existed since time immemorial. Species that are now found among the immediate surroundings were originally proposed to be cave lodgers, for example, bed bugs (Balvin et al., 2012). Many evidences favor the fact suggesting that the insect pest species lived besides our ancestors (Araujo et al., 2009). Camel cricket, belonging to Raphidophoridae, is among the first cave art depiction example (Chopard, 1928; Belles, 1997). Arthropods were recorded from archeological sites of Egypt (dating back from 1353 B.C.E), Israel (Nielsen et al., 2000; Panagiotakopulu, 2001; Panagiotakopulu, 2003; Panagiotakopulu, 2004; Kislev et al., 2004; Kenward & Carrott, 2006), and Europe (The UK, Switzerland and Greenland) with a few species dependent upon livestock and stored food commodities and others representing local insect diversity.

Since the evolution of human society, the insects very rapidly and successfully made use of our resources, both for shelter as well as food. The ability of humans to store food, construct houses, practice agriculture and domesticate animals, accompanied different arthropod species into the daily lives and domiciles of humans.

Today's insect population is a reflection of modernization in the human society, for example, the practice of indoor plumbing has led to less prevalent dung beetles indoors but increased the population of drain moth flies (Psychodidae) inside our homes. Globalization has increased the appearance of household insect species, such as German cockroach, *Blatella germanica* L., Housefly, *Musca domestica* L. and vinegar fly/Fruit fly, *Drosophila melanogaster* Meigen (Legner and McCoy, 1966; Keller, 2007) and have acknowledged the worldwide distribution status.

Particular emphasis on medical pests and pests of commercial importance, such as termites, stored grains, cockroaches, fleas, mosquitoes and bed bugs, has been done for scrutinization on indoor insect pest (Robinson, 2005). Distinguished from the species that are threatening or bothering to humans, there is a limited knowledge about other insect species as most of them remains obtrusive (or, sometimes, unnoticed). Since, the records of insect prevalence and distribution is still ongoing, the true interaction among humans and insects is also relatively unknown.

The complete insect fauna and their significant sampling inside the homes and associated concerns will lead to knowing the role that is played by the indoor microhabitats in determining the ecological dynamics in a huge understudied system (Baz and Monserrat, 1999; Dunn et al., 2013). Despite negative effects of insects, few insects are also valued for various important services rendered to mankind like providing food, that is, edible insects (Lesnik, 2018), Pollination for example, Honey bees (Crane, 1990) and pest control (Van Mele, 2008). Soil insects such as dung beetles and termites help in building and maintaining soil fertility by soil bioturbation (Tuma et al., 2019), plant decomposition (Griffiths et al., 2019) and helping in nutrient recycling (Ulyshen, 2015).

4.3 INSECTS AND STORED GRAINS

Insect pests ravage upon many stored grains/pantry commodities fulfilling their food and shelter requirements, resulting in qualitative as well as quantitative loss. The tropical climate of India exhibits agreeable conditions for these pests throughout the year. A number of insect pests tend to make way inside grain storage at almost any stage of processing of food. Some insect pests start damaging at the mature stage of crops and continue the damage while storage as well. Major infestation is beheld upon old containers, bags, cross over infestation and faulty storage structures (Pruthi and Singh, 1950). Nearly, around 1,000 species of insects have been related with stored grain products in different regions worldwide. These may render the grains rancid and adulterate the rest with nasty flavors and odors. Orders Coleoptera and Lepidoptera share the majority of stored/ pantry insect pests (Khare, 1994).

Based on their feeding behavior, these insects can be sorted as external and internal feeders (Cotton, 1960).

4.3.1 EXTERNAL FEEDERS

These insects feed inside the germplasm and endosperm from outside attacking whole seed leading to damage in the germinal portion, feed on the seeds, which have already been damaged by other insects, or broken mechanically. These insect pests or their stages are generally visible among the seeds for example, Flour beetle, *Cryptolestes* sp., *Oryzaephilus*etc.

4.3.2 INTERNAL FEEDERS

Mostly lay eggs inside or on the surface of grains, spend a part or entire life inside the grains, and emerge as adults. This led to significant loss of germination, which remains undetected from outside for example, rice weevil, pulse beetle, granary weevil, and Angoumois grain moth.

These insects associated with stored grains are of much concern, as they deteriorate the stored commodities inside our homes particularly in the kitchen and storage area and are the most destructive to the stored household commodities and infest stored grains, processed food, spices, dry fruits, flour, pulses, soya products, etc. They cause significant damage to the grains wherein the highest intensity of damage is caused due to the larvae, although, adults also damage in some cases. The damage caused by these insects effect the overall grain production, and in India, the postharvest loss account for 12–16 million metric tons of food grains each year (Rees, 2004).

Worldwide loss due to stored grain insect pests has an estimate between 5 and 10% but surplus loss can reach upto 30% of production. The stored grain insects are accountable for a loss worth a minimum US$200 million per year in USA (Weaver and Petroff, 2004). About 36.92% farmers in Africa (Togola et al., 2013) have adjudged the fact that insects cause considerable damage to the stored rice. In India, due to a typical staple food diet, storage of cereal and pulses becomes extremely important in homes and in godowns. Infestation of grains from field to storage units is a huge matter of concern. Possibly, the reason could be their packing in infested gunny bags, godowns already infested by insects and improper storage (Atwal and Dhaliwal, 1999). Despite India being able to produce million tons of food grains every year, still hunger and poverty prevails because much share of these grains is damaged due to environmental factors accompanied by the infestation of stored grain insect pests and lack of the knowledge of scientific methods of storage adds another advantage in favor of these insects. Therefore, a collective loss of about 25% of food grain in the world is incurred upon by all these associated factors, highest in tropical and sub-tropical areas due to favorable growing conditions. Further, the annual loss of about 8 –10% is witnessed on the store commodities, worldwide (Shankar and Abrol, 2012). Two orders of insects are worldwide known to be economically important postharvest insect pests, that is, Coleoptera and Lepidoptera. A study on storage management in India, by Sharon et al. (2014), deciphered that India incurs about 12–16 million metric tons of postharvest loss which accounts for 10% loss in the country. Food Corporation of India in one of its recent reports (2018) recorded that these pests destroy grains worth Rs. 20,000 crores in India annually.

4.4 INSECTS AS VECTORS

Insects are considered medical pests if they bite, sting, or can act as vectors of diseases to humans. Today, the world is at a great risk from bacteria and viruses that are transmitted through the vectors, such as mosquitoes, ticks, fleas, and other vectors. A vector is any living organism that has the capability of transmitting infectious diseases from among humans or within animals and humans. According to a report by World Health Organisation (2014a), these diseases account for more than 17% of all infectious diseases and cause more than 1 million deaths annually, where mosquito borne diseases hold the major share. Malaria, alone, causes

600,000 deaths annually across the globe and about 2.5 billion people risk themselves from being affected by dengue over the world. These diseases act as a huge burden to the economy of the poverty-stricken countries or the developing countries where health, hygiene and sanitation is still a work under progress. About 70% of the countries affected by vector-borne diseases are low and lower middle-income countries.

Recently, India has witnessed worst cases of malaria and dengue. Malaria led to about 1.06 million deaths in 2006, while dengue was recorded for the first time from Calcutta in 1963, causing severe hemorrhagic fever (Kalra et al., 1997). The major challenge in India posed by vector-borne diseases is that the vectors have increased their geographic distribution and have developed potential of affecting 90% of the population (WHO, 2014b). This is the fact that recently the Zika virus outbreak in India has alarmed the very existence of human beings. Zika virus erupted in the world in 2015, it affected more than 1.5 million people from South America. In India, the first outbreak of Zika virus was recorded in January 2017 from Pune, Maharashtra, which led to a mass awareness program for local people against the disease as a preventive measure, followed by generating National Guidelines and Action plans to prevent the spread of this disease (WHO, 2017). On October 2018, Tribune Medical Express reported the biggest outbreak of Zika virus in India, which confirmed the cases of 80 people, including 22 pregnant women in Western Rajasthan. This is a huge concern for the authorities and the common people to combat the effects of vector-borne diseases, particularly mosquitoes. A bigger concern is that the mosquitoes have increased the infection rate largely that it becomes very tedious to control and monitor the disease cases. Sometimes, it becomes life-threatening to the authorities, which are associated with the prevention and cure of these diseases. Other diseases that are transmitted via mosquitoes include, Chikungunya, filariasis, West Nile fever (WNF), etc. Apart from mosquitoes, other household insects that are nuisance and vectors of various diseases to human beings and other house animals are cockroaches, flies, bugs, etc.

4.5 INSECTS AS NUISANCE PESTS

These insects are found crawling and moving in and around the human dwellings. They live in close association with the human beings, and therefore, are the main reasons of nuisance for the people. Wherever man has gone, these insects have followed him. Almost all of them are cosmopolitan in distribution and affects human life to a significant extent. There is no place in the house where these insects have not registered their presence, preferring warmer places to hide. These are troublesome merely due to their presence inside the house as they cause aesthetic damage to the human property.

4.6 INSECTS AS PREDATORS AND PARASITOIDS

Many insects, in due course of time, have become predators and parasitoids of other insects disorienting various stages of their life cycle. Predators and parasitoids act as natural enemies whose effect ranges from minor to permanent death of the host. First ever successful biological control attempt was undertaken when Vedalia beetle, *Rodolia cardinalis* was used to counter the serious threat posed by the cottony cushion scale, *Icerya purchasi* on the developing citrus industry in California (USA). This resounding success brought about a new revolution in the citrus cultivation in many areas across the globe accompanied with its guaranteed permanency, low-cost high efficiency and opened doors to support for biological control as a solution to other agricultural pest problems as well.

Since then, a lot of other predators and parasitoids have been used to counter the effects of insect pests on various crops. Below mentioned are examples of some of the successful biological control of insect pests in Asian countries.

4.7 ENTOMOLOGICAL THREATS

In the past few decades, there has been a catastrophic decline in the population of insects species and many research papers have authenticated this sudden decline (Fonseca, 2009; Vanbergen, 2013; Hallmann et al., 2017; Sanchez and Wyckhuys, 2019; Seibold et al., 2019). Involvement of human activities has, no doubt, been the most suspicious cause of this decline but question remains whether the change has befallen suddenly, or we have reached the tipping point, or if we have lost too many species and ecosystems permanently? A wider concern over declining insect population was first erupted when a research paper related to the use of Malaise trap to capture flying insects was published (Hallmann et al., 2017). This

TABLE 4.1 Few Examples of Successful Biological Control Against Insect Pests of Various Crops in Asian Continent.

Region/ country	Crop	Insect pest	Natural enemy	Predator or parasitoid	Country from which imported	Year
Kgushu Island (Japan)	Citrus	Spiny black fly *Aleurocanthus spiniferous*	*Encarsia smithi*	Parasite	Japan	1925–1926
Fiji	Coconut	Coconut moth *Levuana iridiscens*	*Bessa remota*	Parasitoid	Malaysia	1925–1929
Assam (India)	Apple	Wooly Apple Aphid *Eriosoma lanigerum*	*Aphelinus mali*	Parasitoid	England	1920s
India	*Acacia decurrens*	*Icerya purchasi*	*Rodolia cardinalis*	Predator	USA	1929–1931
Korea	Apple	Wooly Apple Aphid *Eriosoma lanigerum*	*Aphelinus mali*	Parasitoid	Japan	1934–1935
India	Apple	San Jose Scale *Quadraspidiotus perniciosus*	*Prospaltellaperniciosi*	Parasitoid	China	1958–1960
India	Apple	San Jose Scale *Quadraspidiotus perniciosus*	*Aphytisdiaspidis*	Parasitoid	USA	1960
India	Castor	*Achaea janata*	*Telenomus* sp.	Parasitoid	New Guinea	1964
India	Coconut	*Oryctes rhinoceros*	*Platymeris laevicollis*	Predator	Zanzibar	1965
Korea	Citrus	Wax scale *C. Rubens*	*Anicetus beneficus*	Parasitoid	Japan	1975
Japan	Citrus	Arrowhead scale, *Cinaspis vanonensis*	*Physcus fulvus* *Aphytis yanonensis*	Parasitoid Parasitoid	Taiwan Taiwan	1980–1986

Source: Modified by Sankaran (1974); DeBach and Rosen (1991).

study surpassed 27 years of extensive survey in 63 protected areas in Germany and deciphered an overall decline of 76% flying insect biomass, while the mid-summer biomass decline was recorded to be 82%. This led to criticisms against the method for insect trapping used by the scientists. Studies on the decline of tropical insects remain scanty but still a survey from Puerto Rico from 1976 to 2012 confirmed the decreased arthropod biomass by 4–30 times (Lister and Garcia, 2018). However, the authors focused on the rise in temperature to be the prime factor responsible for the decline, which remains highly debatable for not incorporating human activities particularly disturbances. The group of insects that are most susceptible to extinction are the monophagous insects (feeding on single host plant) due to contraction of their geographical range. Continuous loss of feeding guild and habitat has resulted in an extraordinary loss in the numbers (Fonseca, 2009).

4.8 CAUSES OF DECLINE

Loss of biodiversity is an irreplaceable and irreversible phenomenon much of which is attributed to the human interruptions and there is a very little doubt about that. Even the catastrophic changes in temperature and climate across the globe are the results of human deeds, such as the use of pesticides, pollution, land use change, etc.

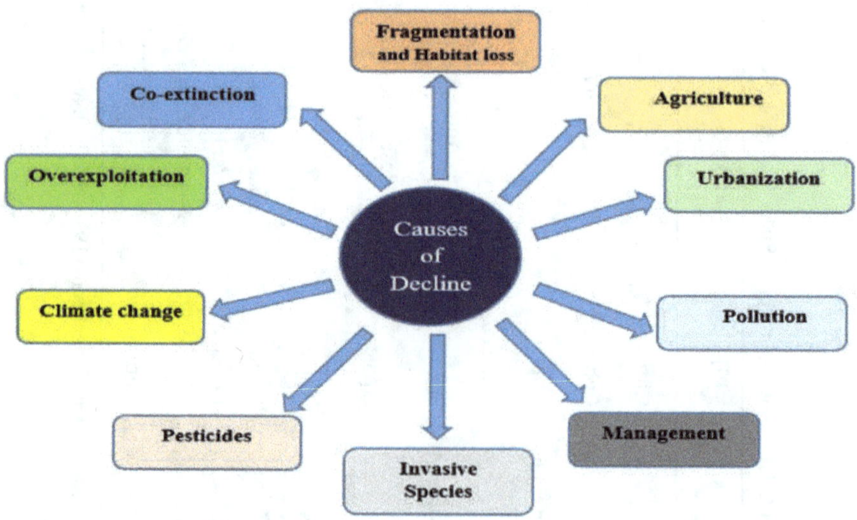

FIGURE 4.1 Causes of decline in insect species

4.8.1 FRAGMENTATION AND HABITAT LOSS

Fragmentation and loss of habitat are the most impactful threats to biodiversity on the whole (Foley et al., 2005; Habel et al., 2019; Cardoso et al., 2020). Loss of habitat and its associated insect species is attributed to deforestation, expansion of agricultural practices, and urban developmental practices (Brook et al., 2003; Basset and Lamarre, 2019). An estimated 22% natural land has been manifested to human use with highest destruction witnessed in the tropical and temperate forests while boreal and tundra habitats being least or relatively unaffected (Hoekstra et al., 2005). Loss of freshwater habitat particularly results from the water regulation and constructing dams and barrages (Dudgeon et al., 2006). Loss of habitat not only uproots the basic mechanism of a working ecosystem but the interrelated species suffer all along. We lose a lot more than just species. Habitat loss and fragmentation leads to decrease in connectivity. The extent and degree of fragmentation vary along wide range of factors. Insects with low mobility tend to survive isolated (Poniatowski et al., 2018) while the ones with higher mobility popularize at suitable habitats with sufficient areas and spatial proximity (Cardoso et al., 2020).

4.8.2 AGRICULTURE

Increase in the agricultural practices during the past few centuries has led to intensive changes in the composition of insects. It has been noted that land area converted to the use of agriculture has increased by 466% from 1,700 to 1,980 (Meyer and Turner, 1992). Reduction of wetlands due to artificial drainage has resulted in the loss of aquatic insect diversity to a significant extent. Aquatic insects can withstand severe and sometimes, often sometimes unpredictable conditions (Polhemus, 1979). The indiscriminate use of inorganic fertilizers to propagate agriculture has led to eutrophic conditions due to surface runoff which highly influences aquatic insects like Ephemeroptera (Boatman et al., 2007). Overgrazing and supplementary feeding has also led to the grassland insect population decline to a considerable extent.

4.8.3 URBANIZATION

Since 1970 and 2000, there has been an estimate 58,000 m^2 rise in global urban area (Seto et al., 2011). This has led to significant change in world insect population for example, decline in beetles and wasps was recorded in West Africa due to urbanization and showed a relative change in complete species abundance in that part (Guenat et al., 2019). The population of Orthoptera severely struck by urbanization in Paris was also reported (Penone et al., 2013). However, a few cases of increase in the insect population have also been reported especially in the case of urban pollinators (Hall et al., 2017).

4.8.4 POLLUTION

Despite reduced pollution in various developed countries, much of the adverse effect of decline in insect population is due to industrial pollution which is known since long time to reduce insect viability (Whittaker, 2001). Industrial pollution has led to significant addition of heavy metals across various water bodies (Sun et al., 2007; Boyd, 2010) and air pollution. The vehicle pollution resulting in emission of nitrous oxides is yet another impending factor in the degradation of insect population worldwide (Campbell and Vallano, 2018). Light pollution is also gaining interest in majority of people these days wherein, especially the nocturnal insects are hindered by changes in natural light/dark conditions. Usually, these insects use natural light from moon and stars to navigate and communicate. Artificial light desynchronizes their activities such as mating, egg laying (Owens and Lewis, 2018), leading to incomplete life cycle thereby reducing the insect population. Soil pollution due to the addition of fertilizers result in soil acidification and eutrophic conditions that are detrimental to the growth and survival of insects (Fox, 2013). Increased use of fertilizers has led to increase in aphid population (kyto et al., 1996) while negatively impacting most insects, thereby, leading to dysfunctional ecosystems.

4.8.5 MANAGEMENT

Regular practice of removing weeds, mowing and removing vegetation litter has led to decrease in many insect species owing to loss of host plants. The practice of tidying up along roadsides and hedgerows in urban and semi-urban areas can reduce the number of plant and insect species (Noordijk et al., 2009).

4.8.6 INVASIVE SPECIES

Alien invasive species are deliberately introduced to areas outside their geographical ranges and have a distinctive ecological or socioeconomic effect. These species pose a direct constant competition to the native species, which are mostly driven away toward extinction (Sala et al., 2000; Havel et al., 2015), due to consumption of habitat and host plant by the alien species. Few renowned examples include extinction of endemic ladybird in Britain due to invasive ladybird, *Harmonia axyridis* Pallas, in 1773 (Comont et al., 2014), large number and predatory nature of invasive ants like the Argentine ants pose grave threat to the native insect fauna. Invasive plants also impact insect communities to a bigger extent. Their monotypic nature reduces the quality and quantity of food thereby limiting essential resources to insects leading to decline in insect population (Severns and Warren, 2008; Preston et al., 2012). Invasive pathogens like *Nosema bombi* (European strain fungal pathogen) resulted in declining North American bumblebee populations (Cameron and Sadd, 2020).

4.8.7 PESTICIDES

Intensive use of pesticides is another reason for considerable decline of insect species across the globe. Pesticides lead to direct toxicity and indirectly through habitat alteration. Biomagnification and bioaccumulation inside food chain further acts as additional stress to insect populations. Such deleterious impacts have been seen on bee where low doses of neonicotinoid pesticides hinder with their navigation (Fischer et al., 2014).

4.8.8 CLIMATE CHANGE

Insects, due to their short life span governed majorly by temperature, are expected to show a greater response toward change in climate. These shifts and overall global temperature increase have led to change in the phenology where we have witnessed some unusual yet predictable phenomenon in certain insects particularly aphids and Lepidoptera (Butterfly and Moth) (Bell et al., 2019). These insects have had earlier emergence as adult in a variety of habitats. Another anticipated effect of climate change is the range shift among various species wherein they have shifted to ranges beyond their present thermal tolerance. Evidences of narrower temperature tolerance of tropical animals than the temperate ones are widely convincing (Eggleton, 2020).

4.8.9 OVEREXPLOITATION

Least emphasized concern is regarding the overexploitation of various insect species as pets, traditional medicine, decorations and food reservoirs. Since time immemorial some insects are known to possess medicinal values (Shrivastava et al., 2009), with the annual commercial value being US$100 million based on medicinal insects. Mealworms, caterpillars, crickets, mantids, ants, etc. are a frequent choice as pets in USA. In Japan, the popularity of insect pests has led to illegal trafficking of various insects in and out of the country (Actman, 2019). Insects have been being preserved for decoration particularly butterflies and beetles due to their colored wings and elytra which are used in jewelry and embroidery (Prasad, 2007; Lokeshwari and Shantibala, 2010). Use of insects as food is another force that drives overexploitation. An inventory enlisted 2,111 edible insect species worldwide (Jongema, 2015), and indiscriminate practice of entomophagy has put aside the sustainable use of edible insects and their habitat.

4.8.10 CO-EXTINCTION

Evolution and specialization have given insects a best hand in becoming co-dependent and therefore jeopardize co-extinction. Although rare, but evidence for it do exist such as co-extinction of dung beetles at continental scales along with declining mammal population due to climate change, habitat disintegration, hunting, etc. (Bogoni et al., 2019).

4.9 CONSERVATION OF INSECTS

The key for the conservation of insects is to create habitat for these mobile species all over the landscapes so that they can find the space to live in. Human beings and insects are entwined with each other. Public parks are highly valuable and can act as a reserve for rare species of insects. The walls of the buildings can be greened that will benefit the insect herbivores and provide the surface for sunning. Setting up sanctuaries will thwart the effect of habitat loss on insects. Parks, gardens, and farm fields can be established that will provide room for the insects to flourish. Conversion of lawns to the least natural habitat could remarkably assist in the conservation of insects. This converted place of the lawn should be kept covered with fallen leaves and twigs that would provide natural home conditions to the insects (Kawahara et al., 2021). Evidence suggests that native plants are more beneficial to native insects as these plants form the source of food and nesting sites for the insects. Appreciation for the benefits of insects should be widespread among the masses. This mechanism is more prevalent in Japan where insects often make their appearance in animated films and popular media, and are rendered as interesting and valuable.

4.10 CONCLUSION

Loss of insect biodiversity has immediate effects on human welfare and folklore due to their close association. Insects act as the driving force of many ecological food chains and food webs. Decline in the number of insect species directly lead to changes in ecology and therefore, manifest the ecological working. It has been a well-known fact that honey bees, alone carry out maximum pollination of plants, thereby increasing food yield. It has been also known that 97% of insects visiting flower blooms are honeybees (Doug, 1999) but domestication of honeybee since a long time ago has rendered them vulnerable in wild, and therefore, added to their peril. Our approach to manage the pest population has always been one-sided, wherein we focus mainly on eradication of pest population by adding pesticides and insecticides keeping aside their possible consequences on soil, water, and other insects. Recently, Integrated Pest Management (IPM) has been keenly followed in order to ensure sustainable pest management. A number of factors must be kept in mind in order to put a halt on the ever-declining insect population.

KEYWORDS

- **conservation**
- **ecosystem**
- **extinction**
- **insect farms**
- **pests**
- **population decline**

REFERENCES

Actman, J. Bug Smuggling Is a Big Business; National Geographic, 2019.

Ali, A. Nuisance Chironomids and Their Control: A Review. *Am. Entomol.* **1980,** *26*, 3–16.

Araujo, A.; Jansen, A. M.; Reinhard, K.; Ferreira, L. F. Palaeoparasitology of Chagas Disease: A Review. *Memorias do Instituto Oswaldo Cruz* **2009,** *104*, 9–16.

Atwal, A. S.; Dhaliwal, G. S. *Agricultural Pests of South Asia and Their Management.* Kalyani Publishers, 1999; pp 377–421.

Balvin, O.; Munclinger, P.; Kratochvil, L.; Vilimova, J. Mitochondrial DNA and Morphology Show Independent Evolutionary Histories of Bedbug *Cimex lectularius* (Hemiptera: Cimicidae) on Bats and Humans. *Parasitol. Res.* **2012,** *111*, 457–469.

Basset, Y.; Lamarre, G. P. A. Toward a World That Values Insects. *Science* **2019,** *364*, 1230–1231.

Baz, A.; Monserrat, V. J. Distribution of Domestic Psocoptera in Madrid Apartments. *Med. Vet. Entomol.* **1999,** *13*, 259–264.

Bebber, D. P.; Holmes, T.; Gurr, S. J. The Global Spread of Crop Pests and Pathogens. *Glob. Ecol. Biogeo.* **2014,** *23*, 1398–1407.

Bell, J. R.; Botham, M. S.; Henrys, P. A.; Leech, D. I.; Pearce-Higgins, J. W.; Shortall, C. R.; Bereton, T. M.; Pickup, J.; Thackeray, S. J. Spatial and Habitat Variation in Aphid, Butterfly, Moth and Bird Phonologies Over the Last Half Century. *Glob. Change Biol.* **2019,** *25*, 1982–1994.

Belles. Los insectos y el hombre prehistorico. *Boletn de la Sociedad Entomologica Aragonesa* **1997,** *20*, 319–325.

Boatman, N.; Parry, H.; Bishop, J.; Cuthberston, A. Impacts of Agricultural Change on Farmland Biodiversity in the UK. *Iss. Environ. Sci. Technol.* **2007,** *25*, 1–32.

Bogoni, J. A.; da Silva, P. G.; Peres, C. A. Co-Declining Mammal-Dung Beetle Faunas Throughout the Atlantic Forest Biome of South America. *Ecography* **2019,** *42*, 1803–1818.

Boyd, R. S. Heavy Metal Pollutants and Chemical Ecology: Exploring New Frontiers. *J. Chem. Ecol.* **2010,** *36*, 46–58.

Brook, B. W.; Sodhi, N. S.; Ng, P. K. L. Catastrophic Extinction Follow Deforestration in Singapore. *Nature* **2003,** *424*, 420–423.

Cameron, S. A.; Sadd, B. M. Global Trends in Bumble Bee Health. *Annu. Rev. Entomol.* **2020,** *65*, 209–232.

Campbell, S. A.; Vallano, D. M. Plant Defences Mediate Interactions Between Herbivory and the Direct Foliar Uptake of Atmospheric Reactive Nitrogen. *Nature Commun.* **2018,** *9*, 1–7.

Cardoso, P.; Barton, PS, Birkhofer, K.; Chichorro, F.; Deacon, C.; Fartmann, T.; Fukushima, C. S.; Gaigher, R.; Haber, J. C.; Hallmann, C. A.; Hill, M. J.; Hochkirch, A.; Kwak, M. L.; Mammola, S.; Noreiga, J. A.; Orfinger, A. B.; Pedraza, F.; Pyrke, J. S.; Roque, F. O.; Settele, J.; Simaika, J. P.; Stork, N. E.; Suhling, F.; Vorster, C.; Samways, M. J. Scientists' Warning to Humanity on Insect Extinctions. *Biol. Conserv.* **2020,** *242*, 1–12.

Chopard, L. Sur une gravure d'insecte de l'epoquemagdalenienne. *ComptesRendus de la Societe de Biogeographie* **1928,** *5*, 64–67.

Comont, R. F.; Roy, H. E.; Harrington, R.; Shortall, C. R.; Purse, B. V. Ecological Correlates of Local Extinction and Colonisation in the British Ladybird Beetles (Coleoptera: Coccinellidae). *Biol. Invasions* **2014,** *16*, 1805–1817.

Condamine, F. L.; Clapham, M. E.; Kergoat, G. J. Global Patterns of Insect Diversification: Towards a Reconciliation of Fossil and Molecular Evidence? *Sci. Rep.* **2016,** *6*, 19208.

Cotton, R. T. *Pest of Stored Grain on Grain Products*; Minneapolis, Burgess Pub. Co., 1960; pp 1–289.

Crane, E. *Bees and Beekeeping. Science, Practice and World Resources*; NCROL. Cornstock Publ.: Los Angeles; Ithaca, 1990; pp 1–593.

DeBach, P.; Rosen, D. *Biological Control by Natural Enemies*, 2nd edn; Cambridge University Press: New York, 1991; pp 1–440.

Dhaliwal, G. S.; Arora, R. *Principles of Insect Pest Management*; Kalyani Publishers, 1998; pp 1–295.

Doug, S. *Pollination of Applesby Honeybees*; NSW Agriculture, 1999; pp 1–4.

Dudgeon, D.; Arthington, A. H.; Gessner, M. O.; Kawabata, Z. I.; Knowler, D. J.; Lévêque, C.; Sullivan, C. A. Freshwater Biodiversity: Importance, Threats, Status and Conservation Challenges. *Biol. Rev*. **2006,** *81*, 163–182.

Dunn, R. R.; Fierer, N.; Henley, J. B.; Leff, J. W.; Menninger, H. L. Home Life: Factors Structuring the Bacterial Diversity Found Within and Between Homes. *PLoS ONE* **2013,** *8*, e1582.

Eggleton, P. The State of the World's Insects. *Annu. Rev. Environ. Resour* **2020,** *46*, 61–82.

Fischer, J.; Muller, T.; Spatz, A. K.; Greggers, U.; Grunewald, B.; Menzel, R. Neonicotinoids Interfere with Specific Components of Navigation in Honeybees. *Plos One* **2014,** *9*, e91364.

Foley, J. A.; DeFries, R.; Asner, G. P.; Barford, C.; Bonan, G.; Carpenter, S. R.; Chapin F. S.; Coe, M. T.; Daily, G. C.; Gibbs, H. K.; Helkowski, J. H.; Holloway, T.; Howard, E. A.; Kucharik, C. J.; Monfreda, C.; Patz, J. A.; Prentice, I. C.; Ramankutty, N.; Snyder, P. K. Global Consequences of Land Use. *Science* **2005,** *309*, 570–574.

Fonseca, C. The Silent Mass Extinction of Insect Herbivores in Biodiversity Hotspots. *Conserv. Biol.* **2009,** *23*, 1507–1515.

Fox, R. The Decline of Moths in Great Britain: A Review of Possible Cause. *Insect Conserv. Divers.* **2013,** *6*, 5–19.

Gaston, K. J. Spatial Patterns of Species Description: How Is Our Knowledge of the Global Insect Fauna Growing? *Biol. Conserv.* **1994,** *67*, 37–40.

Grabowski, N. T.; Klein, G. Bacteria Encountered in Raw Insect, Spider, Scorpion and Centipede Taxa Including Edible Species and Their Significance from the Food Hygiene Point of View. *Trends Food Sci. Technol.* **2017,** *63*, 80–90.

Griffiths, H. M.; Ashton, L. A.; Evans, T.; Parr, C. L.; Eggleton, P. Termites Can Decompose More Than Half of Deadwood in Tropical Rainforest. *Curr. Biol.* **2019,** *29*, R118–R119.

Guenat, S.; Kunin, W. E.; Dougill, A. J.; Dallimer, M. Effects of Urbanisation and Management Practices on Pollinators in Tropical Africa. *J. Appl. Ecol.* **2019,** *56*, 214–224.

Gullan, P. J.; Cranstan, P. S. *The Insects: An Outline of Entomology*; Blackwell Sciences: Malden, 2000; pp 1–470.

Habel, J. C.; Samways, M. J.; Schmitt, T. Mitigating the Precipitous Decline of Terrestrial European Insects: Requirements for a New Strategy. *Biodivers. Conserv.* **2019,** *28*, 1943–1360.

Hall, D. M.; Camilo, G. R.; Tonietto, R. K.; Ollerton, J.; Ahrné, K.; Arduser, M.; Ascher, J. S.; Baldock, K. C.; Fowler, R.; Frankie, G.; Goulson, D.; Gunnarsson, B.; Hanley, M. E.; Jackson, J. I.; Langellotto, G.; Lowenstein, D.; Minor, E. S.; Philpott, S. M.; Potts, S. G.; Sirohi, M. H.; Spevak, E. M.; Stone, G. N.; Threlfall, C. G. The City as a Refuge for Insect Pollinators. *Conserv. Biol.* **2017,** *1*, 24–29.

Hallmann, C. A.; Sorg, M.; Jongejans, E.; Siepel, H.; Hofland, N.; Schwan, H.; Stenmans, W.; Mu¨ller, A.; Sumser, H.; Ho¨rren, T.; Goulson, D.; de Kroon H. More Than 75 Percent Decline Over 27 Years in Total Flying Insect Biomass in Protected Areas. *PLoS One*. **2017,** *10*, e0185809.

Havel,J,E.; Kovalenko, K. E.; Thomaz, S. M.; Amalfitano, S.; Kats, L. B. Aquatic Invasive Species: Challenges for the Future. *Hydrobiologia* **2015,** *750*, 147–170.

Hoekstra, J. M.; Boucher, T. M.; Ricketts, T. H.; Roberts, C. Confronting a Biome Crisis: Global Disparities of Habitat Loss and Protection. *Ecol. Lett.* **2005,** *8*, 23–29.

Jongema, Y. List of Edible Species of the World; Wageningen University, 2015; pp 1–45.

Kalra, N. L.; Kaul, S. M.; Rastogi, R. M. Prevalence of *Aedes aegypti* and *Aedesalbopictus*-Vectors of Dengue Haemorrhagic Fever in North, North-East and Central India. *Dengue Bulletin* **1997,** *21*, 84–92.

Keller, A. *Drosophila melanogaster's* History as a Human Commensal. *Curr. Biol.* **2007,** *17*, R77–R81.

Kenward, H.; Carrott, J. Insect Species Associations Characterise Past Occupation Sites. *J. Archaeol. Sci.* **2006,** *33*, 1452–1473.

Khare, B. P. *Pests of Stored Grain and Their Management*; Kalyani Publishers: New Delhi, 1994; pp 1–304.

Kislev, M. E.; Hartmann, A.; Galili, E. Archaeobotanical and Archaeoentomological Evidence from a Well at Atlit-Yam Indicates Colder, More Humid Climate on the Israeli Coast During the PPNC Period. *J. Archaeol. Sci.* **2004,** *31*, 1301–1310.

Legner, E. F.; McCoy, C. W. The Housefly, *Musca domestica* Linnaeus, as an Exotic Species in the Western Hemisphere Incites Biological Control Studies. *Can. Entomol.* **1966,** *98*, 243–248.

Lesnik, J. J. *Edible Insects and Human Evolution*; Published to University Press Scholarship Online, 2018; pp 1–206.

Lister, B. S.; Garcia, A. Climate-Driven Declines in Arthropod Abundance Restructure a Rainforest Food Web. *PNAS* **2018,** *115*, E10397–E10406.

Lokeshwari, R. K.; Shantibala, T. A Review on the Fascinating World of Insect Resources: Reasons for Thoughts. *Psyche* **2010,** *207570*, 1–12.7.

Lounibos, L. P. Invasions by Insect Vectors of Human Disease. *Annu. Rev. Entomol.* **2002,** *47*, 233–266.

Martin, L. J.; Adams, R. I.; Bateman, A. et al. Evolution of Indoor Biome. *Trends Ecol. Evol.* **2015,** *30*, 223–232.

Meyer, W. B.; Turner, B. L. Human Population Growth and Global Land-Use Cover Change. *Annu. Rev. Ecol. Evol. System.* **1992,** *23*, 39–61.

Nielsen, B. O.; Mahler, V.; Rasmussen, P. An Arthropod Assemblage and the Ecological Conditions in a Byre at the Neolithic Settlement of Weier, Switzerland. *J. Archaeol. Sci.* **2000,** *27*, 209–218.

Noordijk, J.; Delille, K.; Schaffers, A. P.; Sykora, K. V. Optimizing Grassland Management for Flower-Visiting Insects in Roadside Verges. *Biol. Conserv.* **2009,** *142*, 2097–2103.

Owens, A. C. S.; Lewis, S. M. The Impact of Artificial Light at Night on Nocturnal Insects: A Review and Synthesis. *Ecol. Evol.* **2018,** *8*, 11337–11358.

Panagiotakopulu, E. New Records for Ancient Pests: Archaeoentomology in Egypt. *J. Archaeol. Sci.* **2001,** *28*, 1235–1246.

Panagiotakopulu, E. Insect Remains from the Collections in the Egyptian Museum of Turin. *Archaeometry* **2003,** *45*, 355–362.

Panagiotakopulu, E. Dipterous Remains and Archaeological Interpretation. *J. Archaeol. Sci.* **2004,** *31*, 1675–1684.

Penone, C.; Kerbiriu, C.; Julien, J-F.; Julliard, R.; Machon, N.; Le Viol, I. Urbanisation Effects on Orthoptera: Which Scale Matters? *Insect Conserv. Divers.* **2013,** *6*, 319–327.

Polhemus, J. T. Family Naucoridae the Semiaquatic and Aquatic Hemiptera of California (Heteroptera: Hemiptera) *Ed. Menke A. S. Bull California Insect Surv.* **1979,** *21*, 131–138.

Poniatowski, D.; Struhldreher, G.; Loffler, F.; Fartmann, T. Patch Occupancy of Grassland Specialists: Habitat Quality Matters More Than Habitat Connectivity. *Biol. Conserv.* **2018,** *225*, 237–244.

Prasad, B. Insect Diversity and Endemic Bioresources. In: *Endemic Bio-resources in India-Conservation and Sustainable Development with Special Reference to North East India*; Singh, N. I., Ed.; Dehradun, India, 2007, 275–291.

Preston, D. L.; Henderson, J. S.; Johnson, P. T. J. Community Ecology of Invasions: Direct and Indirect Effects of Multiple Invasive Species on Aquatic Communities. *Ecology* **2012,** *93*, 1254–1261.

Pruthi, H. S.; Singh, M. *Pest of Stored Grains and Their Control*, 3rd edn (Rev.), Indian Council of Agriculture Research (ICAR); Government of India Press: Calcutta, 1950.

Rees, D. *Insects of Stored Products*; CSIRO Publishing, 2004; pp 1–146.

Robinson, W. H. *Urban Insects and Arachnids: A Handbook of Urban Entomology*, Vol. 13; Cambridge University Press: Cambridge, 2005; pp 259–264.

Sala, O. F.; Chapin, F. S.; Armesto, J. J.; Berlow, E.; Bloomfield, J.; Dirzo, R. et al. Global Biodiversity Scenario for the Year 2100. *Science* **2000,** *287*, 1770–1774.

Sanakaran, T. Natural Enemies Introduced in Recent Years for Biological Control of Agricultural Pests in India. *Indian J. Agric. Sci.* **1974,** *44*, 425–433.

Sanchez-Bayo, F.; Wyckhuys, K. A. G. Worldwide Decline of the Entomofauna: A Review of Its Drivers. *Biol. Conserv.* **2019,** *232*, 8–27.

Seibold, S.; Gossner, M. M.; Simons, N. K.; Bluthgen, N.; Muller, J. et al. Arthropod Decline in Grasslands and Forests Is Associated with Landscape-Level Drivers. *Nature* **2019,** *574*, 671–674.

Seto, K. C.; Fragkias, M.; Guneralp, B.; Reilly, M. K. A Meta-Analysis of Global Urban Land Expansion. *Plos One* **2011,** *6*, e23777.

Severns, P. M.; Warren, A. D. Selectively Eliminating and Conserving Exotic Plants to Save an Endangered Butterfly from Local Extinction. *Anim. Conserv.* **2008,** *11*, 476–483.

Shankar, U.; Abrol, D. P. Integrated Pest Management in Stored Grains. *Insect Pest Manag.* **2012,** 356–407.

Sharon, M. E. M.; Abirami, C. V. K.; Alagusundaram, K. Grain Storage Management in India. *J. Postharv. Technol.* **2014,** *2*, 12–24.

Shrivastava, S. K.; Babu, N.; Pandey, H. Traditional Insect Bioprospecting—As Human Food and Medicine. *Indian J. Tradit. Knowl.* **2009,** *8*, 485–494.

Stejskal, V.; Hubert, J.; Aulicky, R.; Kucerova, Z. Overview of Present and Past and Pest—Associated Risks in Stored Food and Feed Products: European Perspective. *J. Stored Prod. Res.* **2015,** *34*, 122–132.

Stork, N. E. How Many Species of Insects and Other Terrestrial Arthropods Are There on Earth? *Annu. Rev. Entomol.* **2018,** *63*, 31–45.

Sun, H., Liu, Y., Zhang, G. Effects of Heavy Metal Pollution on Insects. *Acta EntomologicaSinica* **2007,** *50*, 178–185.

Togola, A.; Seck, P. A.; Glitho, I. A.; Diagne, A.; Adda, C.; Toure, A.; Nwilene, F. E. Economic Losses from Insect Pest Infestation on Rice Stored on-farm in Benin. *J. Appl. Sci.* **2013,** *13*, 278–285.

Tribune Medical Xpress. India Reports 80 Zika Cases, 22 Pregnant Women, 2018.

Tuma, J.; Fleiss, S.; Eggleton, P.; Frouz, J.; Klimes, P. et al. Logging of Rainforests and Conversion to Oil Palm Reduces Bioturbator Diversity But Not Levels of Bioturbation. *Appl. Soil Ecol.* **2019,** *144*, 123–133.

Ulyshen, M. D. Insect—Mediated Nitrogen Dynamics in Decomposing Wood. *Ecol. Entomol.* **2015,** *40*, 97–112.

Van Mele, P. A Historical Review of Research on the Weaver Ant Oecophylla in Biological Control. *Agric. Forest Entomol.* **2008,** *10*, 13–22.

Vanbergen, A. J. Threats to an Ecosystem Service: Pressures on Pollinators. *Front. Ecol. Environ.* **2013,** *11*, 251–259.

Weaver; Petroff. Pest Management for Grain Storage and Fumigation. *Montana Depart. Agric*. **2004,** 1–84.

Whittaker, J. B. Presidential Address: Insects and Plants in a Changing Atmosphere. *J. Ecol.* **2001,** *89*, 507–518.

World Health Organisation. Vector-Borne Diseases- Fact Sheet, 2014a.

World Health Organisation. India Vector-Borne Diseases in India: An Analysis from a Health Systems Approach, 2014b.

World Health Organisation Zika Virus Infections—India. Disease Outbreak News, 2017.

CHAPTER 5

Insect-Based Value-Added Products: Current Scenario and Future Scope

ANEELA ASHIQ[1], MUHAMMAD ADNAN BODLAH[2], IMRAN BODLAH[3], AMMARA GUL-E-FAREEN[3], and MUHAMMAD BILAL TAHIR[4]

[1]*Department of Biochemistry, Khwaja Fareed University of Engineering and Information Technology, Rahim Yar Khan, Punjab, Pakistan*

[2]*Fareed Biodiversity Conservation Centre, Department of Agricultural Engineering, Khwaja Fareed University of Engineering and Information Technology, Rahim Yar Khan, Punjab, Pakistan*

[3]*Department of Entomology, Pir Mehr Ali ShahArid Agriculture University, Rawalpindi, Pakistan*

[4]*Department of Physics, Khwaja Fareed University of Engineering and Information Technology, Rahim Yar Khan, Punjab, Pakistan*

ABSTRACT

Over the past few years, insects have received significant attention globally. With the rapidly growing population, it has become challenging to regulate or reduce the consumption of natural resources to balance the ecosystem. The exploitation of natural resources such as fuel consumption has created global energy issues. Scientists have been exploring the benefits of insects for many decades and have always been poorly perceived by humans and regulatory authorities. Insects have greater potential to resolve major issues regarding fuel production, due to their higher fat content also used as raw materials for biodiesel production. In this chapter, we described various examples of insect species recently being used and can

Insect Diversity and Ecosystem Services, Volume 1: Importance, Threats, Conservation and Economic Perspectives. Younis Ahmad Hajam, Sajad Hussain Parey, and Rouf Ahmad Bhat (Eds)

be utilized in near future. The increasing demand for food has focused to explore alternatives to plant and animal food products. Insects are a very promising species on earth for more sustainable food and feed products. Insects hold various properties to ensure large-scale high quality, nutrients rich food and feed sources. Furthermore, insects have the potential nutrition to be consumed as feed for livestock. Advancement in technologies such as nanotechnology and biotechnology enables scientists to design exquisite processing or insect-rearing methods for food production. The following paragraphs presents numerous examples of insect orders and species being used for food and feed production with their beneficial characteristics. Insect-rearing methods and conditions directly influence the yield and quality of food products. Given that context, conditions should be appropriate to obtain a high yield in less time. Lastly, legislation regarding edible insects need more attention for the food industry worldwide according to standard regulations.

5.1 INTRODUCTION

Insects have the largest population on earth and have evolved unique life characteristics, adaptations to the adverse environment, and evolutionary properties. Insects are considered the largest biological resource that are undiscovered by humans. Further exploration of insect's biological characteristics would help us manage insect's resources at a large scale to produce insect-based value-added products in the agricultural and industrial sectors by advanced biological research (Kwon, 2008; Zhang et al, 2008). Previous research exhibit various positive impacts on human life by providing biological resources in the ecosystem (Macadam and Stockan, 2015; Pelini et al, 2015; Prather et al., 2013). Studies of insects have mixed perceptions among the public because most people considered insects negatively (Lockwood, 2013), due to fear and avoidance (Kellert, 1993), particularly insects having any disease (Prokop and Fančovičová, 2010). Lack of awareness results in insect extinction at a large scale and reduces the importance of insect diversity and biomass (Snaddon et al, 2008; Snaddon and Turner, 2007). China has been the top producer of insect-based products such as silk, Chinese gallnuts, insect wax, and honey for many decades worldwide. Insects are used in biodiesel production. In biodiesel production, the bodies of insects are used as raw material and feedstocks (Koonin, 2006). For example, various larval stage insects produce fats very quickly such as *Hermetia illucens* (blacksoldier larvae fly), *Rhynchophorus phoenicis,* and *Gonimbrasia belina* (Li et al, 2011).

Entomophagy—insects' consumption—has been in practice since a long time worldwide (Belluco et al., 2015; Rumpold and Schlüter, 2013). Insects and their consumption can be largely found in the whole world except Antarctica (Lakemond et al., 2017; Costa-Neto and Dunkel, 2016). More than 2000 insect species have been consumed in several parts of the world like Asia, Latin America, and Africa. Major parts of insects are part of human diet due to their gastronomical and cultural importance (Huis et al., 2013). In many countries, insect consumption is considered as habit and demand and development of the insect-based products are minimum. At present entomophagy has become a suitable option in many perspectives. Many established companies provide insect-derived products, and their processing techniques have maintained their nutritional values as per demands for consumers and source history. The processing methods and appropriate ingredients formulations of insect food offer a balanced diet to consumers (Sun-Waterhouse et al., 2016).On the other hand, some industrialized nations like Europe considered entomophagy distasteful (Huis, 2016). Insects are considered as having high nutritional values. Increasing global demand for feed ingredients has shifted attention toward producing food from different resources (animals' products) (Belghit et al., 2018; Dabbou et al., 2018; Lock et al, 2016). From the last few years, insects have become a large source of raw material for animal feed worldwide. Insects fulfill animal's dietary, amino acids, proteins, and nutritional requirements (Makkar, 2018). Mass production of insects have been identified as a promising approach from the environmental point of view due to various factors such as minimum emission of greenhouse gases (van Huis and Oonincx, 2017), production of 1 kg protein in a small area (Oonincx and Boer, 2012), utilization of the small area to produce food to lower food competition (van Huis and Oonincx, 2017), and the ability to convert organic waste to valued protein products. Insects are considered as a remarkable approach to increase the circular economy of the country by the bioconversion of organic waste into highly valued food and animal-feed products (van Huis and Oonincx, 2017).

Insects are also utilized as a large alternative protein source for humans, fish, and livestock (Huis et al., 2013). Due to the overexploitation of insects by utilization of pesticides, insect's reduction increases, so besides collection from the environment, insects rearing occurs in industries (Oonincx and Boer, 2012). With the increasing demand for alternative protein sources, western countries are exploring these types of techniques in the limited areas of livestock (van Huis, 2015). However, it is unclear that the production of insects as a source of alternative protein is more sustainable toward the environment than conventional proteins (Gahukar, 2016; Tabassum-Abbasi et al, 2016). Sustainability exhibit techniques and practices that fulfill the present and future food requirement of the society (Tilman et al, 2002). Insects also contain important vitamins, for example, calcium, magnesium, iron, zinc, and phosphorus, and contain essential nutrients, for example, amino acids which mostly do not found in conventional food products (Banjo et al, 2006). Insects are not only utilized in food and feed products but are also used in medicinal products and additives. Daily used insect products include honey, propolis, royal jelly, food coloring, and many more (Schabel, 2010). Insects-based value-added food products and their products enhance the nutritional quality of food and maintain bioactive components balance and decrease the human efforts by efficient processing technologies. Edible insects consumed worldwide include most common types such as *Coleoptera* (beetles) and *Hymenoptera* (e.g., wasp, bees, and ants), *Diptera* (e.g., flies), *Orthoptera* (e.g., grasshoppers and crickets), *Odonata* (e.g., dragonflies), *Hemiptera* (e.g., cicadas), and *Isoptera* (e.g., termites) (Costa-Neto and Dunkel, 2016). These insect-family consumption depends upon their region, availability, preference, and dish delicacy. In the market, insects are often found as food products such as mealworm cooked with salt and black pepper, crickets with chili lime, and honey mustard roasted seasonal mealworm, like toffee. Processed insect markets contain chocolate-coated lollipops and flavor bars, and crackers and biscuits (La Barbera et al, 2018).

Specific companies provide a large array of high-quality and established insect-based food products to the consumers in the market. In these types of products, ingredients usage is limited due to insect's specific mass is present like in crickets and mealworms. Successful implementation of insect's-based products will depend upon consumer demands, interest, and acceptance (La Barbera et al, 2020). The processing techniques, textures, and flavors will greatly influence the perception of people about insect food products (Schouteten et al., 2016). Entomophagy implementation in societies may be successful by preparing food in such a manner that insects are not visible and placed as such in the product. This approach could reduce the neophobia of people about insects. For example, *T. molitor* larvae powder used to make tortilla having high-fat content, protein, and amino acids level, this type of transformation facilitates the demand of food products in people (Mlcek et al., 2014). Processing methods also affect nutritional balance so methods should be designed to aim at minimal loss of nutrients, proteins, and vitamins from food products (Kinyuru et al., 2010). Safety assessment of edible insects including biological, chemical hazards, and allergens are recently described with help of comparison of two production systems (wild harvested vs. controlled raring) and also analyze distinct processing technique's effect on edible insects (Murefu et al., 2019).

This chapter exhibits an objective overview of insect value-added products and their consumption worldwide. We also describe the human consumption of edible insects and their microbiota. We describe the recent development and utilization of insects in industries to produce essential products. We give a brief overview of how insect industries produce products to facilitate human requirements in terms of food and many others which solve many problems regarding food shortage and adding new flavors. Besides various benefits of insects, environmental hazards are also related to them, so we also describe some hazardous effects of insects and their products.

5.2 BIODIESEL PRODUCTION FROM INSECTS: AN INNOVATIVE APPROACH

For many decades, fossil fuels are used to produce energy which has a hazardous effect on the environment including carbon footprints and distinct types of pollution. In this modern era, the production of sustainable, low-cost, and less hazardous biodiesel is a serious factor worldwide. In the production process, suitable raw material of feedstock is utilized for biodiesel production by optimizing distinct transesterification methods (Ishak and Kamari, 2019a). The problem of clear and sustainable fuel can be solved by the use of biofuels and biodiesels (Youngs and Somerville, 2014). Renewable biofuels are defined as the type of fuel produced from the derived materials of dead plants and animals, or the type of fuels obtained from wastes of different processes. Advancement in technological methods has enabled humans to produce biodiesel from different biomass, still, there is much more to explore (Naik et al, 2010). There are about 1 million insects species belongs to the phylum Arthropoda (Erwin, 2004). Insect usage in biodiesel production has become a key concern due to an important establishing factor metabolic reverse termed as fat bodies. Insects use their fat bodies during the non-feeding period in their life cycle (Tauber et al, 1986; Lees, 1955). Fat bodies have a significant role in the life cycle of insects specifically in the metabolism act as medicine and food for both humans and animals (Fombong and Kinyuru, 2018; Grimaldi et al., 2005). Fat bodies are the largest organ spread throughout the insect body specifically below the integument and surrounds gut and reproductive organ (Dean et al, 1985). These fat bodies are considered as an essential material for biodiesel production and their range in insect bodies is 25–77% (Manzano-Agugliaro et al., 2012).

5.3 INSECT BIOMASS CONTAINING LIPID CONTENT

Insect larvae have been in use effectively for many years in the production of biodiesel as a feedstock or raw material. Tropical countries contain the maximum global biodiversity of insect species 58% and have a more favorable environment for them (Foottit and Adler, 2009). The significant characteristic of insect larvae is to biodegrade organic waste (kitchen waste, agricultural, and municipal wastes) into biodiesel raw material. Lipids are the major contributor to insect development from larvae to adults, and also stored glycogen and protein in hemolymph (Turunen and Chippendale, 1980). Insect larvae use organic waste as their food and absorb all nutrients present in it, and about 30% of lipid content accumulates as biomass in their bodies (Čičková et al., 2015). The selection of insect larvae for biodiesel production is categorized based on four different factors—(a) fat body proportion, (b) reproduction rate and capacity, (c) life cycle period, and (d) requirement for land (Leong et al., 2016; Manzano-Agugliaro et al., 2012). Various species of insect larvae have been utilized for biodiesel production such as *Chrysomya megacephala* (oriental latrine fly, OLF) (Li et al., 2015; Yang and Liu, 2014), *Hermetia illucens* (blacksoldier fly, BSF) (Leong et al., 2016; Li et al., 2015; Zheng et al., 2012; Lin et al., 2011), *tenebrio molitor* (yellow mealwormbeetle, YMB) (Zheng et al., 2013), and *Musca domestica* (house fly, HF) (Yang and Liu, 2014).

Wong et al. (Wong et al., 2018) determined the lipid content of black soldier fly larvae (BSFL) and weevil larvae by using raw coconut endosperm waste, and they exhibited that the lipid yield of BSFL was 32.1% which is higher than the yield obtained by Li et al. (2011), and weevil larvae contain 43.8% of lipid content in dry weight. Various other researches exhibited different lipid content yields from BFSL which is approximately 14.6–33.78% when larvae is fed on the different standard feed treated with different concentrations of glucose and xylose (Tang et al., 2015). Table 5.1 shows the lipid content of various insect larvae to produce biodiesel. According to Li et al., BSFL utilized different organic waste to produce biodiesel including chicken manure to produce a maximum yield of lipid 32.8%, pig manure 20.4%, and cattle manure 12.8%. Biodiesel production yield is observed highest from chicken manure 93%. The biodiesel obtained from this BFSL contains 67.6% saturated fatty acids, the maximum amount reported, which helps to prevent biodiesel from oxidation (Li et al., 2011). Furthermore, Li et al. conducted another study on BFSL and he fed this larva with dairy manure for 21 days and obtained 22.3% crude lipid with the 96% of biodiesel yield and described various properties. Leong et al. conducted a study on *Hermetia illucens* to produce biodiesel. He utilized larvae fed on the three types of organic waste—(a) sewage sludge, (b) fruit waste, and (c) palm decanter cake obtained from palm oil mill. The larval growth rate was observed by determining the change in insect biomass per day. The growth rate of *H. illucens* feed on fruit waste and decanter cake waste was 0.52 ± 0.02 and 0.23 ± 0.09 g d^{-1}, respectively. On the other hand, larvae fed on sewage sludge showed no positive response (-0.04 ± 0.01 g d^{-1}). The fatty acid composition was observed as C12:0, C16:0, and C18:1n9c. Table 5.2 shows biodiesel production from different insect larvae feeding on different manure with their biodiesel properties. C12:0 composition was found to be the most abundant when treated with all three types of organic waste. C12:0 showed 76.13%, 58.31%, and 48.06%, for sewage, fruit, and decanter cake waste, respectively. C16:0 fatty acid composition showed 16.48% and 25.48% for sewage sludge and palm decanter cake, respectively (Leong et al., 2016). Diapausing insects stored a large number of lipids, for example, Diptera adults stored lipid excessively in their bodies (Stoffolano and Matthysse, 1967).

As compared to non-diapausing larvae of *Sarcophaga crassipalis,* the diapausing phase stored a double amount of lipid and proteins in the hemolymph (Adedokun and Denlinger, 1985). Saunders et al. conducted a study on *Calliphora vicina* to determine fat content, the amount of fat in dry weight was observed similarly in both diapause and non-diapause phase (Saunders, 1997). Furthermore, for the same species, 30–40% of fat content was reported reduced during diapause (Saunders, 2000). FAO reported every 100 g dry weight of caterpillar comprises 53 g of proteins, 17% carbohydrates, and 15% lipids content with an energy value of 430 Kcal (Manzano-Agugliaro et al., 2012). Table 5.1 also describes lipid content is according to the different orders of insects and it shows significant variation among orders. *Oryctes boas* from order Coleoptera contain 1.5% lipid content, and *Brachytrupes spp* from order Orthoptera have 2% fat content. Variation in lipid content is not only found in different orders but also observed in the same species. For instance, *Musca domestica* order Diptera showed variation in between 8.1% and 32.6%, because adult insects contain a lesser amount of fat content than a larval phase. Insects-feeding material was found to be the key factor building fat content in insect bodies. For example, *G. mellonella* feeds on beeswax and contains 60% fat. Newton et al. reported that *Hermetiaillucens* larvae that were fed on the fig manure contain 28.0% lipid content and those were fed on poultry manure containing 34.8% lipid content (Newton et al., 2005). Besides the food consumption factor, this heterogeneity is very diverse.

TABLE 5.1 Lipid Content of Different Insect's Larvae and Their Extraction Method to Produce Biodiesel and According to Insect Orders.

Insect's larvae	Extraction method	Lipid yield	References
Zophobas morio	n-hexane	33.8%	Leung et al. (2012)
Hermetia illucens	Petroleum ether	29.1–30.1%	Li et al. (2011)
Musca domestica	Petroleum ether	21.6%	Yang et al. (2014)
Rhynchophorus sp	–	13.8–47.1%	Manzano-Agugliaro et al. (2012)
Hermetia illucens	Petroleum ether	32.1%	Wong et al. (2018)
Rhynchophorus sp	Petroleum ether	43.8%	Wong et al. (2018)

TABLE 5.1 *(Continued)*

Lipid contents according to different insect orders		
Order Coleoptera		
Insect species	**Lipid content (%)**	**References**
Apriona germari	41.5	Liu (2011)
Aplagiognathus spinosus	36.9	Ramos-Elorduy et al. (2006)
Aplagiognathusspinosus	36.0	Ramos-Elorduy et al. (1997)
Aplagiognathusspinosus	37.1	Ramos-Elorduy et al. (2006)
Zophobas morio	38.0	Manzano-Agugliaro et al. (2012)
Zophobas morio	42.0	Finke (2002)
Analeptestrifasciata	18.4	Banjo et al. (2006)
Tenebrio molitor	14.9	Finke (2002)
Tenebrio molitor	55.1	Bernard et al. (1997)
Tenebrio molitor	18.4	Bernard et al. (1997)
Tenebrio molitor	38.2	Ramos-Elorduy et al. (2006)
Tenebrio molitor	36.6	Ramos-Elorduy et al. (2006)
Tenebrio molitor	35.2	Finke (2002)
Tenebrio molitor	30.8	Bernard et al. (1997)
Tenebrio molitor	41.5	Liu (2011)
Tenebrio molitor	30.1	Manzano-Agugliaro et al. (2012)
Tenebrio molitor	32.8	Bernard et al. (1997)
Apomecyna flavovittala	35.2	Liu (2011)
Aromia bungii	35.9	Liu (2011)
Macrodactyluslileaticollis	11.7	Ramos-Elorduy et al. (2006)
Melolontha sp.	18.8	Ramos-Elorduy et al. (2006)
Metamasiusspinolae	25.5	Ramos-Elorduy et al. (2006)
Trichoderespini	36.3	Ramos-Elorduy et al. (2006)
Scyphophorusacupunctatus	50.9	Ramos-Elorduy et al. (2006)
Scyphophorusacupunctatus	52.0	Ramos-Elorduy et al. (1997)
Strataegusaloeus	17.0	Ramos-Elorduy et al. (2006)
Aromia bungii	35.9	Liu (2011)
Pachymerus nucleorum	49.3	Ramos-Elorduy et al. (2006)
Passalusaf. punctiger	44.0	Ramos-Elorduy et al. (1997)
Passalus punctiger.	44.0	Ramos-Elorduy et al. (2006)
Paxillusleachei	47.2	Ramos-Elorduy et al. (2006)
Phyllognathus excavatus	15.9	Manzano-Agugliaro et al. (2012)

TABLE 5.1 *(Continued)*

Insect species	Lipid content (%)	References
Phyllophaga sp.	24.0	Ramos-Elorduy et al. (2006)
Holotrichia oblita	29.8	Liu (2011)
Homolepta sp.	18.0	Ramos-Elorduy et al. (1997)
Homolepta sp.	18.4	Ramos-Elorduy et al. (2006)
Rhantus sp.	6.3	Ramos-Elorduy et al. (2006)
Rhynchophorus ferrugineus	13.8	Manzano-Agugliaro et al. (2012)
Rhynchophorus palmarum	38.5	Cerda et al. (1999)
Rhynchophorus phoenicis	31.4	Banjo et al. (2006)
Rhynchophorus phoenicis	47.1	Oliveira et al. (1976)
Callipogon barbatum	34.0	Ramos-Elorduy et al. (2006)
Callipogon barbatus	34.0	Ramos-Elorduy et al. (1997)
Chalcophora sp.	53.7	Ramos-Elorduy et al. (2006)
Oileusrimator	47.0	Ramos-Elorduy et al. (1997)
Oryctes boas	1.5	Banjo et al. (2006)
	Order Orthoptera	
Locusta migratoria	28.5	Manzano-Agugliaro et al. (2012)
Acheta domestica	13.8	Bernard et al. (1997)
Acheta domestica	14.4	Finke (2002)
Acheta domestica	15.8	Manzano-Agugliaro et al. (2012)
Acheta domestica	22.1	Finke (2002)
Zonocerus variegatus	3.8	Banjo et al. (2006)
Zonocerus variegatus	28.0	Mbah and Elekima (2007)
Zonocerus variegatus	31.0	Mbah and Elekima (2007)
Brachytrypes spp.	2.3	Banjo et al. (2006)
Schistocerasp	17.0	Ramos-Elorduy et al. (1997)
Sphenarium histrio	4.0	Ramos-Elorduy et al. (1997)
Sphenarium purpurascens	11.0	Ramos-Elorduy et al. (1997)
Sphenarium sp	4.0	Ramos-Elorduy et al. (1997)
Sphenarium spp	12.0	Ramos-Elorduy et al. (1997)
Heteracris littoralis	8.8	Manzano-Agugliaro et al. (2012)
Melonoplus mexicanus	7.0	Ramos-Elorduy et al. (1997)
Plectotettia nobilis	7.0	Ramos-Elorduy et al. (1997)
Cytacanthacris naeruginosus	3.5	Liu (2011)

TABLE 5.1 *(Continued)*

Insect species	Lipid content (%)	References
	Order Lepidoptera	
Ostrinia nubilalis	17.0	Bernard et al. (1997)
Ostrinia nubilalis	17.2	Bernard et al. (1997)
Ostrinia nubilalis	46.8	Liu (2011)
Aegialehesperiaris	30.0	Ramos-Elorduy et al. (1997)
Antheraea pernyi	31.3	Liu (2011)
Antheraea pernyi	34.5	Liu (2011)
Anapheinfracta	15.2	Banjo et al. (2006)
Ascalapha odorata	15.0	Ramos-Elorduy et al. (1997)
Anapherecticulata	10.2	Banjo et al. (2006)
Arsenuraarmida	8.0	Ramos-Elorduy et al. (1997)
Anaphe venata	23.1	Banjo et al. (2006)
Xyleutesredtembacheri	48.0	Ramos-Elorduy et al. (1997)
Cirina forda	12.2	Akinnawo and Ketiku (2000)
Cirina forda	14.2	Banjo et al. (2006)
Catastictateutila	19.0	Ramos-Elorduy et al. (1997)
Corcyra cephalonica	43.3	Liu (2011)
Latebrariaamphipyrioides	7.0	Ramos-Elorduy et al. (1997)
Pectinophora gossypiella	49.9	Liu (2011)
Phasus triangularis	77.0	Ramos-Elorduy et al. (1997)
Bombyx mori	35.0	Finke (2002)
Bombyx mori	8.0	Ramos-Elorduy et al. (1997)
Galleria mellonella	46.4	Bernard et al. (1997)
Galleria mellonella	60.0	Finke (2002)
Eucheriasocialis	16.0	Ramos-Elorduy et al. (1997)
Heliothis zea	29.0	Ramos-Elorduy et al. (1997)
Hylesiafrigida	10.0	Ramos-Elorduy et al. (1997)
	Order Hymenoptera	
Brachygastraazteca	22.0	Ramos-Elorduy et al. (1997)
Brachygastramellifica	30.0	Ramos-Elorduy et al. (1997)
Polistes instabilis	62.0	Ramos-Elorduy et al. (1997)
Polybia sp.	13.0	Ramos-Elorduy et al. (1997)
Polybia occidentalis bohemani	19.0	Ramos-Elorduy et al. (1997)

TABLE 5.1 *(Continued)*

Insect species	Lipid content (%)	References
Polybiaparvulina	21.0	Ramos-Elorduy et al. (1997)
Polybia occidentalis nigratella	28.0	Ramos-Elorduy et al. (1997)
Apis mellifera	12.3	Banjo et al. (2006)
Apis mellifera	19.0	Ramos-Elorduy et al. (1997)
Apis mellifera	20.0	Ramos-Elorduy et al. (1997)
Atta cephalotes	31.0	Ramos-Elorduy et al. (1997)
Atta mexicana	39.0	Ramos-Elorduy et al. (1997)
Componotussp	15.9	Mbah and Elekima (2007)
Vespula squamosa	22.0	Ramos-Elorduy et al. (1997)
Melipona beeckeii	41.0	Ramos-Elorduy et al. (1997)
Myrmecosistusmelliger	6.0	Ramos-Elorduy et al. (1997)
Oecophyllalonginoda	41.3	Mbah and Elekima (2007)
Order Diptera		
Protophormiaterraenovae	23.6	Manzano-Agugliaro et al. (2012)
Protophormiaterraenovae	28.3	Manzano-Agugliaro et al. (2012)
Aedes sp.	16.1	Bernard et al. (1997)
Musca domestica	8.1–13.5	Atteh and Ologbenla (1993)
Musca domestica	15.8	Bernard et al. (1997)
Musca domestica	20.0	Bernard et al. (1997)
Musca domestica	20.3	Manzano-Agugliaro et al. (2012)
Musca domestica	20.7–25.3	Ramos-Elorduy et al. (1997)
Musca domestica	15.8	Manzano-Agugliaro et al. (2012)
Musca domestica	20.0	Manzano-Agugliaro et al. (2012)
Chironomus sp.	9.7	Bernard et al. (1997)
Copestylum anna	31.0	Ramos-Elorduy et al. (1997)
Copestylumhaagii	31.0	Ramos-Elorduy et al. (1997)
Hermetiaillucens	28–34.8	Newton et al. (2005)
Hermetiaillucens	15.5	Manzano-Agugliaro et al. (2012)
Hermetiaillucens	18.0	Manzano-Agugliaro et al. (2012)
Hermetiaillucens	18.8	Gutiérrez et al. (2004)
Drosophila melanogaster	10.5	Bernard et al. (1997)
Drosophila melanogaster	12.6	Bernard et al. (1997)
Drosophila melanogaster	29.4	Bernard et al. (1997)

TABLE 5.1 *(Continued)*

Insect species	Lipid content (%)	References
Luciliasericata	16.5	Manzano-Agugliaro et al. (2012)
Luciliasericata	25.7	Manzano-Agugliaro et al. (2012)
Eristalistenax	5.8	Manzano-Agugliaro et al. (2012)
Order Blattodea		
Periploneta americana	21.5	Mbah and Elekima (2007)
Periploneta americana	28.4	Bernard et al. (1997)
Order Homoptera		
Hoplophorionmonograma	14.0	Ramos-Elorduy et al. (1997)
Proarna sp.	4.0	Ramos-Elorduy et al. (1997)
Umboniareclinata	33.0	Ramos-Elorduy et al. (1997)
Order Megaloptera		
Chauliodes sp.	19.5	Bernard et al. (1997)
Order Isoptera		
Macrotermesbellicosus	28.2	Banjo et al. (2006)
Macrotermesbellicosus	36.1	Ekpo and Onigbinde (2007)
Macrotermesnigeriensis	28.3	Mbah and Elekima (2007)
Macrotermesnotalensis	22.5	Banjo et al. (2006)
Microtermessp	17.0	Mbah and Elekima (2007)
Order Hemiptera		
Acantocephaladeclivis	45.0	Ramos-Elorduy et al. (1997)
Corisella mercenaria	9.0	Ramos-Elorduy et al. (1997)
Corisella spp.	7.0	Ramos-Elorduy et al. (1997)
Edessa petersii	42.0	Ramos-Elorduy et al. (1997)
Edessa sp	54.0	Ramos-Elorduy et al. (1997)
Euschistusegglestoni	45.0	Ramos-Elorduy et al. (1997)
Pachilis gigas	19.0	Ramos-Elorduy et al. (1997)
Pachilis gigas	26.0	Ramos-Elorduy et al. (1997)

5.4 CONCERNS ABOUT BIODIESEL PRODUCTION

It must be noted that insects have a unique evolutionary history and potential adaptive strategies toward the environment than mammals, which enables them to exploit distinct ecological niches and characteristics involved in product industrialization. Production of biodiesel from insects

TABLE 5.2 Insect Larvae Consumed Different Manure to Produce Biodiesel.

Insect larvae	Biomass	Biodiesel properties	yield	FAME composition %	References
Blacksoldier fly BFS, *Hermetia Illucens*	Fermented wheat bran	Density: 875 kg/m^3 Water content: 0.03 mg/kg Ester content: 98.3% Acid value: < 0.8 mg KOH/g Viscosity: 5.2 mm^2/s Flash point: 121°C Cetane index: 50	94.14%	Not reported	Nguyen et al. (2018)
Blacksoldier fly BFS, *Hermetia Illucens*	Chicken manure	Density: 885 kg/m^3 Viscosity: 5.8 mm^2/s Flash point: 123°C Cetane number: 53	93%	Lauric acid: 35.5 Oleinic acid: 23.6 Palmitic acid: 14.8	Li et al. (2011)
Blacksoldier fly BFS, *Hermetia Illucens*	Fermented wheat bran	Density: 872 kg/m^3 Water content: 0.03 mg/kg Ester content: 97.2% Acid value: 1.1 mg KOH/g Viscosity at 40°C: 5.8 mm^2/s Flash point: 123°C Cetane index: 53	96.18%	Lauric acid: 58.5 Palmitic acid: 11.2 Myristic acid: 8.8 Oleic acid: 7.2	Nguyen et al. (2018)
Blacksoldier fly BFS, *Hermetia Illucens*	Dairy manure	Density: 872 ± 0.3 kg/m^3 Viscosity: 4.5 ± 0.1 mm^2/s Flash point: 121 ± 2.6°C Water sediment: 300 ± 3.7 mg/kg Acid number: 0.8 ± 0.2 mg KOH/g	96%	Capric acid: 3.1 ± 0.03 Lauric acid: 35.6 ± 0.1 Myristic acid: 7.6 ± 0.1 Pentadecanoic acid: 1.0 ± 0.06 Palmitoleic acid: 3.8 ± 0.2 Palmitic acid: 14.8 ± 0.04 Oleinic acid: 23.6 ± 0.3 Linolenic acid: 2.1 ± 0.3 Stearic acid: 3.6 ± 0.1	Li et al. (2011)

TABLE 5.2 *(Continued)*

Insect larvae	Biomass	Biodiesel properties	yield	FAME composition %	References
Oriental Latrine fly OLF, *Chrysomyamegacephala*	Pig manure	Density: 0.89 g/cm^3 Viscosity: 5.1 mm^2/s Flash point: 138°C Cetane number: 56 Water content: 0.02 % Acid value: 0.28 mg KOH/g	96.6%	Palmitic acid: 36.91 Oleic acid: 27.67 Palmitoleic acid: 10.89 Linoleic acid: 9.49	Yang and Liu (2014)
Housefly, *Musca domestica* L.	Swine manure	Density: 881 kg/m^3 Viscosity: 5.64 mm^2/s Flash point: 145°C Cetane number: 52	96.8%	Palmitic acid: 29.1 Oleic acid: 23.3 Palmitoleic acid: 17.4 Linoleic acid: 17.2	Yang et al. (2014)
Yellow mealworm beetle YMB, *Tenebrio molitor*	Decayed vegetables	Density: 860 kg/m^3 Flash point: 127°C Cetane number: 58 Water content: 300 mg/kg Methanol content: 0.2%	96.8%	Linolenic acid: 19.7 Palmitic acid: 17.6 Linoleic acid: 16.3 Stearic acid: 11.4	Zheng et al. (2013)
Blacksoldier fly larvae BFSL, *Hermetia Illucens*	Restaurant kitchen waste	Cloud point (°C); 5.6 Oxidation stability (h); 5 Water content (mg/kg);387 Flash point (°C); 143 Density at 15°C (kg/m^3);875 Cetane number; 49 Kinematic viscosity at 40°C (mm^2/s); 4.6	97%	Not reported	Ishak and Kamari (2019b)

TABLE 5.2 *(Continued)*

Insect larvae	Biomass	Biodiesel properties	yield	FAME composition %	References
Macrotermesnigeriensis	-	Density at 15°C (kg/m^3); 841±12.62 Viscosity (mm^2.s^{-1}, 40 °C); 2.32 ± 0.05 Flash Point (°C); 125 ± 1.87 Pour Point (°C); -15 ± 0.1 Cloud Point (°C); 0 Cetane Number; 51.4 ± 1.02 HH Value (MJ.kg^{-1}); 41.8 ± 0.80 Acid Value (mg KOH.g^{-1}); 0.44 ± 0.01 Water Content (mg.kg^{-1}); 0.04 ± 0.002 Sulfur Content (wt.%); 0.46 ± 0.02 Sodium Content (wt.%); 0.71 ± 0.02 Potassium Content (wt.%); 0.03 ± 0.001 Carbon Residue (wt.%); 0.05 ± 0.001	86.54%	Undecylic acid methyl ester; 0.72 ± 0.03 Myristic acid methyl ester:0 2.58 ± 0.07 Nervonic acid methyl ester 0.40 ± 0.02 Myristic acid methyl ester 2.58 ± 0.07 Oleinic acid methyl ester 45.88 ± 0.92 Palmitic acid methyl ester 34.70 ± 0.69 Palmitoleic acid methyl ester 5.68 ± 0.17 Stearic acid methyl ester 10.04 ± 0.20 Capric acid methyl ester Lauric acid methyl ester Behenic acid methyl ester Linoleic acid methyl ester	Kalu-Uka et al. (2021)
Zophobasmorio Larva, ZML	–	Cold filter plugging point (°C); 2 Cetane number; 57.3 Flash Point (°C); 159 Density at 20°C (kg/m^3); 871.3 Water and sediments (vol %); <0.03 water content (µg g^{-1}) 454 kinematic viscosity (mm^2 s^{-1}); 4.220 iodine number 73.33 molecular weight 851.29 saponification value 199.88 Copper strip corrosion; 1 Acid Value (mg KOH.g^{-1}); 0.15 Sulphated ash content (%); 0.002	92.35%	palmitic acid (C16:0) 32.74 palmitoleic acid (C16:1) 2.16 heptadecanoic acid (C17:0) 0.58 stearic acid (C18:0) 9.36 oleic acid (C18:1) 29.43 linoleic acid (C18:2) 22.53 linolenic acid (C18:3) 0.85 peroxide value; not detected	Leung et al. (2012)

TABLE 5.2 *(Continued)*

Insect larvae	Biomass	Biodiesel properties	yield	FAME composition %	References
Blacksoldier fly larvae BFSL, *Hermetia Illucens*	Organic waste	Density: 875 kg/m^3 Water content: 0.03 mg/kg Ester content: 99.5% Acid value: < 0.8 mg KOH/g Viscosity at 40°C: 5.4 mm^2/s Flash point: 122°C Cetane index: 49 Sulphur content; not detected	96.97%	Nonadecanoic acid methyl ester (C19:0); 4.74 Capric acid methyl ester (C10:0); 1.62 Linolenic acid methyl ester (C18:3); not detected Lauric acid methyl ester (C12:0); 43.22 Linoleic acid methyl ester (C18:2); 2.33 Myristic acid methyl ester (C14:0); 6.99 Oleic acid methyl ester (C18:1); 13.03 Palmitic acid methyl ester (C16:0); 19.67 Stearic acid methyl ester (18:0); 5.52 Palmitoleic acid methyl ester (C16:1); 2.88	Nguyen et al. (2018)
Blacksoldier fly larvae BFSL, *Hermetia Illucens*	–	Density: 875 kg/m^3 Water content: 300 mg/kg Ester content: 98.7% Acid value: < 0.5 mg KOH/g Viscosity at 40°C: 5.3 mm^2/s Flash point: 121°C Cetane index: 50 Sulphur content wt%; 0.04	–	Nonadecanoic acid methyl ester (C19:0); 0.22 Capric acid methyl ester (C10:0); 0.84 Linolenic acid methyl ester (C18:3); not detected Lauric acid methyl ester (C12:0); 31.15 Linoleic acid methyl ester (C18:2); 4.97 Myristic acid methyl ester (C14:0); 5.64 Oleic acid methyl ester (C18:1); 27.80 Palmitic acid methyl ester (C16:0); 22.63 Stearic acid methyl ester (18:0); 5.51 Palmitoleic acid methyl ester (C16:1); 1.24	Su (2019)

TABLE 5.2 *(Continued)*

Insect larvae	Biomass	Biodiesel properties	yield	FAME composition %	References
Blacksoldier fly larvae BFSL, *Hermetia Illucens*	–	–	32.49%	Capric acid methyl ester (C10:0); 3.1 Lauric acid methyl ester (C12:0); 35.6 Linoleic acid methyl ester (C18:2); 2.1 Myristic acid methyl ester (C14:0); 7.6 Oleic acid methyl ester (C18:1); not detected Palmitic acid methyl ester (C16:0); 14.8 Stearic acid methyl ester (18:0); 3.6 Palmitoleic acid methyl ester (C16:1); 3.8	Wang et al. (2017)
Blacksoldier fly larvae BFSL, *Hermetia Illucens*	Rice straw and restaurant waste	Ester content (%): 96.6 Density (kg/m^3): 895 Viscosity at 40°C (mm^2/s) 5.96 Water content (mg/kg):300 Flash point (closed cup) (°C); 123 Cetane number: 45 Acid number (mg KOH/g): <0.6 Cloud Point (°C): 4.2 Methanol orEthanol (m/m): 0.3% Distillation: 91% at360°C	43.8 g	Capric acid: 3.8 Lauric acid: 27.8 Myristic acid: 8.1 Palmitoleic acid: 4.5 Palmitic acid: 14.2 Oleic acid: 22.5 Linolenic acid: 2.1 Stearic acid: 7.6 Linoleic acid (18:2); 1.8 Pentadecanoic acid: 1.5 Heptadecanoic acid: 0.8 Nonadecanoic acid :1.7	Zheng et al. (2012)

TABLE 5.2 *(Continued)*

Insect larvae	Biomass	Biodiesel properties	yield	FAME composition %	References
Yellow mealworm beetle YMB, *Tenebrio molitor*	Decayed vegetables	Ester content (%): 96.8 Density (kg/m^3): 860 Viscosity at 40°C (mm^2/s) 5.9 Water content (mg/kg): 300 Flash point (closed cup) (°C); 127 Cetane number: 58 Acid number (mg KOH/g): 0.9 Methanol or Ethanol (wt): 0.3% Distillation: 92% at 360°C	34.2g	Capric acid: 1.2 Lauric acid: 1.3 Myristic acid: 8.1 Palmitic acid (C16:0): 17.6 Palmitoleic acid (C16:1): 9.3 Heptadecanoic acid (C17:0): 0.58 Stearic acid (C18:0): 11.4 Oleic acid (C18:1): 1.6 Linoleic acid (C18:2): 16.3 Linolenic acid (C18:3):19.7 Pentadecanoic acid: 1.5 Heptadecanoic acid: 1.7 Nonadecanoic acid: 2.6	Zheng et al. (2013)
Blacksoldier fly larvae BFSL, *Hermetia Illucens*	Restaurant waste	Density (kg/m^3): 860 Viscosity at 40°C (mm^2/s) 4.9 Water content (mg/kg): 0.02 Flash point (closed cup) (°C); 128 Cetane number: 58 Acid number (mg KOH/g): 0.6 Distillation: 92% at 360°C Sulphur content: not detected	93.1%	Capric acid methyl ester (C10:0): 1.8 Lauric acid methyl ester (C12:0): 23.4 Myristic acid methyl ester (C14:1): 3.7 Palmitoleic acid methyl ester (C16:1): 9.4 Palmitic acid methyl ester (C16:0): 18.2 Oleinic acid methyl ester (C18:1): 27.1 Linoleic acid methyl ester (C18:2): 7.5 Stearic acid methyl ester (C18:0): 5.1 Nonadecanoic acid methyl ester (C19:1): 3.3	Zheng et al. (2012)

as feedstock depends upon insect's feeding materials which help to produce insect biomass containing significant fat content. Insects have different feeding behaviors such as filter feeders, necrophagous, predators, suspension feeders, fungivorous, coprophagous, xylophagous, saprophagous, and phytophagous. Adult and larval stages often have different feeding behaviors. Insects have a unique life cycle in which they also go in inactive life stages such as the non-feeding period, diapause phase, prolonged fasting, and pupal phase. Feeding behavior affects biomass building in insects which are the key factor in biodiesel production. Scientists are paying attention to various concerns regarding biodiesel production from insects. There are many important questions about this production that must be considered and because they are relevant to biodiesel production which may affect biodiesel production. Here are few questions which might be particularly significant in biodiesel synthesis—

- Which order or species of insects have the most potential for biodiesel production at a large scale?
- What type of rearing process would be cost-effective and less time-consuming?
- Which insect species have high-fat content?
- What type of organic food is viable for insects on which they can feed and produce biodiesel with maximum yield?
- What type of rearing environment is needed for biodiesel production and how do rearing conditions might affect the yield of biodiesel?
- Are the designed rearing conditions helpful to achieve expected the yield of biodiesel and are these conditions could maintain throughout the process?
- How much does this rearing process affect our ecosystem? And how would insect farming disturb the ecological balance?
- Which type of hazardous chemical is currently being used for lipid extraction from insects to produce biodiesel? And can we use less harmful chemicals and a green approach?

So, there is much more to explore about insects and their enzymes, genes, and microbiomes which can be used in biodiesel production and the science involved in this synthesis. Insect utilization in biodiesel production would help to overcome the global anthropogenic challenges, including economic and sustainable fuel production, greenhouse gases emission, environmental degradation by fossil fuels, and minimizing global conflicts about biofuels production.

5.5 INSECT'S FOOD PRODUCTS EXPLORING NEW HORIZONS

According to United Nations, the world population is estimated to reach about 9 billion up to 2050. So, the demand for food would be increased simultaneously, and adverse climate conditions influence food production (Ksenija, 2018). Having high nutritional values and characteristics, edible insects and their food products have become a highly interesting topic (Garofalo et al., 2019). Entomophagy is getting attention worldwide with the growing interest in insect food products, especially a few western consumers showed changes in their diets by including insects as food (Verbeke, 2015; Tucker, 2014). Western behavior toward entomophagy is quite resistant as compared to Latin, African, and Asian cultures where insect's food products are greatly popular with respect to their taste (Hanboonsong, 2010; Ramos-Elorduy et al., 1997). Various studies exhibited that, in near future, insects can meet the nutritional demand as they are a rich source of fatty acids, proteins, and fibers (Ayensu et al., 2019; Patel et al., 2019; de Castro et al., 2018; Giron et al., 2017). Currently, about 1900 species are being consumed by 2 billion people worldwide (Huis et al., 2013). Edible insects are eaten in different phases such as larvae, pupae, adults, eggs, and nymphs. Insects can be consumed in different forms such as cooked as a whole or mixed in other beverages like alcoholic tequila. Less recognizable forms of insect food are preferred such as dried and paste form or usage of protein, fats, and chitin in chocolates, snacks, crackers, bread, biscuits, sausages, and meatloaves (Ayieko et al, 2010). Farming, wild harvesting, and semi-domestication are three main techniques to obtain edible insects. Geographical distribution influences these contributors (Van Itterbeeck, 2014). Insects nutrients are healthy alternatives to beef, pork, chicken, and fish because they contain reduced fat content and higher protein content, calcium, zinc, and iron compared to conventional meat (Huis et al., 2013). As insects are cold-blooded organisms, they are very convenient and efficient to transform feed into significant proteins. For instance, the feed consumption rate of crickets is 12 times less than cattle, four times lesser than sheep, and half quantity compared to pigs to produce an equal amount of protein. The high metabolic rates of insects enable them to save energy as well as natural resources consumption. Insects require a fewer amount of food and space and consume reduced economic sources. Traditional meat products emit greenhouse gases while insects reported less emissions except for few species as such termites and cockroaches. The rearing process of insects often emits ammonia gas but the amount is quite less than the conventional livestock emission (Huis, 2016).

The insect-rearing process produces a maximum amount of food than other livestock and also consumes less time and land thereby spreading less pollution. Due to these significant factors, entomophagy elevated human livelihood regarding economic and social factors. Insect harvesting provides working opportunities to lower sections of society due to less technological know-hows and low investment methods of rearing. For an efficient rearing process, we just need to understand the biology of insects and also the associated biotic and abiotic factors, production rates and yields, environmental risks, regulatory frameworks, methods, food safety measurements, human behavior and demands toward insects products consumption, production challenges, and basic infrastructure for rearing process (Ramos-Elorduy et al., 2006).

5.5.1 EDIBLE INSECT SPECIES

The insect's species with most consumption are (Coleoptera, 31%) butterflies and moths (Lepidoptera, 18%), bees, wasps, and ants (Hymenoptera, 14%), grasshoppers, locusts, and crickets (Orthoptera, 13%), cicadas, aphids, honeyants, leafhoppers, planthoppers, scale insects and true bugs (Hemiptera, 10%), termites (Isoptera, 3%), dragonflies (Odonata, 3%) and flies (Diptera, 2%) (Jongema, 2015). Insects species are consumed in different life phases such as Coleoptera, Lepidoptera, and Diptera are consumed in the larval stage, while Hemiptera, Orthoptera, Hymenoptera, and Isoptera are eaten in the adult phase (Hanboonsong et al, 2013). Some examples of edible insects are given in Table 5.3.

5.5.1.1 LEPIDOPTERA (CATERPILLARS)

About 27 species of the caterpillar are currently consumed in Chiapas, Mexico (Makkar et al., 2014). In South Africa, *Imbrasiabelina* (mopane caterpillar) is a widely harvested and consumed species reported (Ghazoul, 2006; Nyoka and Fao, 2003). *Omphisafuscidentalis: Pyralidae* (bamboo caterpillar) consumption rate is high in Asia (Yhoung-Aree and Viwatpanich, 2005). Caterpillars' growth and reproduction increase when appropriate conditions are provided like 50–70% humidity, required oxygen level, suitable diets such as organic waste, and 29–31°C temperature (Hardouin and Mahoux, 2003a).

5.5.1.2 ORTHOPTERA (LOCUSTS, GRASSHOPPERS, AND CRICKETS)

Approximately 27,000 species of Orthoptera are found worldwide. These species vary in size, feed selection, microhabitat preferences, reproduction, and environmental behaviors and adaptations. Their body length varies between a minimum of 2 mm (antinquiline crickets, Myrmecophilidae) to 567 mm (*Phobaeticuschani*). Mostly, Orthoptera species feed on the grass while sand groper (Cylindrachetidae) is a wheat crop pest (Vickery and Kevan, 1985). Various species of crickets that survive on different temperature ranges—32–35°C—are available in the commercial market (Kvassay, 2014). Rearing of locust are often done in terrariums with grass and wheat bran. Consumption of orthoptera species is found mostly in Asia, South America, and Africa (Alegbeleye et al., 2012; Cohen et al., 2009; Yhoung-Aree and Viwatpanich, 2005).

TABLE 5.3 Edible Insect Species Worldwide.

Insect species order	Numbers of species	References
Hymenoptera (ants, bees, and wasps)	351	Ramos-Elorduy (2002)
Thysanura (silverfish)	1	Ramos-Elorduy (2002)
Coleoptera (beetles)	468	Ramos-Elorduy (2002)
Anoplura (lice)	3	Ramos-Elorduy (2002)
Diptera (flies, mosquitos)	34	Ramos-Elorduy (2002)
Ephemeroptera (mayflies)	19	Ramos-Elorduy (2002)
Trichoptera (caddis flies)	10	Ramos-Elorduy (2002)
Orthoptera (grasshoppers, cockroaches, crickets)	267	Ramos-Elorduy (2002)
Lepidoptera (butterflies, moths silkworms)	253	Ramos-Elorduy (2002)
Odonata (dragonflies)	29	Ramos-Elorduy (2002)
Neuroptera (dobson flies)	5	Ramos-Elorduy (2002)
Isoptera (termites)	61	Ramos-Elorduy (2002)
Homoptera (cicadas, leafhoppers, mealybugs)	78	Ramos-Elorduy (2002)
Hemiptera (true bugs)	102	Ramos-Elorduy (2002)

5.5.1.3 COLEOPTERA (BEETLES)

Various types of edible beetles are available; aquatic 78 species of Dytiscidae, Gyrinidae, and Hydrophilidae families exist more profoundly (Ramos-Elorduy, Moreno and Camacho, 2009). Beetles are so simple to grow by using plant materials such as wheat, oat, and maize; animal and microbial products like meat waste, feathers and yeast; and fruits, and vegetables such as potatoes, carrots, and lettuce (Veldkamp et al., 2012; Ramos-Elorduy et al., 2002). Larvae of beetles (mealworms) are considered as an efficient protein source, and they require three months for their growth. Mature larvae grow with light yellow-brown color and have body length 20–32 mm and weight 130–160 mg.

5.5.1.4 HYMENOPTERA (WASPS, BEES, AND ANTS)

Hymenoptera varies in size starting from 0.2 mm (*Trichogrammatidae*) to 5 cm (Asian giant hornet), bee's size varies between 2 mm and 39 mm (*Megachilepluto*); furthermore, ants size range is 0.75–60 mm (*Titanomyrmagiganteum*). Hymenoptera has potential adaptative ability toward adverse environmental conditions. Rearing processes and conditions vary from species and developmental phases including temperatures, humidity, feeding materials, and oxygen concentration. Feeding material often used is freshly killed or living insects, fruits, seeds, and nectar. Hymenoptera is mostly consumed in Japan and China (Stabentheiner et al., 2010).

There are a lot more insect species that are widely used in food products such as red weevil *Rhynchophorusferrugineus* worldwide. Homoptera is used as direct food or as a food product with different ingredients (*cicadas, leafhoppers, planthoppers, and scale insects*) (Yen, 2005). Hemiptera is most often consumed in African and Mexican countries and provides cooking oil and processed food products (Ramos-Elorduy et al., 2006). Isoptera also provides oil for cooking and termites are consumed as a side dish in many countries (Kinyuru et al., 2009). Diptera (flies), BFSL, *HermetiaIllucens* are utilized as animal feed, not as human food (Diclaro II and Kaufman, 2009; Hardouin and Mahoux, 2003b). Table 5.4 represents the different insect species and their nutrient content.

TABLE 5.4 Edible Insects and Their Nutrient Content Consumed Worldwide.

Insect specie	Fat and nutrient contents	References
Silkworms	Protein content: 10–17% fresh wt. or 52–76% dry wt. Lipid content: 37% High polyunsaturated fatty acid: linolic acid (18:3) 11–45% Vitamin E Acid detergent fiber 6–12% dry wt. Low Ca	Huis et al. (2013); Tong et al. (2011); Finke (2002)
Weaver ant	Protein content:26–48.5% dry wt Lipid content: 10–25 % fresh wt Cholesterol 4.4–11.7% Ca, Na, Fe, P, Mg, Zn, Vitamin B1, B2, B3 Fibers	Borgohain et al. (2014); Huis et al. (2013); Rumpold and Schlüter (2013); Sribandit et al. (2008)
Termites	Protein content; 32–38% dry wt, fresh wt. 64% rich in Trp; 13–28% fresh wt Lipid content: 49% dry wt Palmitic acid: 30%, oleic acid: 48%, Stearic acid: 9%, linoleic acid: 34–43% fresh wt Fe and Ca	Chakravorty et al. (2013); Huis et al. (2013); Santos et al. (1999)
Melon bug	Protein content: 27.0% dry wt. Lipid content: 45% dry wt Cholesterol 1.4 and sterol 17mg/100g oil, oleic acid: 46.5%, linoleic acid: 44.2%, palmitic acid and linolenic acid: trace quantity Tocophenol: 0.3 mg/100 g oil	Huis et al. (2013); Mariod (2013)
Blacksoldier fly	Protein content: 40–44% dry wt lysine rich Lipid content; 15–49% dry wt Omega 3 fatty acid: 0.2–3%, lauric acid: 21–43%, Oleic acid: 12–32%, Palmitic acid: 11–16% Crude fiber: 7% dry wt Mn, Ca, Zn, P	Huis et al. (2013); Newton et al. (1977); Newton et al. (2005); Arango Gutiérrez et al. (2004)
Stink bug	Protein content: 33–36% dry wt Lipid content: 50–63% dry wt Oleic acid, myristoleate, (Z)-9-hexadeaenoic acid, (Z)-9-octadeaenoic acid, octadeaenoic acid (Z,Z)-9, linoleic acid, 12-octadecadienoic Neutral and acid detergent fiber Luteolin Rutin, quercitrin, apeginin, P, K, Ca, Na, Mg, Fe	Musundire et al. (2016); Huis et al. (2013); Teffo et al. (2007)

TABLE 5.4 *(Continued)*

Insect specie	Fat and nutrient contents	References
Grasshopper	Protein content; 13–28% fresh wt, 57–69% dry wt; low lyc and Met High glutamic acid and glutamine Lipid content: 3–67% dry wt Stearic acid: 12.4%, Palmitoleic acid; 28%, Palmitic acid: 17.2%, linoleic acid: 16–46%, oleic acid: 21%, linolenic acid; 16–25% Acid detergent fiber: 10% Crude fiber: 8.5–12.5%, neutral detergent fiber: 22%, P, Zn, Fe, Cu, Mn, Ca	Ghosh et al. (2016); Chakravorty et al. (2013); Huis et al. (2013); Alegbeleye et al. (2012)
Palm weevil beetle	Protein content: 7–36% fresh wt, 40% dry wt Lipid content: 54% dry wt delicious due to extra fat content Palmitoleic acid: 38%, linoleic acid: 45% Vitamin E	Huis et al. (2013); Fasoranti and Ajiboye (1993)
Ground cricket	Protein content: 48–67% dry wt, low in Lys, Met, Cys, high glutamic acid Lipid content: <5–>20 depends on species type Palmitoleic acid: 20–26%, palmitic acid: 50%, linoleic acid: 36–40%, stearic acid: 9–32% Vitamin B12 Fiber 8.7–18% risk of contamination Zn, Cu, Fe	Chakravorty et al. (2013); Huis et al. (2013); Bukkens (2005); Finke (2002)
Mealworm larvae of beetles	Protein content: 14–25% fresh wt, 47–64% dry wt High glutamic acid and methionine limited Lipid content: 31–43% dry wt Palmitic acid: 21%, oleic acid: 38%, linoleic acid: 27% Vitamin B12, acid detergent fiber: 6.5%, neutral detergent fiber: 12%, Vitamin A (retinol and carotenene), Zn, Fe, Cu, Mn, K, P	Huis et al. (2013); Bukkens (2005); Martins and Pliner (2005); Finke (2002)
Mopane caterpillar	Protein content; 48–80% dry wt Lipid content: 16–20% fresh wt Specific lipid content out of total fat: 40% (may be fresh wt) Fe, Zn, Ca, k, P, Mg	Huis et al. (2013); Bukkens (2005)

5.6 INSECTS UTILIZED AS ANIMAL FEED

Besides human consumption, insects have been in use as animal feed worldwide. According to research, world feed production is 870 million tons, and the manufacturing cost of global commercial feed is US $350 billion (Huis et al., 2013). Silkworms, mealworms, grasshoppers, crickets, BSF, beetles, locusts, and housefly are most profoundly utilized as animal feed (Makkar et al., 2014) (Table 5.5). BSFs are wildly used as animal feed and are mostly found on the poultry and pig manure and also on distinct organic waste, for instance, coffee beans, fruits and vegetable decay, distillers waste, and fish offal. BSF is highly efficient in consuming organic waste so, they rapidly reduce organic waste and manure mass, moisture content, and an adverse odor that is problematic toward a clean environment. They are utilized as feed material to poultry, pig, cattle, and fish (Newton et al., 1977). However, BFS is not associated with human food because they are considered a nuisance; moreover, their high crude fat content is consumed in biodiesel production. For instance, 1000 BFSL grow on distinct manure such as 1 kg of chicken and cattle, and pig manure produce 36 g, 58 g, and 91 g of biodiesel (Li et al., 2011). Common house fly larvae are consumed in tropical regions as animal feed, for example, maggots (larvae of *M domestica*) are commonly used as animal protein for poultry. Maggots can be used as fresh wt but they contain 54% crude protein in 30% dry wt which is more than fresh wt. Storage and transport of maggots are more convenient in dry form. Maggots farming is beneficial in manure decomposition and it is an appropriate alternative to fish meal (Hwangbo et al., 2009). Termites' usage can be found in capturing fish and birds. For example, snouted termites (*Trinervitermes* spp.) are used as fish bait to capture insectivorous birds such as guinea fowl, francolins, quails, and thrushes, but termites farming is highly prohibited due to their difficult rearing conditions and extensive methane gas emission (Hackstein and Stumm, 1994). Mealworms such as *T. molitor* are used as feed to broiler chickens and they have been farming extensively on an industrial scale for feed production. Mealworms grow on low nutritive waste and play a good role as an alternative protein source (Ramos-Elorduy et al., 2002).

5.6.1 A NOVEL APPROACH OF INSECTS REARING/ INDUSTRIALIZATION FOR PRODUCING FOOD AND FEED PRODUCTS

The increasing demand for food creates new techniques and methods for insect industrialization including smart breeding, farming, wild harvesting, post-harvesting, and semi-cultivation technique. In food and

TABLE 5.5 Types of Insect Species Used as Animal Feed, Their Source, and Developmental Phase.

Insect specie	Source	Developmental phase	References
Panchoda marginata (sun beetle)	Farming	Larval stage	Jansson and Berggren, (2015); Oonincx et al. (2010)
Bombyx mori (mulberry silkworm)	Farming	Larvae, pupae	Gahukar (2016); Paul and Dey (2014); Huis et al. (2013)
Hermetiaillucens (blacksoldier fly)	Farming	Larvae	Rumpold and Schlüter (2013)
Tenebrio molitor (yellow mealworm)	Farming	Larvae	Makkar (2018); Gahukar (2016); van Broekhoven et al. (2015); Yang et al. (2015)
Locusta migratoria (locust)	Farming, wild harvesting	Adult, nymphs	Bednářová et al. (2013); Huis et al. (2013); Oonincx and Poel (2011)
Acheta domesticus (house cricket)	Farming	Adult	Gahukar (2016); Lundy and Parrella (2015)
Luciliasericata (green bottlefly)	Semi-cultivation	Larvae	Cortes Ortiz et al. (2016); Makkar et al. (2014); Rueda et al.. (2010)
Gryllusbimacalatus Mediterranean (field cricket)	Farming	Adult	Cortes Ortiz et al. (2016)
Musca domestica (housefly)	Farming	Larvae	Halloran et al. (2016); van Zanten et al. (2015)
Insect specie	Source	Developmental phase	References
Panchoda marginata (sun beetle)	Farming	Larval stage	Jansson and Berggren, (2015); Oonincx et al. (2010)
Bombyx mori (mulberry silkworm)	Farming	Larvae, pupae	Gahukar (2016); Paul and Dey (2014); Huis et al. (2013)
Hermetiaillucens (blacksoldier fly)	Farming	Larvae	Rumpold and Schlüter (2013)
Tenebrio molitor (yellow mealworm)	Farming	Larvae	Makkar (2018); Gahukar (2016); van Broekhoven et al. (2015); Yang et al. (2015)
Locusta migratoria (locust)	Farming, wild harvesting	Adult, nymphs	Bednářová et al. (2013); Huis et al. (2013); Oonincx and Poel (2011)

TABLE 5.5 *(Continued)*

Insect specie	Source	Developmental phase	References
Acheta domesticus (house cricket)	Farming	Adult	Gahukar (2016); Lundy and Parrella (2015)
Luciliasericata (green bottlefly)	Semi-cultivation	Larvae	Cortes Ortiz et al. (2016); Makkar et al. (2014); Rueda et al. (2010)
Gryllusbimacalatus Mediterranean (field cricket)	Farming	Adult	Cortes Ortiz et al. (2016)
Musca domestica (housefly)	Farming	Larvae	Halloran et al. (2016); van Zanten et al. (2015)

feed production from insect biomass, it is important to identify the most suitable and effective techniques in which insect handling, rearing conditions, processing, and packaging are maintained. Advancement in rearing technologies would take food production and development to a new level for the wellness of consumers worldwide. Conventionally, insects are killed followed by freezing, boiling, and drying before cooking and further processing involves removal of legs, wings, and other body parts; however, whole insects frying is sometimes practiced. To attain distinct edible insect dishes, various cooking methods are practiced such as baking, frying, boiling, braising, deep-frying, smoking, steaming, and roasting (Chung, 2010; Fasoranti and Ajiboye, 1993). Caterpillars processing involved initial deglutition and then boiling in salted water to increase shelf life (Stack et al., 2003). Termites are usually eaten as a main dish or side dish in many cities (Kinyuru et al., 2009). A combination of mealworms and chickpeas as spicy Mexican leguminous food snacks are being consumed widely in restaurants and canteens (van Huis, 2015). Another mealworm snack with a mixture of cassava named Crikizz was awarded a culinary innovation prize at the national French contest Eco-trophélia in 2012. Integration of grinding and milling methods to extract proteins and other bioactive compounds from insects by making paste and powder is considered a very effective technique (Costermans, 1955). Edible insects are consumed in a different form for dietary purposes as a whole or in the form of food ingredients to make other dishes such as red colorant carmine (E120) obtained from the cochineal insect, and the melon bug powder often used in culinary. Toxicity, poison, and bacteria are associated with insects, in this case, appropriate handling and processing methods would help to eliminate these toxins (Teffo et al., 2007; Amadi et al., 2005; Gardiner and Gardiner, 2003). For instance, stink bugs can be used after several washing cycles and boiling to eliminate excessive pheromones.

Acanthoplusspiseri Brancsik is eaten after the removal of its head and gut and boiling for at least 5 hours before frying in oil. Appropriate storage and packaging increased shelf life and reduced toxicity (Musundire et al., 2016; van Dijk and Meijerink, 2014; Adamolekun, 1993; Meyer-Rochow, 1975).

For the industrial production of insects, two units are required reproductive and production units. The reproductive unit provides an environment for insects to lay eggs and in the production, unit eggs, growth takes place on a substrate. Larval rearing lasts until the adult larval stage, while trays are used in the case of BFS and mealworms. Substrate added during the rearing process could influence insect growth and chances of attack of predators, microorganisms, and parasitoids may increase (Eilenberg et al., 2015). At the harvest stage, mealworms and prepupae of BFS reached the last larval stage and substrate should remove. The substrate contains chitosan and its derivatives which trigger the plant growth and defense system, so the leftover substrate can be used as fertilizers (Sharp, 2013). In the later stage, larval decontamination and fat removal take place. After the drying process, it could be ground into powder for insect feed, and further protein, fat, and chitosan extraction can be made and utilized for various purposes. Depletion of insect species is increasing due to wild harvesting (Yen, 2015). Therefore, post-saving insect harvesting and processing techniques are required (Kelemu et al., 2015). Published data regarding insect rearing from biomass and agricultural organic waste, and mass technology is promising yet, there is a need to explore more processing techniques for a sustainable insect culture and minimize species depletion. To increase the production level of insects-based products to more sustainable, low-cost, less hazardous, fewer nutrient resources would be required (Cortes Ortiz et al., 2016). So, food biomass waste comprises rich ingredients used in rearing and edible insect production as food (Jansson and Berggren, 2015; Huis et al., 2013; Durst and FAO, 2010) (Table 5.6).

The types of artificial diets used in insect rearing or mass production exist in the form of liquid and solid phases depending on various factors such as nutritional requirements of insects in terms of micro, macronutrients, and minerals. Adaptation strategies regarding feeding are another factor in selecting appropriate diets. The insects containing biting mouthparts are solid feeders and the insect species contain sucking mouthparts are liquid feeders. Furthermore, piercing species pierce the host skin and suck the required liquid (Cortes Ortiz et al., 2016; Morales-Ramos et al., 2014). Artificial diets can be manufactured, for example, liquid material used after encapsulation of specific materials such as paraffin, PVC, polyethylene, and polypropylene to avoid artificial eggs formation (Morales-Ramos et al., 2014). To increase the efficiency of a liquid artificial diet, they can be dried and mixed with other ingredients to make them concentrated for consumption. Semi liquids can be used in the form of the pellet which can use both insects have sucking and biting type of mouthparts (Cortes Ortiz et al., 2016). Solid diets can be used with grinding and mixing other raw materials by pelleting and extrusion. Solid feed is also consumed in

TABLE 5.6 Types of Edible Insect Species and Their Order Reared on Different Food Waste.

Insect specie	Family	Order	Development phase	Reared food material	References
Teleogryllustestaceus (Cambodian field cricket)	*Gryllidae*	Orthoptera	Adult	Waste from mug bean sprout production spent grain	Miech et al. (2016)
Tenebrio molitor L. (mealworm)	*Tenebrionidae*	Coleoptera	Larvae	Spent grain, bear yeast, bread residue, potato peeling waste, biscuits remain, maize distillers' dried grains with soluble	van Broekhoven et al. (2015)
Acheta domesticus (house cricket)	*Gryllidae*	Orthoptera	Adult	Municipal food waste, grocery store food waste after aerobic enzymatic digestion	Lundy and Parrella (2015)
Tenebrio molitor L. (mealworm)	*Tenebrionidae*	Coleoptera	Larvae	Denatured soybean meal, mushroom spent corn stover, spirit distillers' grains	Zhang et al. (2019)
Musca domestica (housefly)	*Muscidae*	Diptera	Larvae	Wheat bran, a mixture of egg contents, hatchery waste	Makkar et al. (2014)
Zophobasatratus Fab. (mealworm)	*Tenebrionidae*	coleoptera	Larvae	Bread remains, biscuits remain, spent grains and beer yeast, potato steam pealing, maize distillers' dried grains with soluble	van Broekhoven et al. (2015)
Hermetiaillucens (blacksoldier fly)	*Stratiomyidae*	Diptera	Larvae	Garden waste, plant tissue waste, food scarps, catering waste, compost tea	Ewusie et al. (2018)
Alphitobiusdiaperinus (mealworm)	*Tenebrionidae*	coleoptera	Larvae	Bread remains, biscuits remain, spent grains and beer yeast, potato steam pealing, maize distillers' dried grains with soluble	van Broekhoven et al. (2015)

the encapsulation phase using coacervation technology in which material is coated with protein and polysaccharides (Xiao et al., 2014).

The selection of appropriate edible insect species is the crucial step in production. Selection depends on various factors such as size, reproduction rate, life cycle, adaptations to survive in high densities, population dynamics, abundance in nature, reduced susceptibility toward diseases and parasitism, survival rate, life cycle, short larval stage, increased biomass, egg hatchability, low feed cost, social behavior, nutrient and bioactive quality, safety, epidemic tendency, high yield of protein content, high and easy storage capacity, market value, and ecological consequences (Rumpold and Schlüter, 2013). Advancement in potential knowledge about edible insect biology and ecology would drive technological improvement in insect breeding and farming. High yield of insect biomass could be produced by selecting the appropriate processing methods and effective habitat management system knowing their biological and ecological characteristics, such as temperature, humidity, oxygen, rearing container, and water facilitation (Tchuinkam et al., 2011).

Usually, insect rearing consumed shorter space, for instance, small cages ($50 \times 44 \times 20.5$ cm^3) are utilized to grow 6000 crickets per day (Parajulee et al., 1993). Various species reared on organic waste such as manure of pigs, cattle, and poultry, vegetables and fruits decay, distillers' grains, food waste, and fish offal, and also human waste (Huis et al., 2013; Newton et al., 1977, 2005). Using these manures for insect rearing solves various environmental issues regarding waste disposal. Harvesting techniques could be updated to attain a high yield of insect biomass. Nanotechnology and biotechnological approaches currently supporting pest management, insect food contamination, and human and animal diseases management. But the success in these technologies involving genetic manipulation, recombinant DNA technology, genomics, fermentation, and enzymatic technology would advance insect rearing, breeding, and processing methods to transform insect biomass into consumer's food, feed, medical products, and diagnostic devices. The nanotechnological approach can support gene and protein manipulation of insects to attain targeted healthy outcomes involving the nutritional and quality requirements of consumers. Nanotechnology could elevate the efficiency of insect-based products and their basic ingredients which ultimately improve the bioabsorption and delivery of bioactive components (Sonkaria et al., 2012; Bruce et al., 1997; Will et al., 1996). Biopolymer's manipulation and altering structures of protein and carbohydrates could increase special properties of food products such as strength quality, coloration, the longer shelf life for packaging, and antimicrobial properties (Fecheyr-Lippens et al., 2015; Igic et al., 2015; Yabutani et al., 2014). Various colors can bring to food products by absorption of different wavelengths by food pigments, or by mixing the pigments and nanoscale structure following interaction with light wavelength (D'Alba et al., 2012; Bagnara et al., 2007).

Different colors variations depend on the angle and intensity of wavelength, for example, in New Zealand Waterhouse group reported gold and silver coloration in *Chrysinaaurigans* and *Chrysinachrysargyrea* beetles, respectively. Pigments and structural components are present in the cuticle layers of beetles which produce fascinating colors by interacting with light (Sun-Waterhouse et al., 2016). Therefore, protein and carbohydrates manipulation not only impart nutritional improvement but also provide potential adaptable colors which give appealing visual impacts to customers. In this regard, the utilization of nanotechnology and biotechnology in insect industrialization will be significant to produce delicate insect food products with high nutrient and bioactive profiles in food sectors.

5.7 BIOACTIVE/BIOLOGICAL COMPONENTS ASSOCIATED WITH INSECTS AND THEIR POTENTIAL APPLICATIONS

Advanced technologies such as genome sequencing, analytical chemistry, and molecular biology promote precise and accurate analysis of micro amounts of biological compounds that exist in insects having efficient interaction with industrial products (Harvey et al., 2015; Johnston et al., 2015; Kuhlisch and Pohnert, 2015; Vizcaino et al., 2014; Zotchev et al., 2012; Davies, 2011) (see Table 5.7). Advanced processing methods and technologies enable maximum retention of nutrients and biological components in insects' food and food products (Kinyuru et al., 2010). Appropriate biological, chemical, and physical processing methods ensure the production of protein and polysaccharides-rich food products, required lipid content according to human health requirements (Steen et al., 2016; Guimei, 2011). Ascorbic acid contains aqueous protein extraction of insects such as *Tenebrio molitor*, *Zophobasmorio*, *Alphitobiusdiaperinus*, *Acheta domesticus,* and *Blapticadubia* following sieving, centrifugation, and freeze-drying produce proteins containing a high level of amino acid (Yi et al., 2013). By modifying the processing conditions such as heating, mixing, centrifugation, sonication, pumping, and filtration can the structure and properties of isolated biopolymers in insects (Sun-Waterhouse et al., 2016). Protein structural changes could be induced by hydrolysis and fermentation process (proteolytic treatment) (Sun-Waterhouse et al., 2016; Zheng et al., 2015). Protein content in mopane caterpillar is increased from 20% (raw) to 48%, 57%, or 37%, via dry roasting, drying, or smoking (Ademolu et al., 2010; Bukkens, 2005). Several methods such as toasting involve stainless steel cooking pan with an open flame for 5 min at 150°C, toasting with drying and solar drying were reported to modify protein structure and digestibility and reduced retinol and riboflavin in termites. These treatments reduced protein digestibility and niacin content in brown and green grasshoppers (Kinyuru et al., 2010). Rheological properties and amino acid-rich proteins for food products from yellow mealworms can obtain via an extraction process with specified conditions (Zhao et al., 2016). Yellow mealworm beetle gives hyperactive antifreeze protein rich in threonine and cysteine by gel exclusion chromatography and reverse-phase high-performance liquid chromatography (Graham et al., 1997). Glutamine-rich proteins obtained from *Tenebrio molitor* exhibit defensive properties and encapsulating functionalities. This protein can encapsulate hemocytes and extracellular hemolymph components (Cho et al., 1999). *Bombyx mori* has been in use in food and drugs for many years in various countries such as Japan, Korea, China, India, and Thailand due to its diabetes-ameliorating, immune-enhancing capacity, and liver-protecting functions found in lipids, proteins, and polysaccharides (Zhou and Han, 2006). Black weaver ants are consumed as various tonics and food products while large black ants increase immune system efficiency in mice (Lirong Shen et al., 2006; Wang et al., 2001).

Industries currently producing food and feed products from insects through mass farming and breeding leads to the accumulation of a large amount of insect frass which creates various environmental issues. To reduce these problems, frass usage should practice establishing sustainable environment, for example, elaboration of biochar (solid product obtained from biomass pyrolysis) (He et al., 2020; Yang et al., 2019). Insect frass could impact nutrient cycle and exchange of nutrients in soils, so, insect frass can be employed to detect the insect diversity in soil (Sweetapple and Barron, 2016). Insect frass can be used as fish feed due to its high-nutrient qualities, for example, the frass of *Hermetiaillucens* is used as feed for channel catfish *(Ictalurus punctatus)* (Yildirim-Aksoy et al., 2020). Plants often lack nitrogen content which is significant for plant growth leading to denitrification, microbial competition, and soil erosion (Goulding et al., 1998). Insects are an integral source of nitrogen as they impart in the nitrogen cycle in soil, therefore, they provide nitrogen content to plants (Behie and Bidochka, 2013). Insect frass is the most studied fertilizer as a nitrogen source to plants. Field experiment reported high efficiency of this fertilizer to provide nitrogen content to plants, for example, cabbage moth *(Mamestrabrassicae)* provides nitrogen content to cabbage plants (*Brassica rapa var. perviridis*) and support accumulation of inorganic nitrogen and ammonium content in plant leaves on interaction with fertilizer (Kagata and Ohgushi, 2012, 2013).

In this regard, various studies reported about defoliating insects whose dropping contribute as nitrogen supply to plants they feed on, for instance, *Chorthippuscurtipennis and Melanoplus borealis* provides nitrogen to soil while feeding grass (Fielding et al., 2013) (see Table 5.8). Insect frass production of the eucalyptus-defoliating beetle *Paropsisatomaria* and the lepidopteran *Doratiferaquadriguttata* is approximately 160 and 270 kg/ha in 2 years and nitrogen deposition is 2–4 kg (Gherlenda et al., 2016). This high contribution of nitrogen in the soil promotes plant growth to maximum level by accumulating in plant tissues, which is described in distinct studies such as *Trirhabda virgata* on goldenrod (Uriarte, 2000), the defoliation moths *Malacosoma americanum,* and *Orgyialeucostigma* on red oak (Frost and Hunter, 2007), the gypsy moth (*Lymantria dispar*) and the forest tent caterpillar (*Malacosoma disstria*) on trembling aspen (Madritch et al., 2007).

The defoliation moth *Lymantria monacha* on *Pinus sylvestris* enhances nitrogen concentration in soil but trees often do not respond in high nitrogen concentration and deposition by insect frass, therefore, they accumulate this nitrogen as amino acid and protein form as a defensive strategy (Grüning et al., 2017). Nitrogen content also improves roots growth via amides. Searching economical use of plastic reconversion in the agriculture field, *Zophobasmorio* beetle while feeding on polystyrene produces insect frass rich in amides, enhances plant growth in terms of root and shoot length in *(Hylocereusundatus)* (Gan et al., 2021). Insect frass not only provides nitrogen content to the soil but also acts as a reservoir for other essential nutrients such as the defoliation moth *M.*

TABLE 5.7 Biological Compounds Isolated from Different Insect Species and Their Applications.

Genus	Family	Biological components	Functionality	References
Lepidoptera (butterflies and moths)				
Bombyx mori (silkmoth)	*Bombycidae*	20-Hydroxyecdysone	Analgesic effects	Tulp and Bohlin (2004)
Byasapolyeuctestermessa (Taiwan butterfly)	*Papilionidae*	Papilistatin	Antiproliferative	Pettit et al. (2010)
Hymenoptera (bees, wasps, ants, and sawflies)				
Polybiapaulista (wasp)	*Vespidae*	Polybioside	Neuroactive	Saidemberg et al. (2010)
Pergidae (sawfly)	*Symphyta*	Macrocarpal; grandinol	Antimicrobial	Vihakas et al. (2010)
Polyrhachis dives (black ant)	*Formicidae*	(±)-polyrhadopamine A, (±)-polyrhadopamine B, polyrhadopamines C–E; Troline; and dopamine derivatives	Anti-inflammatory, reno protective, anti-proliferative	Tang et al. (2015); Tang et al. (2014); Cope et al. (2007)
Tetramoriumsp. (red ant)	*Formicidae*	Coumarins, N- (2-hydro-xyethyl)-benzamide, alkaloid N-phenyl-3,4,5-trihydroxybenzamide	Antimicrobial	Song et al. (2012)
Tetraponerarufonigra, (iron ant)	*Formicidae*	Tetraponerines and its derivatives	Antimicrobial	Bosque et al. (2014); Rouchaud and Braekman (2009)
Solenopsisinvicta (fire ant)	*Formicidae*	Solenopsin	Antiangiogenic	El-Tantawy (2015); Alonso-Castro (2014); Heinen and Gorini da Veiga (2011); Arbiser et al. (2007)
Solenopsis germinate (fire ant)	*Formicidae*	Akaloids	–	Yu et al. (2014); Chen et al. (2012)
Orthoptera (grasshopper and crickets)				
Brachystola magna (Texas grasshopper)	*Romaleidae*	Pancratistatin, narciclasine, ungeremine	Antiproliferative	Pettit et al. (2005)
Coleoptera (beetles)				

TABLE 5.7 *(Continued)*

Genus	Family	Biological components	Functionality	References
Blapsjapanensis (stink beetle)	*Tenebrionidae*	Blapsols and other dopamine dimers	Anti-inflammatory	Yan et al. (2015)
–	Synthetic	N-benzylcantharidinamide	Antiproliferative	Lee et al. (2014)
–	Synthetic	Synthetic cantharidin and norcantharidin derivatives	Antiproliferative	Shimizu et al. (2012); Tarleton et al. (2012)
–	Synthetic	N-farnesyloxy-norcantharimide; N-farnesyl-norcantharimide	Antiproliferative	Wu et al. (2014)
–	Synthetic	Synthetic cantharidin and norcantharidin derivatives (bis-trichloroethyl and diphenyl)	Antiproliferative	Robertson et al. (2011)
Catharsiusmolossus (dung beetle)		Actyldopamine dimers (Molossusamides) and dopamine dimer	Anti-inflammatory	Lu et al. (2015)
Bruchidius dorsalis (seed beetle)	*Bruchidae*	Antioxidant lipids (dorsamin)	Antioxidant	Hirose et al. (2013)
Hycleus (blister beetle)	*Meloidae*	Monoterpene cantharidin (CTD) (R)- (±)-palasonin	–	Mebs et al. (2009); Dettner et al. (2003)
Hycleuslunata (blister beetle)	*Meloidae*	Cantharimide, andpalasonimide	–	Dettner et al. (2003)
Various sp. (blister beetle)	*Meloidae*	Cantharidin	Antiproliferative; immunomodulatory	Alzoubi et al. (2015); Wang et al. (2015); Hsia et al. (2014); Hsiao et al. (2014); Zhang et al. (2014); Kim et al. (2013); Twu et al. (2012); Verma and Prasad (2012); Hsieh et al. (2011); Mebs et al. (2009)

TABLE 5.7 *(Continued)*

Genus	Family	Biological components	Functionality	References
Mylabrisphalerate (blister beetle)	*Meloidae*	5-hydroxycantharimide and norcantharidin (NCTD)	–	Puerto Galvis et al. (2013)
Hemiptera (true bugs, aphids, cicadas, and scale insects)				
Kerrialacca (lac insect)	*Kerriidae*	Shellolic acid,	Antimicrobial	Lu et al. (2014)
Kerrialacca (lac insect)	*Kerriidae*	Sesquiterpene acids	–	Lu et al. (2014)
Aspongopus chinensis (Chinese stinkbug)	*Pentatomidae*	Aspongamide	Active against kidney diseases	Yan et al. (2014)
Aspongopus chinensis (Chinese stinkbug)	*Pentatomidae*	Aspongopusamides	Anti-inflammatory	Shi et al. (2014)
Aspongopus chinensis (Chinese stinkbug)	*Pentatomidae*	Aspongnoids, asponglactam	–	Shi et al. (2014)
Aspongopus chinensis (Chinese stinkbug)	*Pentatomidae*	Sesquiterpenoids and other small molecules	Promote neutral stem cell proliferation	Di et al. (2015)
Blattodea (cockroaches and termites)				
Polyphagaplancyi (Chinese cockroach)	*Corydiidae*	Plancyamides, plancypyrazine, plancyols	Antiproliferative	Zhu et al. (2016)
Periplaneta americana (American cockroach)	*Blattidae*	Isocoumarinsperiplatins	Antiproliferative	Luo et al. (2014); Zhang et al. (2013)
Eupolyphaga sinensis (Chinese cockroach)	*Corydiidae*	Isocoumarins and an alkaloid	Antiproliferative	Jiang et al. (2012)

*americanum*provides carbon, total nitrogen, and ammonium content to red oak from insect frass also enhance anti-microbial activity in soil (Frost and Hunter, 2004). Various insects feed on dead wood and provide nutrients to the soil ecosystem in return, for example, *Imbrasiabelina* feeds on *Colophospermum mopane* and produces insect frass rich in nitrogen, phosphorus, and potassium (Swardt et al., 2018). *T. molitor* also produces insect frass rich in these nutrients. Poveda et al. described an association between feed used by insects and frass containing nutrients diversity. By changing insects feeding material, nutrients variety changed in the organic fertilizer form. He obtained insect frass from mealworms with NPK content of 3-2-2 (g/100 g) and iron content balance of 140mg/kg in pots which increases the growth of plants in terms of increased chlorophyll content, stem length and width, and a massive weight of aerial parts of plants (Poveda et al., 2019). Furthermore, it has been concluded that frass obtained from mealworms is an essential NPK fertilizer to various crops such as barely due to its significant nutrients range and capacity in readily available form leading to increased growth and microbial activity (Houben et al., 2020). Frass application as fertilizer with NPK balance 5-2-2 in barely field increase nutrients contents in soil ultimately elevate plant biomass. Subsequently, earthworms (*Lumbricusterrestris*) frass were observed highly rich in nutrient content promoting plant growth in pots (Dulaurent et al., 2020). It has been reported that insect frass not only provides nutrients to plants but also gives other biological components for plant development. For example, larvae of *Chlorophorusannularis* produce frass rich in sugars, alkaloids, and phenols while feeding on dead twigs of *Acacia stenophylla,* parallel to beneficial fungus *Trichoderma hamatum*in soil (Khan et al., 2016).

Enzymes are an exquisite product of nature with exclusive complexities which can be utilized in various fields. Enzymes have been using in many fields involving cheese making, bakery, extraction of fruit juices, and starch processing for years. Enzymes can increase the nutritional quality, texture, and aroma of products. Enzymes isolated from insects increase the possibilities to produce food products of high quality. Cytochrome p450 and aldehyde oxidase are common enzymes isolated from insects to produce intermediate compounds. Natural selection could impart efficiency of metabolites and enzymes obtained from insects has been improved for many years. These intermediate compounds have been tested regarding safety and hazardous issues in eukaryotic hosts

TABLE 5.8 Types of Insect Species Frass Used as Fertilizers in the Agricultural Field and Their Applications.

Insect species	Crop/plants	Applications	References
		Lepidoptera	
AgrotisIpsilon	Potato	Reduce the oviposition of *Phthorimaeaoperculella*, phenols, and flavonoids presence	Ahmed et al. (2013)
Parasa Consocia	*Salix gilgiana*	Nitrogen supply to soil	Kagata and Ohgushi (2012)
Doratiferaquadriguttata	Eucalyptus	Nitrogen supply to soil	Gherlenda et al. (2016)
Orgyialeucostigma	Red oak	Increase nitrogen content in plants tissues	Frost and Hunter (2004)
Imbrasiabelina	–	Nutrients supply to soil	Swardt et al. (2018)
Orgyialeucostigma	–	Carbon and nitrogen supply to soil increase antimicrobial activity	Hillstrom et al. (2010)
Lymantria dispar	*Trembling aspen*	Nitrogen supply to soil	Madritch et al. (2007)
Lymantria dispar	*Pinus densiflora, Quercus acutissima*	Nitrogen supply to soil	Kagata and Ohgushi (2013)
Malacosoma disstria	*Trembling aspen*	Increase nitrogen content in plants tissues	Madritch et al. (2007)
Lymantria monacha	*Pinus sylvestris*	Increase nitrogen and defensive capacity	Grüning et al. (2017)
Lymantria monacha	*Pinus sylvestris*	Nitrogen supply to soil	Grüning et al. (2018)
Malacosoma americanum	Red oak	Carbon and nitrogen supply to soil increase antimicrobial activity	Frost and Hunter (2004)
Malacosoma americanum	Red oak	Increase nitrogen content in plants tissues	Frost and Hunter (2007)
		Blattodea	
Zootermopsisnevadensis	–	Nitrogen supply to soil	Chen and Forschler (2016)
Reticulitermes sp.	–	Nitrogen supply to soil	Chen and Forschler (2016)
		Diptera	
Hermetiaillucens	Cow pea	Control Fusarium wilt disease	Quilliam et al. (2020)

TABLE 5.8 *(Continued)*

Insect species	Crop/plants	Applications	References
Orthoptera			
Melanoplus borealis	–	Nitrogen supply to soil	Fielding et al. (2013)
Chorthippuscurtipennis	–	Nitrogen supply to soil	Fielding et al. (2013)
		Coleoptera	
Trirhabda virgata	Goldenrod	Increased plant growth	Uriarte (2000)
Anoplophoraglabripennis	*Salix babylonica*	Reduce insect attack	Wei et al. (2013)
Paropsisatomaria	Eucalyptus	Nitrogen supply to soil	Gherlenda et al. (2016)

and increase possibilities for their applications in various products for human use [253]. Cytochrome p450 isolated from bark beetle (CYP9T3) can catalyze hydroxylation of β-myrcene to produce ipsdienol, the main component of aggregation pheromone, and an intermediate compound of the antimalarial drug via biotransformation to produce polar antimalarial 1, 2, 4-trioxanes (Griesbeck et al., 2006). Mountain pine beetle's cytochrome p450 is used in antifungal drugs (Dai et al., 2015). Aldehyde oxidase is a pheromone degrading enzyme isolated from insects that can oxidize a variety of substrates associated with pyrimidine and purine bases (Table 5.9). Aldehyde oxidase involves the intermediary metabolism of various compounds that have been involved in drug metabolism and pharmacokinetics (Sanoh et al., 2015). The fruit fly, silkworm, and moth reported having aldehyde oxidase capacity to produce intermediate compounds used in drugs synthesis. Various types of hydrolases isolated from grain beetles, red flour beetle, and wheat weevil can degrade carbohydrates and proteins. This enzyme hydrolyzes celiac disease-associate proteins (Mika et al., 2013). This gluten degrading enzyme can be used to produce gluten-free food. Chitinase can be used in preserving food items and in the agriculture industry. Insect cuticles comprise chitin monomer, N-acetyl-β-D-glucosamine, and sclerotized proteins (Stoykov et al., 2015; Merzendorfer and Zimoch, 2003). Chitosan and chitin prohibit degradation and this property is utilized in crop defense and food preservation (Muzzarelli et al., 2012). Wood-feeding insects such as red palm weevil, termites, and wood-boring beetle have cellulolytic enzymes which can degrade lignin in plants (Scharf et al., 2011; Khiyami and Alyamani, 2008). This ability of insects enables them to produce a variety of enzymes that play the role of biocatalyst to transform lignocellulosic components into biofuels and pulping industry (Wang et al., 2017). Cellulase and xylanase from the insect gut are natural enzymes to produce biofuels.

5.8 AN INSIGHT INTO FOOD SAFETY MEASUREMENT, LEGISLATIONS, AND HAZARDOUS EFFECTS RELATED TO INSECTS-BASED FOOD AND FEED PRODUCTS

The legislation strategies regarding insect-based food and feed products vary worldwide. The current European legislation is very strict, with the application of two rules. (i) Regulation 2015/2283 (European Food Safety

TABLE 5.9 Enzymes Obtained from Insect Species and Their Functions.

Insect specie	Enzymes	Functionality	References
Eurygasterintegriceps (sun pest)	Ascorbate peroxidase protolytic enzyme	Utilized in blood clothing medicines	Bandani et al. (2009)
Helicoverpazea (corn earworm)	Ascorbate peroxidase protolytic enzyme	Used as antioxidant	Fountain et al. (2014)
Drosophila melanogaster (CYP450)	Cytochrome P450	CNS drugs	Sanoh et al. (2015)
Eastern Ips pini (CYP9T3)	Cytochrome P450	Antimalarial drugs	Griesbeck et al. (2006)
Mountain pine beetle (MPB-CPR)	Cytochrome P450	Antifungal drugs	Dai et al. (2015)
Bombyx mori (domestic silkworm)	Aldehyde oxidase	Antibacterial drugs	Sanoh et al. (2015); Pryde et al. (2010)
Drosophila melanogaster	Aldehyde oxidase	Antibacterial drugs	Sanoh et al. (2015); Pryde et al. (2010)
Manduca sexta (moth)	Aldehyde oxidase	Antibacterial drugs	Sanoh et al. (2015); Pryde et al. (2010)
Antheraea polyphemus (polyphemus silkmoth)	Aldehyde oxidase	Antibacterial drugs	Sanoh et al. (2015); Pryde et al. (2010)
Mamestrabrassicae (cabbage armyworm)	Aldehyde oxidase	Antibacterial drugs	(Sanoh et al. (2015); Pryde et al. (2010)
Bombyx mori (silkworm)	Peroxidase	Degradation of lignin	Angzzas et al. (2016); Shil et al. (2014); Scharf et al. (2011); Nakashima et al. (2002)
Eurycanthacalcarata (walking stick)	Peroxidase	Modifying or degradation of lignin	
Oxyavelox (rice pest)	Cellulase 1-β Glucosidase 1-α Glucosidase	Modifying or degradation of cellulose and hemicellulose	
Rhynchophorusferrugineus (sago worm)	Laccase	Modifying or degradation of lignin	
Aularches miliaris (grasshopper)	Cellulase 1-β Glucosidase 1-α Glucosidase	Modifying or degradation of cellulose and hemicellulose	
Coptotermusformosanus	Laccase	Modifying or degradation of lignin	
Reticulitermes flavipes (termites)	Hemicellulase - Xylanase	Modifying or degradation of cellulose and hemicellulose	
Manduca sexta (tobacco hornworm)	Chitinase	Dietary ingredient, preservative, and crop protective agent	Stoykov et al. (2015); Hamid et al. (2013); Muzzarelli et al. (2012)

TABLE 5.9 *(Continued)*

Insect specie	Enzymes	Functionality	References
Phthorimaeaoperculella (Zeller)	Protease	Food additives bakery	Mansouri et al. (2013); Prem Anand et al. (2010); Alarcón et al. (2002)
Bombyx mori (silkworm)	Protease	Food additives bakery	
Rhynchophorusferrugineus (red palm weevil)	Protease	Food additives bakery	
Bombyx mori (silkworm)	cellulase	Used in gluten-free food and animal feedstock	Prem Anand et al. (2010); Alarcón et al. (2002)
Rhynchophorusferrugineus (red palm weevil)	Xylanase	Used in gluten-free food, food additive, and animal feedstock	Prem Anand et al. (2010)
Rhynchophorusferrugineus (red palm weevil)	Pectinase	Used for extraction of juices	Prem Anand et al. (2010)
Rhizoperthadominica	Prolyl-specific peptidase	Used in food and medical industry	Mika et al. (2013)
Spilosomaobliqua	Alkaline protease	Addictive in detergents	Anwar and Saleemuddin (2000)
Neotermskoshunensis	B-glucosidase Laccase	Production of bioethanol, middle-density fiberboard, and dyes	Uchima et al. (2011); Dittmer et al. (2009)
Cuelxquinquefasciatus	Trypsin-like serine peptidase	Pest control	Borges-Veloso et al. (2012)
Bombyx mori	Pupal gut serine protease	Additives in detergents, medical, and chemical industry	Kannan et al. (2017)
Aedes albopictus	Trypsin like serine peptidase	Pest control	Saboia-Vahia et al. (2014)
Helicoverpaarmigera	Alkaline serine protease	Addictive in detergents	Akbar and Sharma (2017)
Helicoverpaarmigera	Trypsin like protease	Pest control	Grover et al. (2018)
Periplaneta americana	Serine alkaline protease	Used in additives in detergent, medical, and chemical synthesis	Sanatan et al. (2013)
Choreutisnemorana	Gut protease	Proteolytic activity	Chitgar et al. (2013)

Authority, EFSA) describes edible insects as food. However, before they were termed as novel food (Parliament and Union, 2015), because before 1997 insects were not utilized as food. In the modified regulation, edible insects do not mention as novel foods. They authorized consumption of insect-derived proteins from the following insect species—common Housefly (*Musca domestica*), BSF (*Hermetiaillucens*), yellow mealworm (*Tenebrio molitor*), lesser mealworm (*Alphitobiusdiaperinus*), house cricket (*Acheta domesticus*), banded cricket (*Gryllodessigillatus*), and field cricket (*Gryllusassimilis*), as feed-in aquaculture (Regulation, 2015). Authorization from EFSA is required for the marketing sale of these edible insect products. (ii) Second rule, Regulation EU 999/2001 exhibit edible insects as feed ("Regulation (EC) No. 999/2001 laying down rules for the prevention, control and eradication of certain transmissible spongiform encephalopathies.", n.d.). International Platform on Insects for Food and Feed (IPIFF) regulated that only purified insect fat and hydrolyzed insect proteins are permitted to be consumed as feed for livestock. Non-hydrolyzed proteins can only be used for pet food and fur animal feed; however, insect-derived proteins cannot be used for pig and poultry feed ("Insects as Feed EU Legislation – Aquaculture, Poultry & Pig Species", n.d.). Furthermore, many countries are seeking approval for new insects' species for feed and food products (Marone, 2016). The Netherlands marketed edible insects by providing the "List of edible insects of the world" (Jongema, 2015). In Belgium, an analysis on the sale of edible insects (mealworms, crickets, and locusts) as novel food is provided by AgenceFédérale pour la Sécurité de la ChaîneAlimentaire (AFSCA) (Marone, 2016; Halloran et al., 2015). According to EU regulation, in Germany, only processed animal protein is allowed to consume, while insect proteins are not allowed as feed because they are not slaughtered, but this regulation does not imply live insects. In this regard, a plan is being made and presented to Deutsche Landwirtschaftsgesellschaft (DLG) to refer to a list of edible insects as direct animal feed. This rule is quite flexible in the United Kingdom in which insects are sold as food but this regulation is going to be modified soon with specific application requirements for the classification of the edible insect as food to be sold. In Switzerland, the Federal Office of Food Safety and Veterinary Services (FFSVO) authorizes edible insect species to be sold in the market; currently, FFSVO is following Belgium's policy in referring insect species as food (Marone, 2016).

5.9 CONCLUSION AND FUTURE PERSPECTIVE

This chapter summarizes biodiesel production and food, and feed production from various insect species. Biodiesel production depends on the biomass of the insects containing lipid contents. Higher biomass can produce higher biodiesel. We exhibited numerous insect orders including Orthoptera, Coleoptera, Lepidoptera, Isoptera, Hymenoptera, Hemiptera, Blattodea, Megaloptera, and Diptera, and their lipid content used in biodiesel production. Insects have potential properties to transform or biodegrade organic wastes into biodiesel raw material. Insect selection for biodiesel synthesis involves fat bodies percentage, reproduction rate, and life cycle of the specific insect species. Insects feeding material is a major contributor to their biomass building. We also give examples of organic wastes on which insects feed and produce their biomass containing lipid contents. The properties of lipid contents obtained from insects fed on different types of manure are given in Table 5.2. Then we discussed some concerns about insect consumption in biodiesel production and we also raised some questions regarding synthesis, yield, and conditions. This chapter gives potential knowledge about biodiesel production, raw material, and suitable insects species being utilized for the future.

Currently, we are facing challenges with the human population, food demand, reduction of environmental sources, and growth problems. The emergence of the insect-based food industry is a revolution in the food sector. Many countries consumed insects as their food due to their nutritional, and medicinal benefits. Industrialized nations have planned regulations for edible insects and their market sale according to the customer's requirement and health demand. Insects have been replacing many plants and animal food products as alternative food sources. Given that industries are focusing to develop new varieties of food products from insects to meet the global food demand. We discussed various insect species consumed like food products. We also described types of edible insects' species and their properties. Some insects are not edible so they can utilize as feed for livestock. Various insect-rearing methods have been developed to produce edible insects on large scale. Insects have potential biological components that are used in other industries such as agriculture, pulp, and pharmaceutical industries. We also give knowledge about these industries and their applications associated with biological components. Furthermore, legislation or regulatory framework are given by food authorities to minimize potential hazards associated with insects. Engineers need to develop more advanced technologies and design methods to produce significant products from insects. These products should be more sustainable, low-cost, and environmentally friendly. The food sector needs study on insects' ecology, history, and biology, nature of insect's food, nutritional content, functions, organoleptic properties, insect-based food ingredients, and processing methods. Marketing policy-makers should advertise insect-based products with their nutritional contents labeled on food products. To produce desirable and efficient biodiesel, various constraints must be overcome, such as high-cost raw materials, costly processing methods, difficult rearing conditions, less-hazardous fuel, and large area requirements for production. There is a need to develop multiple products to analyze their beneficial and hazardous aspects to launch the appropriate product in the market. To promote the insect industry there is a need for:

- Greater support and attention worldwide
- Funding and focus from policymakers, government, agencies, and industries
- Establishment of efficient laboratories
- Promotion of insect-based products from lab to industry and markets
- International societies to promote food and feed products
- Advanced regulations regarding edible insects
- Storage and transfer means for insect products to market
- Industrial associations to promote and support food products
- Funding for industries to design advanced methods to produce insect products
- Knowledge about insect-based products to societies

This will lead insect industries to develop at international forums and will explore new horizons to produce alternatives for food and energy sources.

KEYWORDS

- **insects**
- **products**
- **biomass**
- **biodiesel**
- **food products**
- **edible**

REFERENCES

Adamolekun, B. Anaphevenata Entomophagy and Seasonal Ataxic Syndrome in Southwest Nigeria. *Lancet* **1993,** *341*, 629.

Adedokun, T. A.; Denlinger, D. L. Metabolic Reserves Associated with Pupal Diapause in the Flesh Fly, Sarcophagacrassipalpis. *J. Insect Physiol.* **1985,** *31*, 229–233.

Ademolu, K. O.; Idowu, A. B.; Olatunde, G. O. Nutritional Value Assessment of Variegated Grasshopper, *Zonocerus variegatus* (L.) (Acridoidea: Pygomorphidae), During Post-Embryonic Development : Research Article. *Afr. Entomol.* **2010,** *18*, 360–364.

Ahmed, A. A. I.; Hashem, M. Y.; Mohamed, S. M.; Khalil, S. H. S. Protection of Potato Crop Against Phthorimaeaoperculella (Zeller) Infestation Using Frass Extract of Two Noctuid Insect Pests Under Laboratory and Storage Simulation Conditions. *Arch. Phytopathol. Plant Prot.* **2013,** *46*, 2409–2419.

Akbar, S. M.; Sharma, H. C. Alkaline Serine Proteases from Helicoverpaarmigera: Potential Candidates for Industrial Applications. *Arch. Insect Biochem. Physiol.* **2017,** *94*, e21367.

Akinnawo, O.; Ketiku, A. O. Chemical Composition and Fatty Acid Profile of Edible Larva of Cirinaforda (Westwood). *Afr. J. Biomed. Res.* **2000,** *3*, 93–96.

Alarcón, F. J.; Martínez, T. F.; Barranco, P.; Cabello, T.; Díaz, M.; Moyano, F. J. Digestive Proteases During Development of Larvae of Red Palm Weevil, Rhynchophorusferrugineus (Olivier, 1790) (Coleoptera: Curculionidae). *Insect Biochem. Mol. Biol.***2002,** *32*, 265–274.

Alegbeleye, W. O.; Obasa, S. O.; Olude, O. O.; Otubu, K.; Jimoh, W. Preliminary Evaluation of the Nutritive Value of the Variegated Grasshopper (*Zonocerus variegatus* L.) for African Catfish Clariasgariepinus (Burchell. 1822) Fingerlings. *Aquac. Res.* **2012,** *43*, 412–420.

Alonso-Castro, A. J. Use of Medicinal Fauna in Mexican Traditional Medicine. *J. Ethnopharmacol.* **2014,** *152*, 53–70.

Alzoubi, K.; Egler, J.; Briglia, M.; Fazio, A.; Faggio, C.; Lang, F. Induction of Suicidal Erythrocyte Death by Cantharidin. *Toxins* **2015,** *7*, 2822–2834.

Amadi, E. N.; Ogbalu, O. K.; Barimalaa, I. S.; Pius, M. Microbiology and Nutritional Composition of an Edible Larva (bunaeaAlcinoe Stoll) of the Niger Delta. *J. Food Saf.* **2005,** *25*, 193–197.

Angzzas, S. M. K.; Ashuvila, M. A.; Dayang, N. F. A. Z. Potential Lignin Degraders Isolated from the Gut of Rhynchophorusferrugineus. In: *International Conference on Mechanics, Materials and Structural Engineering (ICMMSE 2016)*; Atlantis Press, 2016; pp 124–130.

Anwar, A.; Saleemuddin, M. Alkaline Protease from Spilosomaobliqua: Potential Applications in Bio-Formulations. *Biotechnol. Appl. Biochem.* **2000,** *31*, 85–89.

Arango Gutiérrez, G. P.; Vergara Ruiz, R. A.; Mejía Vélez, H. Analisis Composicional, Microbiológico Y Digestibilidad De La Proteína De La Harina De Larvas De Hermetiailluscens L (Diptera:Stratiomyiidae) En Angelópolis-Antioquia, Colombia. *Rev. Fac. Nac. Agron. Med.* **2004,** *57*, 2491–2500.

Arbiser, J. L.; Kau, T.; Konar, M.; Narra, K.; Ramchandran, R.; Summers, S. A.; Vlahos, C. J. Ye, K. Solenopsin, the Alkaloidal Component of the Fire Ant (Solenopsisinvicta), Is a Naturally Occurring Inhibitor of Phosphatidylinositol-3-Kinase Signaling and Angiogenesis. *Blood.* **2007,** *109*, 560–565.

Atteh, J. O.; Ologbenla, F. D. Replacement of Fish Meal with Maggots in Broiler Diets: Effects on Performance and Nutrient Retention. *Niger. J. Anim. Prod.* **1993,** *20*, 44–49.

Ayensu, J.; Annan, R. A.; Edusei, A.; Lutterodt, H. Beyond Nutrients, Health Effects of Entomophagy: A Systematic Review. *Nutr. Food Sci.* **2019.**

Ayieko, M.; Oriaro, V.; Nyambuga, I. A. Processed Products of Termites and Lake Flies: Improving Entomophagy for Food Security Within the Lake Victoria Region. *Afr. J. Food Agric. Nutr. Dev.* **2010,** *10.*

Bagnara, J. T.; Fernandez, P. J.; Fujii, R. On the Blue Coloration of Vertebrates. *Pigment Cell Res.* **2007,** *20*, 14–26.

Bandani, A. R.; Kazzazi, M.; Mehrabadi, M. Purification and Characterization of Midgut α-Amylases of Eurygasterintegriceps. *Entomol. Sci.* **2009,** *12*, 25–32.

Banjo, A. D.; Lawal, O. A.; Songonuga, E. A. The Nutritional Value of Fourteen Species of Edible Insects in Southwestern Nigeria. *Afr. J. Biotechnol.* **2006,** *5*, 298–301.

Bednářová, M.; Borkovcová, M.; Mlček, J.; Rop, O.; Zeman, L. Edible Insects—Species Suitable for Entomophagy Under Condition of Czech Republic. *Acta Univ. Agric. et Silvic. Mendelianae Brun.* **2013,** *61*, 587–593.

Behie, S. W.; Bidochka, M. J. Insects as a Nitrogen Source for Plants. *Insects.* **2013,** *4*, 413–424.

Belghit, I.; Liland, N. S.; Waagbø, R.; Biancarosa, I.; Pelusio, N.; Li, Y.; Krogdahl, Å.; Lock, E. J. Potential of Insect-Based Diets for Atlantic salmon (Salmo salar). *Aquac.* **2018,** *491*, 72–81.

Belluco, S.; Losasso, C.; Maggioletti, M.; Alonzi, C.; Ricci; Paoletti, M. G. Edible Insects: A Food Security Solution or a Food Safety Concern? *Anim. Front.* **2015,** *5*, 25–30

Berasategui, A.; Shukla, S.; Salem, H.; Kaltenpoth, M. Potential Applications of Insect Symbionts in Biotechnology. *Appl. Microbiol. Biotechnol.* **2016,** *100*, 1567–1577.

Bernard, J. B.; Allen, M. E.; Ullrey, D. E. Feeding Captive Insectivorous Animals: Nutritional Aspects of Insects as Food. *Nut. Adv. Group Handbook, Fact Sheet.* **1997,** *3*, 1–7.

Borges Veloso, A.; SaboiaVahia, L.; Cuervo, P.; Pires, R. C.; Britto, C.; Fernandes, N.; d'Avila-Levy, C. M.; De Jesus, J. B. Proteolytic Profiling and Comparative Analyses of Active Trypsin-Like Serine Peptidases in Preimaginal Stages of Culex Quinquefasciatus. *Parasites Vectors* **2012,** *5*, 123.

Borgohain, M.; Borkotoki, A.; Mahanta, R. Total Lipid, Triglyceride, and Cholesterol Contents in *Oecophyllasmaragdina Fabricius* Consumed in Upper Assam of Northeast India. *Int. J. Sci. Res. Publ.* **2014,** *4*, 1–7.

Bosque, I.; Gonzalez-Gomez, J. C.; Loza, M. I.; Brea, J. Natural Tetraponerines: A General Synthesis and Antiproliferative Activity. *J. Org. Chem.* **2014,** *79*, 3982–3991.

Bruce, M. E.; Will, R. G.; Ironside, J. W.; McConnell, I.; Drummond, D.; Suttie, A.; McCardle, L.; Chree, A. Transmissions to Mice Indicate That 'New Variant' CJD Is Caused by the BSE Agent. *Nature* **1997,** *389*, 498–501.

Bukkens, S. G. F. Insects in the Human Diet: Nutritional Aspects. In: *Ecological Implications of Minilivestock: Potential of Insects, Rodents, Frogs and Snails*; 2005; pp 545–577.

Cerda, H.; Martínez, R.; Briceño, N.; Pizzoferrato, L.; Hermoso, D.; Paoletti, M. Cría, análisisnutricional y sensorial del picudo del cocoteroRhynchophorus palmarum (Coleoptera: Curculionidae). *insecto de la dietatradicionalindígenaamazónica*, 1999.

Chakravorty, J.; Ghosh, S.; Meyer-Rochow, V. B. Comparative Survey of Entomophagy and Entomotherapeutic Practices in Six Tribes of Eastern Arunachal Pradesh (India). *J. Ethnobiol. Ethnomedicine*. **2013,** *9*, 50.

Chen, L.; Lu, Y. Y.; Hu, Q. B.; Fadamiro, H. Y. Similarity in Venom Alkaloid Chemistry of Alate Queens of Imported Fire Ants: Implication for Hybridization Between Solenopsisrichteri and *S. invicta* in the Southern United States. *Chem. Biodivers*. **2012,** *9*, 702–713.

Chen, Y.; Forschler, B. T. Elemental Concentrations in the Frass of Saproxylic Insects Suggest a Role in Micronutrient Cycling. *Ecosphere* **2016,** *7*, e01300.

Chitgar, M. G.; Ghadamyari, M.; Sharifi, M. Identification and Characterisation of Gut Proteases in the Fig Tree Skeletoniser Moth, *ChoreutisnemoranaHübner* (Lepidoptera: Choreutidae). *Plant Prot. Sci.* **2013,** *49*, 19–26.

Cho, M. Y.; Choi, H. W.; Moon, G. Y.; Kim, M. H.; Kwon, T. H.; Homma, K.; Natori, S.; Lee, B. L. An 86 kDa Diapause Protein 1-Like Protein Is a Component of Early-Staged Encapsulation-Relating Proteins in Coleopteran Insect, Tenebrio Molitor Larvae. *FEBS Lett.* **1999,** *451*, 303–307.

Chung, A. Y. C. Edible Insects and Entomophagy in Borneo. In *Forest Insects as Food: Humans Bite Back. Proceedings of a Workshop on Asia-Pacific Resources and Their Potential for Development, Chiang Mai, Thailand, 19–21 February, 2008*; 2010; pp 141–150.

Čičková, H.; Newton, G. L.; Lacy, R. C.; Kozánek, M. The Use of Fly Larvae for Organic Waste Treatment. *Waste Manag.* **2015,** *35*, 68–80.

Cohen, J. H.; Sáánchez, N. D. M.; Montiel-Ishino, F. Chapulines and Food Choices in Rural. *Oaxaca Gastronomica* **2009,** *9*, 61–65.

Cope, A. P.; Schulze-Koops, H.; Aringer, M. The Central Role of T Cells in Rheumatoid Arthritis. *Clin. Exp. Rheumatol.* **2007,** *25*, S4–11.

Cortes Ortiz, J. A.; Ruiz, A. T.; Morales-Ramos, J. A.; Thomas, M.; Rojas, M. G.; Tomberlin, J. K.; Yi, L.; Han, R. et al. Chapter 6—Insect Mass Production Technologies.

In Dossey, A. T., Morales-Ramos, J. A., Rojas, M. G., Eds.; Academic Press: San Diego, 2016; pp 153–201.

Costa-Neto, E. M.; Dunkel, F. V. Chapter 2—Insects as Food: History, Culture, and Modern Use around the World. In: Dossey, A. T., Morales-Ramos, J. A., Rojas, M. G., Eds.; 2016; pp 29–60.

Costermans, J. B. Het termieten-stokenbij de Logo-Avokaya (vervolg). *Aequat.* **1955,** *18*, 50–55.

D'Alba, L.; Kieffer, L.; Shawkey, M. D. Relative Contributions of Pigments and Biophotonic Nanostructures to Natural Color Production: A Case Study in Budgerigar (*Melopsittacus undulatus*) Feathers. *J. Exp. Biol.* **2012,** *215*, 1272–1277.

Dabbou, S.; Gai, F.; Biasato, I.; Capucchio, M. T.; Biasibetti, E.; Dezzutto, D.; Meneguz, M.; Plachà, I. Black Soldier Fly Defatted Meal as a Dietary Protein Source for Broiler Chickens: Effects on Growth Performance, Blood Traits, Gut Morphology and Histological Features. *J. Anim. Sci. Biotechnol.* **2018,** *9*, 49.

Dai, L.; Li, Z.; Yu, J.; Ma, M.; Zhang, R.; Chen, H.; Pham, T. The CYP51F1 Gene of Leptographiumqinlingensis: Sequence Characteristic, Phylogeny and Transcript Levels. *Int. J. Mol. Sci.* **2015,** *16*, 12014–12034.

Davies, J. How to Discover New Antibiotics: Harvesting the Parvome. *Curr. Opin. Chem. Biol.* **2011,** *15*, 5–10.

de Castro, R. J. S.; Ohara, A.; Aguilar, J. G.; dos, S.; Domingues, M. A. F. Nutritional, Functional and Biological Properties of Insect Proteins: Processes for Obtaining, Consumption and Future Challenges. *Trends Food Sci. Technol.* **2018,** *76*, 82–89.

Dean, R. L.; Locke, M.; Collins, J. V.; Kerkut, G. A.; Gilbert, L. I. Structure of the fat body Comparative insect physiology. *Biochem. Pharmacol.* **1985,** 155–209.

Dettner, K.; Schramm, S.; Seidl, V.; Klemm, K.; Gäde, G.; Fietz, O.; Boland, W. Occurrence of Terpene Anhydride Palasonin and Palasoninimide in Blister Beetle Hycleuslunata (Coleoptera: Meloidae). *Biochem. Syst. Ecol.* **2003,** *31*, 203–205.

Di, L.; Shi, Y. N.; Yan, Y. M.; Jiang, L. P.; Hou, B.; Wang, X. L.; Zuo, Z. L.; Chen, Y-B. Nonpeptide Small Molecules from the Insect *Aspongopus chinensis* and Their Neural Stem Cell Proliferation Stimulating Properties. *RSC Adv.* **2015,** *5*, 70985–70991.

Diclaro II, J. W.; Kaufman, P. E. *Black Soldier Fly Hermetiaillucenslinnaeus (Insecta: Diptera: Stratiomyidae)*; EDIS, 2009.

Dittmer, N. T.; Gorman, M. J. Kanost, M. R. Characterization of Endogenous and Recombinant Forms of Laccase-2, a Multicopper Oxidase from the Tobacco Hornworm, *Manduca sexta. Insect Biochem. Mol. Biol.* **2009,** *39*, 596–606.

Dulaurent, A. M.; Daoulas, G.; Faucon, M. P.; Houben, D. Earthworms (Lumbricusterrestris L.) Mediate the Fertilizing Effect of Frass. *Agron.* **2020,** *10*, 783.

Durst, P. B.; F. A. O., Eds. Forest Insects as Food: Humans Bite Back ; Proceedings of a Workshop on Asia-Pacific Resources and Their Potential for Development. In: *Food and Agriculture Organization of the United Nations, Regional Office for Asia and the Pacific Bangkok, Thailand, 19–21 February* 2008, *Chiang Mai, Thailand* (RAP Publication no. 2010/02). *Bangkok, Thailand,* 2010.

Eilenberg, J.; Vlak, J. M.; Nielsen-LeRoux, C.; Cappellozza, S.; Jensen, A. B. Diseases in Insects Produced for Food and Feed. *J. Insects Food Feed.* **2015,** *1*, 87–102.

Ekpo, K. E.; Onigbinde, A. O. Characterization of Lipids in Winged Reproductives of the Termite Macrotermisbellicosus. *Pak. J. Nutr.* **2007,** *6*, 247–251.

El-Tantawy, N. L. Helminthes and Insects: Maladies or Therapies. *Parasitol. Res.* **2015,** *114*, 359–377.

Erwin, T. L. The Biodiversity Question: How Many Species of Terrestrial Arthropods are There. *Forest Canopies* **2004,** *10*, 259–269.

Ewusie, E.; Kwapong, P. et al. Development of Black Soldier Fly, Hermetiaillucens (Diptera: Stratiomyidae) in Selected Organic Market Waste Fractions in Accra, Ghana. *Asian J. Biotechnol. Bioresour. Tech.* **2018,** *4*, 1–16.

Fasoranti, J. O.; Ajiboye, D. O. Some Edible Insects of Kwara State, Nigeria. *Am. Entomol.* **1993,** *39*, 113–116.

Fecheyr-Lippens, D. C.; Igic, B.; D'Alba, L.; Hanley, D.; Verdes, A.; Holford, M.; Waterhouse, G. I. N.; Grim, T. The Cuticle Modulates Ultraviolet Reflectance of Avian Eggshells. *Biol. Open.* **2015,** *4*, 753–759.

Fielding, D. J.; Trainor, E.; Zhang, M. Diet Influences Rates of Carbon and Nitrogen Mineralization from Decomposing Grasshopper Frass and Cadavers. *Biol. Fertil. Soils* **2013,** *49*, 537–544.

Finke, M. D. Complete Nutrient Composition of Commercially Raised Invertebrates Used as Food for Insectivores. *Zoo. Biol.* **2002,** *21*, 269–285.

Fombong, F. T.; Kinyuru, J. N. Termites as Food in Africa. In: Khan, Md. A., Ahmad, W., Eds.; Springer International Publishing: Cham, 2018; pp 217–240.

Foottit, R. G.; Adler, P. H. *Insect Biodiversity: Science and Society*; John Wiley & Sons, 2009.

Fountain, J.; Scully, B.; Ni, X.; Kemerait, R.; Lee, D.; Chen, Z. Y.; G. U.O, B. Environmental Influences on Maize—*Aspergillus flavus* Interactions and Aflatoxin Production. *Front. Microbiol.* **2014,** *5*, 40.

Frost, C. J. Hunter, M. D. Insect Canopy Herbivory and Frass Deposition Affect Soil Nutrient Dynamics and Export in Oak. *Ecology* **2004,** *85*, 3335–3347.

Frost, C. J.; Hunter, M. D. Recycling of Nitrogen in Herbivore Feces: Plant Recovery, Herbivore Assimilation, Soil Retention, and Leaching Losses. *Oecologia* **2007,** *151*, 42–53.

Gahukar, R. T. Chapter 4 - Edible Insects Farming: Efficiency and Impact on Family Livelihood, Food Security, and Environment Compared With Livestock and Crops. In: Dossey, A. T., Morales, R., Rojas, M. G.; 2016; pp 85–111.

Gan, S. K. E.; Phua, S. X.; Yeo, J. Y.; Heng, Z. S. L.; Xing, Z. Method for Zero-Waste Circular Economy Using Worms for Plastic Agriculture: Augmenting Polystyrene Consumption and Plant Growth, 2021; pp 3–11 (Preprint).

Gardiner, A. J.; Gardiner, E. M. Edible Insects, Part 1. Preparation of Species from Mushumbi Pools, Zimbabwe. *Afr. Entomol.* **2003,** *11*, 125–127.

Garofalo, C.; Milanović, V.; Cardinali, F.; Aquilanti, L.; Clementi, F.; Osimani, A. Current Knowledge on the Microbiota of Edible Insects Intended for Human Consumption: A State-of-the-Art Review. *Food Res. Inter.* **2019,** *125*, 108527.

Ghazoul, J. *Mopane Woodlands and the Mopane Worm: Enhancing Rural Livelihoods and Resource Sustainability*; Final Technical Report, 2006.

Gherlenda, A. N.; Crous, K. Y.; Moore, B. D.; Haigh, A. M.; Johnson, S. N. Riegler, M. Precipitation, Not CO_2 Enrichment, Drives Insect Herbivore Frass Deposition and Subsequent Nutrient Dynamics in a Mature Eucalyptus Woodland. *Plant Soil* **2016,** *399*, 29–39.

Ghosh, S.; Haldar, P.; Mandal, D. Evaluation of Nutrient Quality of a Short Horned Grasshopper, OxyahylahylaServille (Orthoptera: Acrididae) in Search of New Protein Source. *J. Entomol. Zool. Stud.* **2016,** *4*.

Giron, R. J. C.; Hidalgo, G. G.; Garcia, J. E. B.; Pino Hernández, E. J. G.; Villa, P. M. Exploring the Food and Nutritional Potential of Three Edibles Amazonian Arthropods, 2017.

Goulding, K. W. T.; Bailey, NJ.; Bradbury, N. J.; Hargreaves, P.; Howe, M.; Murphy, D. V.; Poulton, P. R.; Willison, T. W. Nitrogen Deposition and Its Contribution to Nitrogen Cycling and Associated Soil Processes *New Phytol.* **1998,** *139*, 49–58.

Graham, L. A.; Liou, Y. C.; Walker, V. K.; Davies, P. L. Hyperactive Antifreeze Protein from Beetles. *Nature*. **1997,** *388*, 727–728.

Griesbeck, A. G.; Hoinck, O.; Lex, J.; Synthetic Approaches to Polar Antimalarial 1,2,4-Trioxanes from C5-Aldehyde and Ipsdienol. *Lett. Org. Chem.* **2006,** *3*, 247–249.

Grimaldi, D.; Engel, M. S.; Engel, M. S.; Engel, S. C. et al. *Evolution of the Insects*; Cambridge University Press, 2005.

Grover, S.; Kaur, S.; Gupta, A. K.; Taggar, G. K.; Kaur, J. Characterization of Trypsin Like Protease from Helicoverpaarmigera (Hubner) and Its Potential Inhibitors. *Proc. Natl. Acad. Sci. India. Sect. B. Biol. Sci.* **2018,** *88*, 49–56.

Grüning, M. M.; Germeshausen, F.; Thies, C.; Arnold, A. Increased Forest Soil CO_2 and N_2O Emissions During Insect Infestation. *Forests* **2018,** *9*, 612.

Grüning, M. M.; Simon, J.; Rennenberg, H.; Arnold, A. Defoliating Insect Mass Outbreak Affects Soil N Fluxes and Tree N Nutrition in Scots Pine Forests. *Front. Plant Sci.* **2017,** *8*, 954.

Guimei, H. Study on Extraction and Antioxidant Activity Analysis of Barley Pest Protein, 2011.

Gutiérrez, G. P. A.; Ruiz, R. A. V.; Vélez, H. M. Análisiscomposicional, microbiológico y digestibilidad de la proteína de la harina de larvas de hermetiailluscens L (diptera:stratiomyiidae) enangelópolis-Antioquia, Colombia. *Rev. Fac. Nac. Agron. Medellin.* **2004,** *57*, 2491–2499.

Hackstein, J. H.; Stumm, C. K. Methane Production in Terrestrial Arthropods. *Proc. Natl. Acad. Sci.* **1994,** *91*, 5441–5445.

Halloran, A.; Roos, N.; Eilenberg, J.; Cerutti, A.; Bruun, S. Life Cycle Assessment of Edible Insects for Food Protein: A Review. *Agron. Sustain. Dev*. **2016,** *36*, 57.

Halloran, A.; Vantomme, P.; Hanboonsong, Y.; Ekesi, S. Regulating Edible Insects: The Challenge of Addressing Food Security, Nature Conservation, and the Erosion of Traditional Food Culture. *Food Secur.* **2015,** *7*, 739–746.

Hamid, R.; Khan, M. A.; Ahmad, M.; Ahmad, M. M.; Abdin, M. Z.; Musarrat, J.; Javed, S. Chitinases: An Update. *J. Pharm. Bioallied. Sci.* **2013,** *5*, 21–29.

Hanboonsong, Y. Edible Insects and Associated Food Habits in Thailand in Forest Insects as Food: Humans Bite Back. In: *Proceedings of a Workshop on Asia-Pacific Resources and Their Potential for Development*; FAO Regional Office for Asia and the Pacific: Bangkok, 2010; pp 173–182.

Hanboonsong, Y.; Jamjanya, T.; Durst, P. B. *Six-Legged Livestock: Edible Insect Farming, Collection and Marketing in Thailand*; RAP Publication, 2013; p 3.

Hardouin, J.; Mahoux, G. Zootechnied'insectes-Elevage et utilisation au bénéfice de l'homme et de certainsanimaux, 2003.

Harvey, A. L.; Edrada-Ebel, R.; Quinn, R. J. The Re-emergence of Natural Products for Drug Discovery in the Genomics Era. *Nat. Rev. Drug Discov*. **2015,** *14*, 111–129.

He, L.; Yang, S. S.; Bai, S. W.; Pang, J. W.; Liu, G. S. Cao, G. L.; Zhao, L.; Feng, X. C. Fabrication and Environmental Assessment of Photo-Assisted Fenton-Like Fe/FBC Catalyst Utilizing Mealworm Frass Waste. *J. Clean. Prod*. **2020,** *256*, 120259.

Heinen, T. E.; Gorini da Veiga, A. B. Arthropod Venoms and Cancer. *Toxicon* **2011,** *57*, 497–511.

Hillstrom, M.; Meehan, T. D.; Kelly, K.; Lindroth, R. L. Soil Carbon and Nitrogen Mineralization Following Deposition of Insect Frass and Greenfall from Forests Under Elevated CO_2 and O_3. *Plant Soil* **2010,** *336*, 75–85.

Hirose, Y.; Ohta, E.; Kawai, Y.; Ohta, S. Dorsamin-A's, Glycerolipids Carrying a Dehydrophenylalanine Ester Moiety from the Seed-Eating Larvae of the Bruchid Beetle *Bruchidius dorsalis*. *J. Nat. Prod*. **2013,** *76*, 554–558.

Houben, D.; Daoulas, G.; Faucon, M. P.; Dulaurent, A. M. Potential Use of Mealworm Frass as a Fertilizer: Impact on Crop Growth and Soil Properties. *Sci. Rep*. **2020,** *10*, 4659.

Hsia, T. C.; Yu, C. C.; Hsu, S. C.; Tang, N. Y.; Lu, H. F.; Huang, Y. P.; Wu, S. H.; Lin, J. G. Cantharidin Induces Apoptosis of H460 Human Lung Cancer Cells Through Mitochondria-Dependent Pathways. *Int. J. Oncol*. **2014,** *45*, 245–254.

Hsiao, Y. P.; Tsai, C. H.; Wu, P. P.; Hsu, S. C.; Liu, H. C.; Huang, Y. P.; Yang, J.; Chung, J. G. Cantharidin Induces G2/M Phase Arrest by Inhibition of Cdc25c and Cyclin A and Triggers Apoptosis Through Reactive Oxygen Species and the Mitochondria-Dependent Pathways of A375.S2 Human Melanoma Cells. *Int. J. Oncol*. **2014,** *45*, 2393–2402.

Hsieh, C. H.; Huang, Y. C.; Tsai, T. H.; Chen, Y. J. Cantharidin Modulates Development of Human Monocyte-Derived Dendritic Cells. *Toxicol. In Vitro* **2011,** *25*, 1740–1747.

Hwangbo, J.; Hong, E. C.; Jang, A.; Kang, H. K.; Oh, J. S.; Kim, B. W.; Park, B. S. Utilization of House Fly-Maggots, a Feed Supplement in the Production of Broiler Chickens. *J. Environ. Biol*. **2009,** *30*, 609–614.

Igic, B.; Fecheyr-Lippens, D.; Xiao, M.; Chan, A.; Hanley, D.; Brennan, P. R. L.; Grim, T.; Waterhouse, G. I. N. A Nanostructural Basis for Gloss of Avian Eggshells. *J. R. Soc. Interface* **2015,** *12*, 20141210.

Insects as Feed EU Legislation—Aquaculture, Poultry & Pig Species. https://ipiff.org/insects-eu-legislation/

Ishak, S.; Kamari, A. A Review of Optimum Conditions of Transesterification Process for Biodiesel Production from Various Feedstocks. *Int. J. Environ. Sci. Technol*. **2019,** *16*, 2481–2502.

Ishak, S.; Kamari, A. Biodiesel from Black Soldier Fly Larvae Grown on Restaurant Kitchen Waste. *Environ. Chem. Lett*. **2019,** *17*, 1143–1150.

Jansson, A.; Berggren, Å. *Insects as Food—Something for the Future?* (Report), 2015.

Jiang, H. L.; Luo, X. H.; Wang, X. Z.; Yang, J. L.; Yao, X. J.; Crews, P.; Valeriote, F. A.; Wu, Q. X. New Isocoumarins and Alkaloid from Chinese Insect Medicine, *Eupolyphaga sinensis* Walker. *Fitoterapia*. **2012,** *83*, 1275–1280.

Johnston, C. W.; Skinnider, M. A.; Wyatt, M. A.; Li, X.; Ranieri, M. R. M.; Yang, L.; Zechel, D. L.; Ma, B. An Automated Genomes-to-Natural Products Platform (GNP) for the Discovery of Modular Natural Products. *Nat. Commun*. **2015,** *6*, 8421.

Jongema, Y. *List of Edible Insect Species of the World*; Laboratory of Entomology: Wageningen; Wageningen University: The Netherlands, 2014.

Kagata, H.; Ohgushi, T. Home-Field Advantage in Decomposition of Leaf Litter and Insect Frass. *Pop. Ecol.* **2013,** *55*, 69–76.

Kagata, H.; Ohgushi, T. Non-Additive Effects of Leaf Litter and Insect Frass Mixture on Decomposition Processes. *Ecol. Res.* **2012,** *27*, 69–75.

Kalu-Uka, G. M.; Kumar, S.; Kalu-Uka, A. C.; Vikram, S.; Okorafor, O. O.; Kigozi, M.; Ihekweme, G. O.; Onwualu, A. P. Prospects for Biodiesel Production from Macrotermesnigeriensis: Process Optimization and Characterization of Biodiesel Properties. *Biomass Bioenergy* **2021,** *146*, 105980.

Kannan, M.; Ramya, T.; Anbalagan, S.; Suriya, J.; Krishnan, M. Proteomic Analysis of Pupal Gut Serine Protease of Silkworm, *Bombyx mori*: Partial Purification and Biochemical Characterization. *Biocatal. Agric. Biotechnol.* **2017,** *12*, 159–165.

Kelemu, S.; Niassy, S.; Torto, B.; Fiaboe, K.; Affognon, H.; Tonnang, H.; Maniania, N.k.; Ekesi, S. African Edible Insects for Food and Feed: Inventory, Diversity, Commonalities and Contribution to Food Security. *J. Insects Food Feed.* **2015,** *1*, 103–119.

Kellert, S. R. Values and Perceptions of Invertebrates. *Conserv. Biol.* **1993,** *7*, 845–855.

Khan, D.; Sahito, Z.; Dawar, S.; Zaki, M. Frass of Saproxylic-Cerambycid Larvae From Dead Twigs of Acacia Stenophylla A. Cunn. Ex. Benth. and Its Effects on Germination and Seedling Growth of Lactuca Sativa L. Var. Grand Rapids. *Int. J. Biol. Biotech.* **2016,** *13*, 461–470.

Khiyami, M.; Alyamani, E. Aerobic and Facultative Anaerobic Bacteria from Gut of Red Palm Weevil (Rhynchophorusferrugineus). *Afr. J. Biotechnol.* **2008,** *7*, 1432–1437.

Kim, Y. M.; Ku, M. J.; Son, Y. J.; Yun, J. M.; Kim, S. H.; Lee, S. Y. Anti-Metastatic Effect of Cantharidin in A549 Human Lung Cancer Cells. *Arch. Pharm. Res.* **2013,** *36*, 479–484.

Kinyuru, J. N.; Kenji, G. M.; Njoroge, M. S. Process Development, Nutrition and Sensory Qualities of Wheat Buns Enriched with Edible Termites (Macrotermessubhylanus) from Lake Victoria Region, Kenya. *Afr. J. Food Agric. Nutr. Dev.* **2009,** *9*, 8.

Kinyuru, J. N.; Kenji, G. M.; Njoroge, S. M.; Ayieko, M. Effect of Processing Methods on the In Vitro Protein Digestibility and Vitamin Content of Edible Winged Termite (Macrotermessubhylanus) and Grasshopper (Ruspoliadifferens). *Food BioprocTech.* **2010,** *3*, 778–782.

Koonin, S. E. Getting Serious About Biofuels. *Science* **2006,** *311*, 435.

Ksenija, N. Mycotoxins–Climate Impact and Steps to Prevention Based on Prediction. *Acta. Veterinaria.* **2018,** *68*, 1–15.

Kuhlisch, C.; Pohnert, G. Metabolomics in Chemical Ecology. *Nat. Prod. Rep.* **2015,** *32*, 937–955.

Kvassay, G. The Complete Cricket Breeding Manual: Revolutionary New Cricket Breeding Systems Zega Enterprises, 2014.

Kwon, Y. J. Bombiculture: A Fascinating Insect Industry for Crop Pollination in Korea. *Entomol. Res.* **2008,** *38*, S66–S70, 1748–5967.

La Barbera, F.; Verneau, F.; Amato, M.; Grunert, K. Understanding Westerners' Disgust for the Eating of Insects: The Role of Food Neophobia and Implicit Associations. *Food Qual Prefer.* **2018,** *64*, 120–125.

La Barbera, F.; Verneau, F.; Videbæk, P. N.; Amato, M.; Grunert, K. G. A Self-Report Measure of Attitudes Toward the Eating of Insects: Construction and Validation of the Entomophagy Attitude Questionnaire. *Food Qual. Prefer.* **2020,** *79*, 103757.

Lee, J. Y.; Chung, T. W.; Choi, H. J.; Lee, C. H.; Eun, J. S.; Han, YT.; Choi, J. Y.; Kim, S. Y. A Novel Cantharidin Analog *N*-Benzylcantharidinamide Reduces the Expression of MMP-9 and Invasive Potentials of Hep3B via Inhibiting Cytosolic Translocation of HuR. *Biochem. Biophys. Res. Commun.* **2014,** *447*, 371–377.

Lees, A. D. *The Physiology of Diapause in Arthropods*; CUP Archive, 1955.

Leong, S. Y.; Kutty, S. R. M.; Malakahmad, A.; Tan, C. K. Feasibility Study of Biodiesel Production Using Lipids of Hermetiaillucens Larva Fed with Organic Waste. *Waste Manag.* **2016,** *47*, 84–90.

Leung, D.; Yang, D.; Li, Z.; Zhao, Z.; Chen, J.; Zhu, L. Biodiesel from Zophobasmorio Larva Oil: Process Optimization and FAME Characterization. *Ind. Eng. Chem. Res.* **2012,** *51*, 1036–1040.

Li, Q.; Zheng, L.; Cai, H.; Garza, E.; Yu, Z.; Zhou, S. From organic waste to biodiesel: Black Soldier Fly, Hermetiaillucens, Makes It Feasible. *Fuel.* **2011,** *90*, 1545–1548.

Li, Q.; Zheng, L.; Hou, Y.; Yang, S.; Yu, Z. Insect Fat, a Promising Resource for Biodiesel. *J. Pet. Environ. Biotechnol.* **2011,** *2*, 2–4.

Li, Q.; Zheng, L.; Qiu, N.; Cai, H.; Tomberlin, J. K.; Yu, Z. Bioconversion of Dairy Manure by Black Soldier Fly (Diptera: Stratiomyidae) for Biodiesel and Sugar Production. *Waste Manag.* **2011,** *31*, 1316–1320.

Li, W.; Li, Q.; Zheng, L.; Wang, Y.; Zhang, J.; Yu, Z.; Zhang, Y. Potential Biodiesel and Biogas Production from Corncob by Anaerobic Fermentation and Black Soldier Fly. *Bioresour. Technol.* **2015,** *194*, 276–282.

Lin, L.; Cunshan, Z.; Vittayapadung, S.; Xiangqian, S.; Mingdong, D. Opportunities and Challenges for Biodiesel Fuel. *Appl. Energy*. **2011,** *88*, 1020–1031.

Lirong Shen, Z. U.; Duo Li, Z. U.; Fengqin Feng, Z. U.; Yucui; Ren, Z. U. Nutritional Composition of Polyrhachisvicina Roger (Edible Chinese Black Ant) *Warasan Songkhla Nakharin (Sakha Witthayasatlae Technology).***2006.**

Liu, Y. The Feasibility Study of the New Kind Crude Fat Material of Biodiesel from Insects. *Mater. Renew. Sustain. Energy* **2011,** *1*, 271–279.

Lock, E. R.; Arsiwalla, T.; Waagbø, R. Insect Larvae Meal as an Alternative Source of Nutrients in the Diet of *Atlantic salmon* (*Salmo salar*) Postsmolt. *Aquacult. Nutr.* **2016,** *22*, 1202–1213.

Lockwood, J. The infested mind: Why humans fear, loathe, and love insects. *Oxford University Press*. **2013**

Lu, J.; Sun, Q.; Tu, Z. C.; Lv, Q.; Shui, P. X.; Cheng, Y. X. Identification of *N*-Acetyldopamine Dimers from the Dung Beetle *Catharsius molossus* and Their COX-1 and COX-2 Inhibitory. *Activities Mols*. **2015,** *20*, 15589–15596.

Lu, J.; Wang, H.; Huang, J.; Li, G.; Wang, Q.; Xu, W.; Chen, Y.; Zhang, K. Sesquiterpene Acids from Shellac and Their Bioactivities Evaluation. *Fitoterapia* **2014,** *97*, 64–70.

Lundy, M. E.; Parrella, M. P. Crickets are Not a Free Lunch: Protein Capture from Scalable Organic Side-Streams via High-Density Populations of *Acheta domesticus*. *PloS One*. **2015**. *10*, e0118785.

Luo, S. L.; Huang, X. J.; Wang, Y.; Jiang, R. W.; Wang, L.; Bai, L. L.; Peng, Q. L.; Song, C. L. Isocoumarins from American Cockroach (*Periplaneta americana*) and Their Cytotoxic Activities. *Fitoterapia* **2014,** *95*, 115–120.

Macadam, C. R.; Stockan, J. A. More Than Just Fish Food: Ecosystem Services Provided by Freshwater Insects. *Ecol. Entomol.* **2015,** *40*, 113–123.

Madritch, M. D.; Donaldson, J. R.; Lindroth, R. L. Canopy Herbivory Can Mediate the Influence of Plant Genotype on Soil Processes Through Frass Deposition. *Soil Biol. Biochem.* **2007,** *39*, 1192–1201.

Makkar, H. P. S. Review: Feed Demand Landscape and Implications of Food-Not Feed Strategy for Food Security and Climate Change. *Animal* **2018,** *12*, 1744–1754.

Makkar, H. P. S.; Tran, G.; Heuzé, V.; Ankers, P. State-of-the-Art on Use of Insects as Animal Feed. *Anim. Feed Sci. Technol.* **2014,** *197*, 1–33.

Mansouri, S. M.; Ganbalani, G. N.; Fathi, S. A. A.; Naseri, B.; Razmjou, J. Nutritional Indices and Midgut Enzymatic Activity of Phthorimaeaoperculella (Lepidoptera: Gelechiidae) Larvae Fed Different Potato Germplasms. *J. Econ. Entomol.* **2013,** *106*, 1018–1024.

Manzano-Agugliaro, F.; Sanchez-Muros, M. J.; Barroso, F. G.; Martínez-Sánchez, A.; Rojo, S.; Pérez-Bañón, C. Insects for Biodiesel Production. *Renew. Sust. Energ. Rev.* **2012,** *16*, 3744–3753.

Mariod, A. A. Insect Oil and Protein: Biochemistry, Food and Other Uses: Review. *Agric. Sci.* **2013,** *4*, 9.

Marone, P. A. Chapter 7—Food Safety and Regulatory Concerns. In: Dossey, A. T., Morales-Ramos, J. A., Rojas, M. G., Eds.; Academic Press: San Diego, 2016; pp 203–221.

Martins, Y.; Pliner, P. Human Food Choices: An Examination of the Factors Underlying Acceptance/Rejection of Novel and Familiar Animal and Nonanimal Foods. *Appetite.* **2005,** *45*, 214–224.

Mbah, C. E.; Elekima, G. O. V. Nutrient Composition of Some Terrestrial Insects in Ahmadu Bello University, Samaru Zaria Nigeria. *Sci. World J.* **2007,** 2.

Mebs, D.; Pogoda, W.; Schneider, M.; Kauert, G. Cantharidin and Demethylcantharidin (Palasonin) Content of Blister Beetles (Coleoptera: Meloidae) from Southern Africa. *Toxicon* **2009,** *53*, 466–468.

Meneguz, M.; Schiavone, A,.; Gai, F,.; Dama, A.; Lussiana, C.; Renna, M.; Gasco, L. Effect of Rearing Substrate on Growth Performance, Waste Reduction Efficiency and Chemical Composition of Black Soldier Fly (Hermetiaillucens) Larvae. *J. Sci. Food Agric.* **2018,** *98*, 5776–5784.

Merzendorfer, H.; Zimoch, L. Chitin Metabolism in Insects: Structure, Function and Regulation of Chitin Synthases and Chitinases. *J. Exp. Biol.* **2003,** *206*, 4393–4412.

Meyer-Rochow, V. Can Insects Help to Ease the Problem of World Food Shortage? *Search* **1975,** *6*, 261–262.

Miech, P.; Berggren, Å.; Lindberg, J. E.; Chhay, T.; Khieu, B.; Jansson, A. Growth and Survival of Reared Cambodian Field Crickets (Teleogryllustestaceus) Fed Weeds, Agricultural and Food Industry By-Products. *J. Insects Food Feed.* **2016,** *2*, 285–292.

Mika, N.; Zorn, H.; Rühl, M. Insect-Derived Enzymes: A Treasure for Industrial Biotechnology and Food Biotechnology. *Yellow Biotechnol. II.* **2013,** 1–17.

Mlcek, J.; Rop, O.; Borkovcova, M.; Bednarova, M. A Comprehensive Look at the Possibilities of Edible Insects as Food in Europe—A Review. *Polish J. Food Nutr. Sci.* **2014,** *64*, 147–157.

Morales-Ramos, J. A.; Rojas, M. G.; Coudron, T. A. Chapter 7 - Artificial Diet Development for Entomophagous Arthropods. In: Morales-Ramos, J. A.; Rojas, M. G.; Shapiro-Ilan, D. I., Eds.; Academic Press: San Diego, 2014; pp 203–240.

Murefu, T. R.; Macheka, L.; Musundire, R.; Manditsera, F. A. Safety of Wild Harvested and Reared Edible Insects: A Review. *Food Cont.* **2019,** *101*, 209–224.

Musundire, R.; Osuga, I. M.; Cheseto, X.; Irungu, J.; Torto, B. Aflatoxin Contamination Detected in Nutrient and Anti-Oxidant Rich Edible Stink Bug Stored in Recycled Grain Containers. *Plos One* **2016,** *11*, e0145914.

Muzzarelli, R. A. A.; Boudrant, J.; Meyer, D.; Manno, N.; DeMarchis, M.; Paoletti, M. G. Current Views on Fungal Chitin/Chitosan, Human Chitinases, Food Preservation, Glucans, Pectins and Inulin: A Tribute to Henri Braconnot, Precursor of the Carbohydrate Polymers Science, on the Chitin Bicentennial. *Carbohydr. Polym.* **2012,** *87*, 995–1012.

N. N. F. Regulation, *Regulation (EU) 2015/2283 of the European Parliament and of the Council of 25 November 2015 on Novel Foods, Amending Regulation (EU) No 1169/2011 of the European Parliament and of the Council and Repealing Regulation (EC) No 258/97 of the European Parliament and of the Council and Commission Regulation (EC) No 1852/2001.*

Naik, S. N.; Goud, V. V.; Rout, P. K.; Dalai, A. K. Production of First and Second Generation Biofuels: A Comprehensive Review. *Renew. Sust. Energ. Rev.* **2010,** *14*, 578–597.

Nakashima, K.; Watanabe, H.; Saitoh, H.; Tokuda, G.; Azuma, J. I. Dual Cellulose-Digesting System of the Wood-Feeding Termite, CoptotermesformosanusShiraki. *Insect Biochem. Mol. Biol.* **2002,** *32*, 777–784.

Newton, G. L.; Booram, C. V.; Barker, R. W.; Hale, O. M. Dried HermetiaIllucens Larvae Meal as a Supplement for Swine. *J. Anim. Sci.* **1977,** *44*, 395–400.

Newton, L.; Sheppard, C.; Watson, D. W.; Burtle, G.; Dove, R. Using the Black Soldier Fly, Hermetiaillucens, as a Value-Added Tool for the Management of Swine Manure. *Anim. Poultry Waste Manag. Center, North Carolina State Univ.* **2005,** *17*, 18.

Nguyen, H. C.; Liang, S. H.; Chen, S. S.; Su, C. H.; Lin, J. H.; Chien, C. C. Enzymatic Production of Biodiesel from Insect Fat Using Methyl Acetate as an Acyl Acceptor: Optimization by Using Response Surface Methodology. *Energy Convers. Manag.* **2018,** *158*, 168–175.

Nguyen, H. C.; Liang, S. H.; Li, S. Y.; Su, C. H.; Chien, C. C.; Chen, Y. J.; Huong, D. T. M. Direct Transesterification of Black Soldier Fly Larvae (Hermetiaillucens) for Biodiesel Production. *J. Taiwan Inst. Chem. Eng.* **2018,** *85*, 165–169.

Nyoka, B. I.; Fao, R. *State of Forest and Tree Genetic Resources in Dry Zone Southern Africa Development Community Countries*; Rome (Italy), 2003.

Oliveira, J. F. S.; de Carvalho, J. P.; de Sousa, R. F. X. B.; Simão, M. M. The Nutritional Value of Four Species of Insects Consumed in Angola. *Ecol. Food Nutr.* **1976,** *5*, 91–97.

Oonincx, D. G. A. B.; Poel, A. F. B. van der. Effects of Diet on the Chemical Composition of Migratory Locusts (*Locusta migratoria*). *Zoo. Biol.* **2011,** *30*, 9–16.

Oonincx, D. G. A. B.; Boer, I. J. M. de. Environmental Impact of the Production of Mealworms as a Protein Source for Humans—A Life Cycle Assessment. *Plos One* **2012,** *7*, e51145.

Ooninex, D. G. A. B.; van Itterbeeck, J.; Heetkamp, M. J. W.; Brand, H.; van den Loon; J. J. A.; Huis, A, van. An Exploration on Greenhouse Gas and Ammonia Production by Insect Species Suitable for Animal or Human Consumption. *Plos One* **2010,** *5*, e14445.

Parajulee, M. N.; Defoliart, G. R.; Hogg, D. B. Model for Use in Mass-Production of *Acheta domesticus* (Orthoptera: Gryllidae) as Food. *J. Econ. Entomol.* **1993,** *86*, 1424–1428.

Parliament, E.; Union, C. E. Regulation (EU) 2015/2283 of the European Parliament and of the Council of 25 November 2015 on Novel Foods, Amending Regulation (EU) No 1169/2011 of the European Parliament and of the Council and Repealing Regulation (EC) No 258/97 of the European Parliament and of the Council and Commission Regulation (EC) No 1852/2001. *Off. J. Eur. Union.* **2015,** *327*, 1–22.

Patel, S.; Suleria, H. A. R.; Rauf, A. Edible Insects as Innovative Foods: Nutritional and Functional Assessments. *Trends Food Sci. Technol.* **2019,** *86*, 352–359.

Paul, D.; Dey, S. Essential Amino Acids, Lipid Profile and Fat-Soluble Vitamins of the Edible Silkworm *Bombyx mori* (Lepidoptera: Bombycidae). *Int. J. Trop. Insect Sci.* **2014,** *34*, 239–247.

Pelini, S. L.; Maran, A. M.; Chen, A. R.; Kaseman, J.; Crowther, T. W. Higher Trophic Levels Overwhelm Climate Change Impacts on Terrestrial Ecosystem Functioning. *Plos One* **2015,** 10.

Pettit, G. R.; Meng, Y.; Herald, D. L.; Knight, J. C.; Day, J. F. Antineoplastic Agents: The Texas Grasshopper *Brachystola magna*. *J. Nat. Prod.* **2005,** *68*, 1256–1258.

Pettit, G. R.; Ye, Q.; Herald, D. L.; Hogan, F.; Pettit, R. K. Antineoplastic Agents: Isolation and Structure of Papilistatin from the Papilionid Butterfly Byasapolyeuctestermessa. *J. Nat. Prod.* **2010,** *73*, 164–166.

Poveda, J.; Jiménez Gómez, A.; Saati Santamaría, Z.; Usategui Martín, R.; Rivas, R.; García Fraile, P. Mealworm Frass as a Potential Biofertilizer and Abiotic Stress Tolerance-Inductor in Plants. *Appl. Soil Ecol.* **2019,** *142*, 110–122.

Prather, C. M.; Pelini, S. L.; Laws, A.; Rivest, E.; Woltz, M.; Bloch, C. P.; Toro, I. D.; Ho, C. K. et al. Invertebrates, Ecosystem Services and Climate Change. *Biol. Rev.* **2013,** *88*, 327–348.

Prem Anand, A. A.; Vennison, S. J.; Sankar, S. G.; Gilwax Prabhu, D. I.; Vasan, P. T.; Raghuraman, T.; Jerome Geoffrey, C.; Vendan, S. E. Isolation and Characterization of Bacteria from the Gut of *Bombyx mori* That Degrade Cellulose, Xylan, Pectin and Starch and Their Impact on Digestion. *J. Insect Sci.* **2010,** *10*, 107.

Prokop, P.; Fančovičová, J. The Association Between Disgust, Danger and Fear of Macroparasites and Human Behaviour. *Acta Ethologica* **2010,** *13*, 57–62.

Pryde, D. C.; Dalvie, D.; Hu, Q.; Jones, P.; Obach, R. S.; Tran, T. D. Aldehyde Oxidase: An Enzyme of Emerging Importance in Drug Discovery. *J. Med. Chem.* **2010,** *53*, 8441–8460.

Puerto Galvis, C. E.; Vargas Méndez, L. Y.; Kouznetsov, V. V. Cantharidin-Based Small Molecules as Potential Therapeutic Agents. *Chem. Biol. Drug Des.* **2013,** *82*, 477–499.

Quilliam, R.s.; NukuAdeku, C.; Maquart, P.; Little, D.; Newton, R.; Murray, F. Integrating Insect Frass Biofertilisers Into Sustainable Peri-Urban Agro-Food Systems. *J. Insects Food Feed.* **2020,** *6*, 315–322.

Ramos-Elorduy, J. Edible Insects of Chiapas. *Mexico Ecol. Food Nutr.* **2002,** *41*, 271–299.

Ramos-Elorduy, J.; González, E. A.; Hernández, A. R.; Pino, J. M. Use of Tenebrio molitor (Coleoptera: Tenebrionidae) to Recycle Organic Wastes and as Feed for Broiler Chickens. *J. Ethnobiol. Ethnomed.* **2002,** *95*, 214–220.

Ramos-Elorduy, J.; Medeiros Costa Neto, E.; Ferreira dos Santos, J.; Pino Moreno, J. M.; Landero-Torres, I.; Ángeles Campos, S. C.; García Pérez, Á. Estudiocomparativo del valor nutritivo de varioscoleoptera comestibles de México y Pachymerusnucleorum (FABRICIUS, 1792) (BRUCHIDAE) de Brasil. *Interciencia.* **2006,** *31*, 512–516.

Ramos-Elorduy, J.; Moreno, J. M. P.; Camacho, V. H. M. Edible aquatic Coleoptera of the world with an emphasis on Mexico. *J. Ethnobiol. Ethnomed.* **2009,** *5*, 11.

Ramos-Elorduy, J.; Moreno, J. M. P.; Prado, EE.; Perez, M. A.; Otero, J. L.; de Guevara, O. L. Nutritional Value of Edible Insects from the State of Oaxaca, Mexico. *J. Food. Compost. Anal.* **1997,** *10*, 142–157.

Regulation (EC) No. 999/2001 Laying down Rules for the Prevention, Control and Eradication of Certain Transmissible Spongiform Encephalopathies. https://www.ecolex.org/details/legislation/regulation-ec-no-9992001-laying-down-rules-for-the-prevention-control-and-eradication-of-certain-transmissible-spongiformencephalopathies-lex-fao-c030438/.

Robertson, M. J.; Gordon, C. P.; Gilbert, J.; McCluskey, A.; Sakoff, J. A. Norcantharimide Analogues Possessing Terminal Phosphate Esters and Their Anti-Cancer Activity. *Bioorg. Med. Chem.* **2011,** *19*, 5734–5741.

Rouchaud, A.; Braekman, J. C. Synthesis of New Analogues of the Tetraponerines. *Eur. J. Org. Chem.* **2009,** 2666–2674.

Rueda, L. C.; Ortega, L. G.; Segura, N. A.; Acero, V. M.; Bello, F. Luciliasericata strain from Colombia: Experimental Colonization, Life Tables and Evaluation of Two Artificial Diets of the Blowfly Luciliasericata (Meigen) (Diptera: Calliphoridae), Bogotá, Colombia strain. *Biol. Res.* **2010,** *43*, 197–203.

Rumpold, B. A.; Schlüter, O. K. Nutritional Composition and Safety Aspects of Edible Insects. *Mol. Nutr. Food Res.* **2013,** *57*, 802–823.

SaboiaVahia, L.; Cuervo, P.; Borges Veloso, A.; de Souza, N. P.; Britto, C.; Dias Lopes, G.; De Jesus, J. B. The Midgut of *Aedes albopictus* Females Expresses Active Trypsin-Like Serine Peptidases. *Parasites Vectors* **2014,** *7*, 253.

Saidemberg, DM.; da Silva-Filho, L. C.; Tognoli, L. M. M. C.; Tormena, C. F.; Palma, M. S.; Polybioside, a Neuroactive Compound from the Venom of the Social Wasp Polybiapaulista. *J. Nat. Prod.* **2010,** *73*, 527–531.

Sanatan, P. T.; Lomate, P. R.; Giri, A. P.; Hivrale, V. K. Characterization of a Chemostable Serine Alkaline Protease from *Periplaneta americana. BMC Biochem.* **2013,** *14*, 32.

Sanoh, S.; Tayama, Y.; Sugihara, K.; Kitamura, S.; Ohta, S. Significance of Aldehyde Oxidase During Drug Development: Effects on Drug Metabolism, Pharmacokinetics, Toxicity, and Efficacy. *Drug Metab. Pharmacokinet.* **2015,** *30*, 52–63.

Santos, A. B. R.; Chapman, M. D.; Aalberse, R. C.; Vailes, L. D.; Ferriani, V. P.; Oliver, C.; Rizzo, M. C.; Naspitz, C. K. Cockroach Allergens and Asthma in Brazil: Identification of Tropomyosin as a Major Allergen with Potential Cross-Reactivity with Mite and Shrimp Allergens. *J. Allergy Clin. Immunol.* **1999,** *104*, 329–337.

Saunders, D. S. Larval Diapause Duration and Fat Metabolism in Three Geographical Strains of the Blow Fly, Calliphoravicina. *J. Insect Physiol.* **2000,** *46*, 509–517.

Saunders, D. S. Under-Sized Larvae from Short-Day Adults of the Blow Fly, Calliphoravicina, Side-Step the Diapause Programme. *Physiol. Entomol.* **1997,** *22*, 249–255.

Schabel, H. G. Forest Insects as Food: A Global Review. *Forest insects Food: Humans Bite Back* **2010,** *37*.

Scharf, M. E.; Karl, Z. J.; Sethi, A.; Boucias, D. G. Multiple Levels of Synergistic Collaboration in Termite Lignocellulose Digestion. *Plos One* **2011,** *6*, e21709.

Schouteten, J. J.; De Steur, H.; De Pelsmaeker, S.; Lagast, S.; Juvinal, J. G.; De Bourdeaudhuij, I.; Verbeke, W.; Gellynck, X. Emotional and Sensory Profiling of Insect-, Plant- and Meat-Based Burgers Under Blind, Expected and Informed Conditions. *Food Qual. Prefer.* **2016,** *52*, 27–31.

Sharp, R. G. A Review of the Applications of Chitin and Its Derivatives in Agriculture to Modify Plant-Microbial Interactions and Improve Crop Yields. *Agron.* **2013,** *3*, 757–793.

Shi, Y. N.; Tu, Z. C.; Wang, X. L.; Yan, Y. M.; Fang, P.; Zuo, Z. L.; Hou, B.; Yang, T. H. Bioactive Compounds from the Insect *Aspongopus chinensis*. *Bioorg. Med. Chem. Lett.* **2014,** *24*, 5164–5169.

Shil, R. K.; Mojumder, S.; Sadida, F. F.; Uddin, M.; Sikdar, D. Isolation and Identification of Cellulolytic Bacteria from the Gut of Three Phytophagus Insect Species. *Braz. Arch. Biol. Technol.* **2014,** *57*, 927–932.

Shimizu, T.; Iizuka, M.; Matsukura, H.; Hashizume, D.; Sodeoka, M. Synthesis of Optically Pure Norcantharidin Analogue NCA-01, a Highly Selective Protein Phosphatase 2B Inhibitor, and its Derivatives. *Chem. Asian J.* **2012,** *7*, 1221–1230.

Snaddon, J. L.; Turner, E. C. A Child's Eye View of the Insect World: Perceptions of Insect Diversity. *Environ. Conserv.* **2007,** *34*, 33–35

Snaddon, J. L.; Turner, E. C.; Foster, W. A. Children's Perceptions of Rainforest Biodiversity: Which Animals Have the Lion's Share of Environmental Awareness? *Plos One* **2008,** *3*, e2579.

Song, Z. W.; Liu, P.; Yin, W. P.; Jiang, Y. L.; Ren, Y. Isolation and Identification of Antibacterial Neo-Compounds from the Red Ants of ChangBai Mountain, Tetramorium sp. *Bioorg. Med. Chem. Lett.* **2012,** *22*, 2175–2181.

Sonkaria, S.; Ahn, S. H.; Khare, V. Nanotechnology and its Impact on Food and Nutrition: A Review. *Recent Pat. food, Nutr. Agric.* **2012,** 4, 8–18.

Sribandit, W.; Wiwatwitaya, D.; Suksard, S. The Importance of Weaver Ant (OecophyllasmaragdinaFabricus) Harvest to a Local Community in Northeastern Thailand. *Asian Myrmecol.* **2008,** 129–138.

Stabentheiner, A.; Kovac, H.; Brodschneider, R. Honeybee Colony Thermoregulation—Regulatory Mechanisms and Contribution of Individuals in Dependence on Age, Location and Thermal Stress. *Plos One* **2010,** *5*, e8967.

Stack, J.; Dorward, A..Gondo, T.; Frost, P.; Taylor, F.; Kurebgaseka, N. Mopane Worm Utilisation and Rural Livelihoods in Southern Africa. *Undefined*, 2003.

Steen, L.; Glorieux, S.; Goemaere, O.; Brijs, K.; Paelinck, H.; Foubert, I.;Fraeye, I. Functional Properties of Pork Liver Protein Fractions. *Food Bioproc. Tech.* **2016,** *9*, 970–980.

Stoffolano, J. G. Jr.; Matthysse, J. G. Influence of Photoperiod and Temperature on Diapause in the Face Fly, Musca autumnalis (Diptera: Muscidae). *Ann. Entomol. Soc. Am.* **1967,** *60*, 1242–1246.

Stoykov, Y. M. Pavlov, A. I. Krastanov, A. I. Chitinase Biotechnology: Production, Purification, and Application. *Eng. Life Sci.* **2015,** *15*, 30–38.

Su, C. H.; Nguyen, H. C.; Bui, T. L.; Huang, D. L. Enzyme-Assisted Extraction of Insect Fat for Biodiesel Production. *J. Clean. Prod.* **2019,** *223*, 436–444.

Sun-Waterhouse, D.; Waterhouse, G. I. N.; You, L.; Zhang, J.; Liu, Y.; Ma, L.; Gao, J.; Dong, Y. Transforming Insect Biomass into Consumer Wellness Foods: A Review. *Int. Food Res. J.* **2016,** *89*, 129–151.

Swardt, D. B. de; Wigley Coetsee, C.; O'Connor, T. G. Insect Outbreaks Alter Nutrient Dynamics in a Southern African Savanna: Patchy Defoliation of Colophospermum Mopane Savanna by Imbrasiabelina Larvae. *Biotropica* **2018,** *50*, 789–796.

Sweetapple, P.; Barron, M. Frass Drop for Monitoring Relative Abundance of Large Arboreal Invertebrates in a New Zealand Mixed Beech Forest. *N. Z. J. Ecol.* **2016,** *40*, 321–329.

Tabassum, A.; Abbasi, T.; Abbasi, S. A. Reducing the Global Environmental Impact of Livestock Production: The Minilivestock Option. *J. Clean. Prod.* **2016,** *112*, 1754–1766.

Tang, J. J.; Fang, P.; Xia, H. L.; Tu, Z. C.; Hou, B. Y.; Yan, Y. M.; Di, L.; Zhang, L. Constituents from the Edible Chinese Black Ants (*Polyrhachis dives*) Showing Protective Effect on Rat Mesangial Cells and Anti-Inflammatory Activity. *Int. Food Res. J.* **2015,** *67*, 163–168.

Tang, J. J.; Zhang, L.; Jiang, L. P.; Di, L.; Yan, Y. M.; Tu, Z. C.; Yang, C. P.; Zuo, Z. L. Dopamine Derivatives from the Insect *Polyrhachis dives* as Inhibitors of ROCK1/2 and Stimulators of Neural Stem Cell Proliferation. *Tetrahedron* **2014,** *70*, 8852–8857.

Tang, X.; da Costa Sousa, L.; Jin, M.; Chundawat, S. P.; Chambliss, C. K.; Lau, M. W.; Xiao, Z.; Dale, B. E. Designer Synthetic Media for Studying Microbial-Catalyzed Biofuel Production. *Biotechnol. Biofuels*. **2015,** *8*, 1.

Tarleton, M.; Gilbert, J.; Sakoff, J. A.; McCluskey, A. Synthesis and Anticancer Activity of a Series of Norcantharidin Analogues. *Eur. J. Med. Chem.* **2012,** *54*, 573–581.

Tauber, M. J.; Tauber, C. A.; Masaki, S. *Seasonal Adaptations of Insects*; Oxford University Press, 1986.

Tchuinkam, T.; Mpoame, M.; Make-Mveinhya, B.; Simard, F.; Lélé-Defo, E.; Zébazé-Togouet, S.; Tateng-Ngouateu, A.; Awono-Ambéné, H. P. Optimization of Breeding Output for Larval Stage of *Anopheles gambiae* (Diptera: Culicidae): Prospects for the Creation and Maintenance of Laboratory Colony from Wild Isolates. *Bull. Entomol. Res.* **2011,** *101*, 259–269.

Teffo, L. S.; Toms, R. B.; Eloff, J. N. Preliminary Data on the Nutritional Composition of the Edible Stink-Bug, EncosternumdelegorgueiSpinola, Consumed in Limpopo Province, South Africa. *S. Afr. J. Sci.* **2007,** *103*, 434–436.

Tilman, D.; Cassman, K. G.; Matson, P. A.; Naylor, R.; Polasky, S. Agricultural Sustainability and Intensive Production Practices. *Nature.* **2002,** *418*, 671–677.

Tong, L.; Yu, X.; Liu, H. Insect Food for Astronauts: Gas Exchange in Silkworms Fed on Mulberry and Lettuce and the Nutritional Value of These Insects for Human Consumption During Deep Space. *Bull. Entomol. Res.* **2011,** *101*, 613–622.

Tucker, C. A. The Significance of Sensory Appeal for Reduced Meat Consumption. *Appetite* **2014,** *81*, 168–179.

Tulp, M.; Bohlin, L. Unconventional Natural Sources for Future Drug Discovery. *Drug Discov.* **2004,** *9*, 450–458.

Turunen, S.; Chippendale, G. M. Fat Body Protein Associated with the Larval Diapause of the Southwestern Corn Borer, Diatraeagrandiosella: Synthesis and Characteristics. *Comp. Biochem. Physiol.* **1980,** *65*, 595–603

Twu, N. F.; Srinivasan, R.; Chou, C. H.; Wu, L. S.; Chiu, C. H. Cantharidin and Norcantharidin Inhibit Caprine Luteal Cell Steroidogenesis In Vitro. *Exp. Toxicol. Pathol.* **2012,** *64*, 37–44.

Uchima, C. A.; Tokuda, G.; Watanabe, H.; Kitamoto, K.; Arioka, M. Heterologous Expression and Characterization of a Glucose-Stimulated β-Glucosidase from the Termite Neotermeskoshunensis in *Aspergillus oryzae*. *Appl. Microbiol. Biotechnol.* **2011,** *89*, 1761–1771.

Uriarte, M. Interactions Between Goldenrod (*Solidago altissima* L.) and Its Insect Herbivore (*Trirhabda virgata*) Over the Course of Succession. *Oecologia* **2000,** *122*, 521–528.

van Broekhoven, S.; Oonincx, D. G. A. B.; van Huis, A.; van Loon, J. J. A. Growth Performance and Feed Conversion Efficiency of Three Edible Mealworm Species (Coleoptera: Tenebrionidae) on Diets Composed of Organic By-Products. *J. Insect Physiol.* **2015,** *73*, 1–10.

van Dijk, M.; Meijerink, G. W. A Review of Global Food Security Scenario and Assessment Studies: Results, Gaps and Research Priorities. *Glob. Food Sec.* **2014,** *3*, 227–238.

van Itterbeeck, J. *Semi-Cultivating Edible Insects: A Historical Perspective and Future Prospects*; Book of Abstracts of Conference on Insects to Feed the World: The Netherlands, 2014; pp 14–17.

van Huis, A.; van Itterbeeck, J.; Klunder, H.; Mertens, E.; Halloran, A.; Muir, G.; Vantomme, P. *Edible Insects: Future Prospects for Food and Feed Security*; FAO Forestry Paper, 2013.

van Huis, A. Edible Insects Contributing to Food Security? *Agric. Food Secur.* **2015,** *4*, 20.

van Huis, A. Edible Insects Are the Future? *Proc. Nutr. Soc.* **2016,** *75*, 294–305.

van Huis, A.; Oonincx, D. G. A. B. The Environmental Sustainability of Insects as Food and Feed: A Review. *Agron. Sustain. Dev. Agron. Sustain. Dev.* **2017,** *37*, 43.

van Zanten, H. H. E.; Mollenhorst, H.; Oonincx, D. G. A. B.; Bikker, P.; Meerburg, B. G.; de Boer, I. J. M. From Environmental Nuisance to Environmental Opportunity: Housefly Larvae Convert Waste to Livestock Feed. *J. Clean. Prod.* **2015,** *102*, 362–369.

Veldkamp, T.; Duinkerken, G.; van Huis.; A.; Lakemond, C. M. M.; Ottevanger, E.; Bosch, G.; Boekel, T. Insects as a Sustainable Feed Ingredient in Pig and Poultry Diets : A Feasibility Study = Insectenalsduurzamediervoedergrondstof in varkens-enpluimveevoeders : eenhaalbaarheidsstudie (638) Lelystad. *Wageningen UR Livestock*, 2012.

Verbeke, W. Profiling Consumers Who Are Ready to Adopt Insects as a Meat Substitute in a Western Society. *Food Qual. Prefer*. **2015,** *39*, 147–155.

Verma, A. K.; Prasad, S. B. Bioactive Component, Cantharidin from Mylabriscichorii and Its Antitumor Activity Against *Ehrlich ascites* Carcinoma. *Cell Biol. Toxicol.* **2012,** *28*, 133–147.

Vickery, V. R.; Kevan, D. K. The Grasshoppers, Crickets, and Related Insects of Canada and Adjacent Regions. Ulonata: Dermaptera, Cheleutoptera, Notoptera, Dictuoptera, Grylloptera, and Orthoptera. (no. Part 14). Agriculture Canada, 1985.

Vihakas, M. A.; Kapari, L.; Salminen, J. P.; New Types of Flavonol Oligoglycosides Accumulate in the Hemolymph of Birch-Feeding Sawfly Larvae. *J. Chem. Ecol.* **2010,** *36*, 864–872.

Vizcaino, M. I.; Guo, X.; Crawford, J. M. Merging Chemical Ecology with Bacterial Genome Mining for Secondary Metabolite Discovery. *J. Ind. Microbiol. Biotechnol.* **2014,** *41*, 285–299.

Wang, C.; Qian, L.; Wang, W.; Wang, T.; Deng, Z.; Yang, F.; Xiong, J.; Feng, W. Exploring the Potential of Lipids from Black Soldier Fly: New Paradigm for Biodiesel Production (I). *Rene. Energy*. **2017,** *111*, 749–756.

Wang, T.; Liu, J.; Xiao, X. Q. Cantharidin Inhibits Angiogenesis by Suppressing VEGF-Induced JAK1/STAT3, ERK and AKT Signaling Pathways. *Arch. Pharm. Res.* **2015,** *38*, 282–289.

Wang, Y.; Liu, J.; Ruan, H.; Wu, K.; Gao, M.; Yu, H.; Huang, J. The Immunological Regulation Effect of Water Extract from Polyrhachisvicina Roger on Normal and Immune Deficient Mic. **2001,** *V*. 26.

Wei, J. R. Lu, X. P.; Jiang, L. Monoterpenes from Larval Frass of Two Cerambycids as Chemical Cues for a Parasitoid, Dastarcushelophoroides. *J. Insect Sci*. **2013,** *13*.

Will, R. G.; Ironside, J. W.; Zeidler, M.; Estibeiro, K.; Cousens, S. N.; Smith, P. G.; Alperovitch, A.; Poser, S. A New Variant of Creutzfeldt-Jakob Disease in the UK. *Lancet* **1996,** *347*, 921–925.

Wong, C. Y.; Lim, J. W.; Uemura, Y.; Chong, F. K.; Yeong, Y. F.; Mohamad, M.; Hermansyah, H. Insect-Based Lipid for Biodiesel Production, 2018.

Wu, J. Y.; Kuo, C. D.; Chu, C. Y.; Chen, M. S.; Lin, J. H.; Chen, Y. J.; Liao, H. F. Synthesis of Novel Lipophilic N-Substituted Norcantharimide Derivatives and Evaluation of Their Anticancer Activities. *Mols* **2014,** *19*, 6911–6928.

Xiao, Z.; Liu, W.; Zhu, G.; Zhou, R.; Niu, Y. A Review of the Preparation and Application of Flavour and Essential Oils Microcapsules Based on Complex Coacervation Technology. *J. Sci. Food Agric*. **2014,** *94*, 1482–1494.

Yabutani, T.; Waterhouse, G. I. N.; Sun-Waterhouse, D.; Metson, J. B.; Iinuma, A.; Thuy, L. T. X.; Yamada, Y.; Takayanagi, T. Facile Synthesis of Platinum Nanoparticle-Containing Porous Carbons, and Their Application to Amperometric Glucose Biosensing. *Microchim. Acta*. **2014,** *181*, 1871–1878.

Yan, Y. M.; Ai, J.; Shi, Y. N.; Zuo, Z. L.; Hou, B.; Luo, J.; Cheng, Y. X. (±)-Aspongamide A, an N-Acetyldopamine Trimer Isolated from the Insect *Aspongopus chinensis*, Is an Inhibitor of p-Smad3. *Org. Lett.* **2014,** *16*, 532–535.

Yan, Y. M.; Li, L. J.; Qin, X. C.; Lu, Q.; Tu, Z. C.; Cheng, Y. X. Compounds from the Insect Blapsjapanensis with COX-1 and COX-2 Inhibitory Activities. *Bioorg. Med. Chem. Lett.* **2015,** *25*, 2469–2472.

Yang, S.; Li, Q.; Gao, Y.; Zheng, L.; Liu, Z. Biodiesel Production from Swine Manure via Housefly Larvae (*Musca domestica* L.). *Renew. Energy* **2014,** *66*, 222–227.

Yang, S.; Liu, Z. Pilot-Scale Biodegradation of Swine Manure via Chrysomyamegacephala (Fabricius) for Biodiesel Production. *Appl. Energy*. **2014,** *113*, 385–391.

Yang, S. S.; Chen, Y.; Kang, J. H.; Xie, T. R.; He, L.; Xing, D. F.; Ren, N. Q.; Ho, S. H. Generation of High-Efficient Biochar for Dye Adsorption Using Frass of Yellow Mealworms (Larvae of Tenebrio molitor Linnaeus) Fed with Wheat Straw for Insect Biomass Production. *J. Clean. Prod.* **2019,** *227*, 33–47.

Yang, Y.; Yang, J.; Wu, WM.; Zhao, J.; Song, Y.; Gao, L.; Yang, R.; Jiang, L. Biodegradation and Mineralization of Polystyrene by Plastic-Eating Mealworms: Part 1. Chemical and Physical Characterization and Isotopic Tests. *Environ. Sci. Technol.***2015,** *49*, 12080–12086.

Yen, A. L. Insect and Other Invertebrate Foods of the Australian Aborigines. In: *Ecological Implications of Minilivestock: Potential of Insects, Rodents, Frogs and Snails*, 2005; pp 367–387.

Yen, A. L. Insects as Food and Feed in the Asia Pacific Region: Current Perspectives and Future Directions. *J. Insects Food Feed.* **2015,** *1*, 33–55.

Yhoung-Aree, J.; Viwatpanich, K. Edible Insects in the Laos PDR, Myanmar, Thailand, and Vietnam. In: *Ecological Implications of Minilivestock: Potential of Insects, Rodents, Frogs and Snails* 2005; pp 415–440.

Yi, L.; Lakemond, C. M. M.; Sagis, L. M. C.; Eisner-Schadler, V.; van Huis, A.; van Boekel, M. A. J. S. Extraction and Characterisation of Protein Fractions from Five Insect Species. *Food Chem.* **2013,** *141*, 3341–3348.

Yi, L.; Van Boekel.; M. A. J. S.; Lakemond, C. M. M. Extracting Tenebrio Molitor Protein While Preventing Browning: Effect of pH and NaCl on Protein Yield. *J. Insects Food Feed.* **2017,** *3*, 21–31.

Yildirim-Aksoy, M.; Eljack, R.; Schrimsher, C.; Beck, B. H. Use of dietary frass from black Soldier Fly Larvae, Hermetiaillucens, in Hybrid Tilapia (Nile x Mozambique, Oreocromisniloticus x O. mozambique) Diets Improves Growth and Resistance to Bacterial Diseases. *Aquac. Rep.* **2020,** *17*, 100373.

Youngs, H.; Somerville, C. Best Practices for Biofuels. *Sciences* **2014,** *344*, 1095–1096.

Yu, Y. T.; Wei, H. Y.; Fadamiro, H. Y.; Chen, L. Quantitative Analysis of Alkaloidal Constituents in Imported Fire Ants By Gas Chromatography. *J. Agric. Food Chem.* **2014,** *62*, 5907–5915.

Zhang, C.; Chen, Z.; Zhou, X.; Xu, W.; Wang, G.; Tang, X.; Luo, L.; Tu, J. Cantharidin induces G2/M Phase Arrest and Apoptosis in Human Gastric Cancer SGC–7901 and BGC–823 Cells. *Oncol. Lett.* **2014,** *8*, 2721–2726.

Zhang, C. X.; Tang, X. D.; Cheng, J. A. The Utilization and Industrialization of Insect Resources in China. *Entomol. Res.* **2008,** *38*, S38–S47, 1748–5967.

Zhang, H.; Wei, L.; Zhang, Z.; Liu, S.; Zhao, G.; Zhang, J.; Hu, Y. Protective Effect of Periplaneta Americana Extract on Intestinal Mucosal Barrier Function in Patients with Sepsis. *J. Tradit. Chin. Med.* **2013,** *33*, 70–73.

Zhang, X.; Tang, H.; Chen, G.; Qiao, L.; Li, J.; Liu, B.; Liu, Z.; Li, M. Growth Performance and Nutritional Profile of Mealworms Reared on Corn Stover, Soybean Meal, and Distillers' Grains. *Eur. Food Res. Technol.* **2019,** *245*, 2631–2640.

Zhao, X.; Vázquez-Gutiérrez, J. L.; Johansson, D. P.; Landberg, R.; Langton, M. Yellow Mealworm Protein for Food Purposes—Extraction and Functional Properties. *Plos One* **2016,** *11*, e0147791.

Zheng, L.; Hou, Y.; Li, W.; Yang, S.; Li, Q.; Yu, Z. Biodiesel Production from Rice Straw and Restaurant Waste Employing Black Soldier Fly Assisted by Microbes. *Energy.* **2012,** *47*, 225–229.

Zheng, L.; Hou, Y.; Li, W.; Yang, S.; Li, Q.; Yu, Z. Exploring the Potential of Grease from Yellow Mealworm Beetle (Tenebrio molitor) as a Novel Biodiesel Feedstock. *Appl. Energy* **2013,** *101*, 618–621.

Zheng, L.; Li, Q.; Zhang, J.; Yu, Z. Double the Biodiesel Yield: Rearing Black Soldier Fly Larvae, Hermetiaillucens, on Solid Residual Fraction of Restaurant Waste After Grease Extraction for Biodiesel Production. *Renew. Energy* **2012,** *41*, 75–79.

Zheng, L.; Zhao, Y.; Xiao, C.; Sun-Waterhouse, D.; Zhao, M.; Su, G. Mechanism of the Discrepancy in the Enzymatic Hydrolysis Efficiency Between Defatted Peanut Flour and Peanut Protein Isolate by Flavorzyme. *Food Chem.* **2015,** *168*, 100–106.

Zhou, J.; Han, D. Safety Evaluation of Protein of Silkworm (*Antheraea pernyi*) Pupae. *Food Chem. Toxicol.* **2006,** *44*, 1123–1130.

Zhu, H. J.; Yan, Y. M.; Tu, Z. C.; Luo, J. F.; Liang, R.; Yang, T. H.; Cheng, Y. X.; Wang, S. M. Compounds from Polyphagaplancyi and Their Inhibitory Activities Against JAK3 and DDR1 Kinases. *Fitoterapia* **2016,** *114*, 163–167.

Zotchev, S. B.; Sekurova, O. N.; Katz, L. Genome-Based Bioprospecting of Microbes for New Therapeutics. *Curr. Opin. Biotechnol.* **2012,** *23*, 941–947.

CHAPTER 6

Caddisfly (Trichoptera): Diversity and Ecosystem Services

MANPREET S. PANDHER[1], DEEPTI GARIMA[1,2], and SIMARJIT KAUR[1,3]

[1]*Zoological Survey of India, Kolkata, West Bengal, India*

[2]*Department of Zoology, University of Calcutta, Kolkata, West Bengal, India*

[3]*Post Graduate Government College for Girls, Chandigarh, India*

ABSTRACT

Trichoptera are important insect order which are aquatic and are one of the most diverse. These are distributed throughout the world and are present in all continents except Antarctica. In India, they are more diverse in the Himalayan region than the rest other parts of the country. Trichoptera provide diverse ecosystem services like including their essential roles in food webs, in biological monitoring of water quality, as food for fish and other predators. Their larvae are well known for their case-making properties and clearing of the organic matter from water bodies during the process. Rapid change in climate is affecting their diversity and habitats drastically which needs much attention as this group of insects play many important roles in our ecosystem.

6.1 INTRODUCTION

Trichoptera (Caddisflies) are aquatic insects which are the seventh most speciose insect order. These are holometabolus insects whose larval forms are exclusively aquatic. However, the adult flies are terrestrial. The

Insect Diversity and Ecosystem Services, Volume 1: Importance, Threats, Conservation and Economic Perspectives. Younis Ahmad Hajam, Sajad Hussain Parey, and Rouf Ahmad Bhat (Eds)

characteristic features of these insects are the presence of hairy wings and while sitting, the wings are folded to form a conical roof/tent-like covering over the abdomen. The adult body size ranges from 1.5 mm to 400 mm and the color in general has the variation of dull yellow, grey, brown, tan, and black. They have well-developed maxillary palps and prominent long antennae, but they always lack coiled proboscis which is a characteristic feature of Lepidoptera—their sister taxon under the Amphiesmenoptera.

Larvae belonging to different taxa feed in a wide variety of ways. They are either free living or construct different kinds of stationary retreats and portable cases. When they are mature, the larvae construct stationary shelters or modify their portable cases into stationary shelters and prepare to pupate. They often spin different kinds of cocoons and then moult to the pupal form. After a few weeks, the pupation is completed which leads to emerging of the adults. The emergence of adults occurs often at specific times of the day or night as well as at predictable times of the year depending upon daylight changes, temperature, or other environmental conditions.

6.2 HISTORICAL RESUME

Primarily, Walker (1852), McLachlan (1875), Morton (1900, 1902), Banks (1911, 1913, 1939), Martynov (1926, 1935, 1936), Mosely (1935), Kimmins (1950, 1957, 1963), Schmid (1963, 1968a, 1968b, 1970, 1971, 1972, 1982, 1983, 1984, 1987, 1991, 1994a, 1994b, 1995a, 1995b, 1995c), and Malicky (2005, 2009) contributed to the major work on systematics of Caddisflies. But the Indian Trichoptera was not dealt fairly by any systematics. Though largest among primary aquatic orders, it remained ignored and finds no mention in the "Fauna of British India." In recent years, Saini et al. (2010, 2011, 2012, 2013) initiated the major work on Trichoptera from India, followed by Saini and Pandher (2011), Saini and Parey (2011, 2012), Pandher and Saini (2011, 2012, 2013a, 2013b, 2014, 2015), Pandheret al. (2012, 2013, 2014, 2017, 2018a, 2018b, 2018c, 2020a, 2020b, 2020c, 2020d, 2020e), Pandher and Parey (2018, 2019), Pandher (2018, 2019a, 2019b, 2020), Pandher and Chandra (2019), Parey (2015), Parey and Pandher (2016, 2019), Parey et al. (2013, 2016), and Kaur et al. (2020).

6.3 DIVERSITY

Caddisflies are present in all the continents except the Antarctica (Morse, 2021). At present, 16,267 species of Trichoptera positioned under 632 genera and 63 families have been reported throughout the world. 1301 species in 102 genera and 28 families are reported from India (Morse, 2021).The entire India was classified into 10 biogeographic zones by Rodgers et al. (2002): Trans Himalaya; The Himalaya; Desert; Semi-Arid, Western Ghats, Deccan Peninsula; Gangetic Plains; The North-east; Islands; and Coasts (Fig. 6.1). Among these biogeographic zones, Caddisflies are most diverse within the Himalaya biogeographic zones. This is because of the high altitudinal range and snow fed streams of the Indian Himalaya, which makes a favorable habitat for the caddisflies. The maximum number of Trichoptera has been reported from the Himalaya Biogeographic zone (762 spp.) followed by the Northeast biogeographic zone (389 spp.). About 159 species are reported from the Deccan Peninsula biogeographic zone and 116 species are on record from the Gangetic Plains biogeographic zone based on available literature. Only 20 species are reported from the Island biogeographic zone, whereas 53 species are recorded from the Coastal biogeographic zone of India. No species have been reported from the Desert Biogeographic zone (Table 6.1).

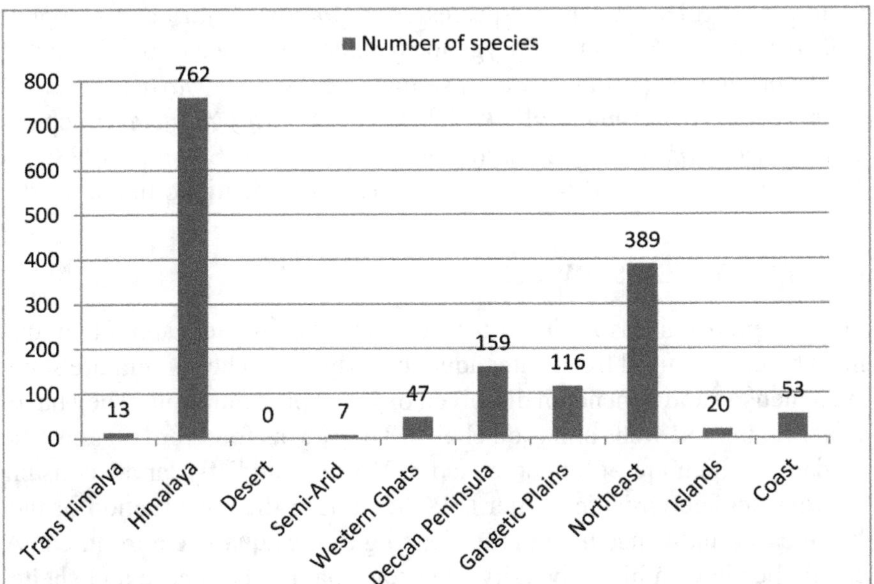

FIGURE 6.1 Graph showing the total number of Trichoptera species in different biogeographic zones of India.

TABLE 6.1 Species of Trichoptera in Different Biogeographic Zones of India.

Biogeographic zones	Number of species
Trans Himalaya	13
Himalaya	762
Desert	0
Semi-Arid	7
Western Ghats	47
Deccan Peninsula	159
Gangetic Plains	116
Northeast	389
Islands	20
Coast	53

Among all the 29 families of Trichoptera reported from India, the family Leptoceridae is the largest with 293 species followed by Rhyachophilidae with 193 species, Philopotamidae with 153 species, and Hydropsychidae with 135 species. A single species *Phryganopsychelatipennis* (Banks, 1906) represents the family Phryganopsychidae. Six species of Trichoptera are uncertain in superfamily Sericostomatoidea, viz, *Asahayaasambaddha* Schmid, 1991; *Karomanadidinki* Schmid, 1993; *Mpugalafcadio* Schmid, 1993; *Mpugalafcadiello* Schmid, 1993; *Ngoyaelwe* Schmid, 1993; and *Ngoyaolwe* Schmid, 1994. All these species are endemic to India.

6.4 ECOSYSTEM SERVICES

Trichoptera plays various important roles in the ecosystem and is abundant in all types of natural freshwater aquatic ecosystem. Their immature stages are apneustic and depend on dissolved oxygen for respiration. They have a vital role in food webs and food chains. They act as food for fish and other predators which are of human concern. Various caddisfly larvae consume Fine Particulate Organic Matter (FPOM) during the construction of their larval cases and hence help in the cleaning of the aquatic environment. As caddisflies have a high diversity of species having both case and shelter-constructing larvae, they are useful indicators of organic pollution. The high species diversity is correlated with an extraordinarily broad range of ecological specialization (Morse, 2019).

They are important processors of organic matter and act as a keystone taxon in the development of the river continuum concept (RCC) (Vannote et al., 1980). Their larvae, pupae, and adults form a vital link in the food chain, and they have also been used extensively by trout fishing enthusiasts as models for "flies" (McCafferty, 1981). Trichoptera has been used extensively in biomonitoring assays with indicator species. Trichoptera, in association with other aquatic insect orders, have also been used to assess aquatic biodiversity EPT (Ephemeroptera, Plecoptera, and Trichoptera) or ETS (Ephemeroptera, Trichoptera, Simuliidae) and habitat diversity (Hannaford and Resh, 1995; Hewlett, 2000; de Moor, 2002).

The silk which are produced by the caddisfly larvae have adhering properties and are employed in the self-healing synthetic materials, fibers, and fabrics, especially suited to underwater and wet-field biomaterial applications (Morse, 2019). The larval cases of caddisflies are used as a model in the making of jewelleries (Prommi, 2018). Trichoptera have some negative impacts also. The net-spinning caddisflies block water intake pipes of hydro-electric power plants. Sometimes, caddisflies also destroy manmade wooden materials. Some people are also allergic to the hair of adult caddisflies (Morse, 2019).

The order Trichoptera has been divided into three suborders—Annulipalpia, Integripalpia, and Spicipalpia which is a basal lineage of Integripalpia. The suborder Annulipalpia has larvae which are fixed retreat makers. They built stationary shelters of silk and pieces of plant materials and mineral particles. These retreats are attached to stones, woody debris, and aquatic plants of moving waters or lakes. The larva of Hydropsychidae constructs nets of silk threads and hence is named net spinning caddisflies. The larvae of family Philopotamidae occupy all of a single sac-like or finger-like net and hence they have no separate retreat. Stenopsychidae larvae make loosely woven retreats/capture nets in fast-flowing water. This helps them in making a cover for themselves, a primary feeding mesh, and a pocket for collecting minute living and nonliving. Fine particulate organic matter (FPOM) is suspended in moving water. Larvae of two families Xiphocentronidae and Psychomyiidae, which are sister families, are called "nettubecaddisflies." They usually construct "lairs" that hide them from potential predators as they feed. Larvae of polycentropodidae family of genus *Holocentropus* and *Plectrocnemia* are predators having silken retreats suspended on aquatic plants and have many irregularly radiating silken threads to make capture nets which functions very like those of terrestrial web-spinning spiders.

Larvae of the suborder Integripalpia make portable cases essentially of tubular design. The cases are of a wide variety of shapes and are constructed with different materials and in several arrangements that are generally distinct for different taxon. A number of the Integripalpia larvae are also filter feeders. Larvae of Helicopsychidae shred and consume detritus, mosses, and liverworts and also feed on the attached algae, fungi, microbes, and detritus adhering to underwater surfaces.

The suborder Spicipalpia is a basal lineage of the suborder Integripalpia. Larvae of this suborder are usually free living as in the case of the families Rhyacophilidae and Hydrobiosidae (collectively called "free-living caddisflies"). They feed as predators that pursue their prey. The final instar of a microcaddisfly larva usually feeds in a genus-specific manner, by shredding living algae or liverworts, gathering deposited FPOM, scraping periphyton, or piercing and sucking contents of individual cells of filamentous algae. Some species are parasitic and a few entrain suspended FPOM in silken shelters spun over their cases. Larvae of Glossossomatidae ("saddle-case caddisflies") are scrapers of periphyton growing on the tops of rocks.

6.5 CONCLUSION

The species diversity of Trichoptera is greater than that of all other primarily aquatic insect orders combined, surpassing that of aquatic beetles and equating to about one-third of that of aquatic flies (Morse, 2019). Their contribution to the ecosystem is also non-negligible. But besides having so much of importance, this group of insects have not gained much attention. All the information is available only through sporadic publications. Also, most of the works have been done considering the adults and thus ignoring the larvae.

Therefore, much research work is needed for this group in order to discover species association, that is, adult–larval association for a better understanding of their biology as well as their origin. The study on the larvae should also be encouraged along with the adults. Also, due to climate change, caddisflies are becoming extinct at an alarming rate. Because of the importance of Trichoptera in freshwater ecosystems, there is a necessity to increase our knowledge of the diversity and ecosystem services of caddisflies.

KEYWORDS

- **Caddisfly**
- **trichoptera**
- **biogeographic zones**
- **biomonitoring**
- **services**

REFERENCES

Banks, N. Notes on Indian Neuropteroid Insects. *Proc. Entomol. Soc. Wash.* **1911,** *13*, 99–106.

Banks, N. Synopses and Descriptions of Exotic Neuroptera. *Trans. Am. Entomol. Soc.* **1913,** *39*, 201–242.

Banks, N. Notes and descriptions of Oriental Oestropsychidae (Trichoptera). *Psyche.***1939**, *47*, 52–61

de Moor, F. C. Shortcomings and Advantages of Using Rapid Biological Assessment Techniques for the Purpose of Characterising Rivers in South Africa. Verhandlungen Internationale Vereinigung fur theoretische und angewandte Limnologie. *Stuttgart* **2002,** *28*, 651–662.

Hannaford, M. J.; Resh, V. H. Variability in Macroinvertebrate Rapid-Bioassessment Surveys and Habitat Assessments in a Northern California Stream. *J. North Am. Benthol. Soc.* **1995,** *14* (*3*), 430–439.

Hewlett, R. Implications of Taxonomic Resolution and Sample Habitat for Stream Classification at a Broad Geographic Scale. *J. North Amer. Benthological Soc.* **2000,** *19*, 352–361.

Kaur, S.; Garima, D.; Pandher, M. S. New Species and New Records of Caddisflies (Insecta: Trichoptera) from India. *Zootaxa* **2020**, *4747* (2), 350–360.

Kimmins, D. E. Indian Caddisflies (Trichoptera) I. New Species of the Genus *Limnocentropus* Ulmer. *Ann. Mag. Nat. Hist.* **1950**, *12*, 590–603.

Kimmins, D. E. Entomological Results from the Swedish Expedition 1934 to Burma and British India. Trichoptera. The genus *Chimarra* Stephens (Fam. Philopotamidae). *Arkiv. Zool.* **1957,** *11*, 53–75.

Kimmins, D. E. On the Leptocerinae of the Indian Sub-Continent and North-East Burma (Trichoptera). *Bull. Br. Mus. Nat. Entomol.***1963,** *14 (6)*, 261–316.

Malicky, H. Beiträgezur Kenntnisasiatischer *Oecetis* (Trichoptera, Leptoceridae). *Linz. Biol. Beitr.* **2005,** *37*, 605–669

Malicky, H. Ubersichtuber die Gattung Pseudoneureclipsis (Trichoptera, Polycentropodidae), mitBeschreibung von neuenArten. *Linz. Biol. Beitr.* **2009,** *41* (1), 709–735.

Martynov, A. V. On the Family Stenopsychidae Mart., with a Revision of the Genus *Stenopsyche*McLachl. (Trichopt.). *Eos-Revista Española de Entomología* **1926,** *2*, 281–308.

Martynov, A. V. On a Collection of Trichoptera from the Indian Museum Part I. Annulipalpia. *Rec. Ind. Mus.* **1935,** *37*, 93–209.

Martynov, A. V. On a Collection of Trichoptera from the Indian Museum Part Integripalpia. *Rec. Ind. Mus.* **1936,** *38*, 239–306.

Mc Cafferty, W. P. In *Aquatic Entomology: The Fishermen's and Ecologists' Illustrated Guide to Insects and Their Relatives*; Jones & Bartlett Publishers, Inc.: Boston, 1981; p 448.

McLachlan, R. Descriptions de Plusieures Névroptères-Plannipenneset Trichoptères nouveau de l'île de Célèbes et de quelquesespècesnouvelles de *Dipseudopsis*, avec considérationssurce genre". *TijdschriftvoorEntomologie* **1875,** *18*, 1–21.

Morse, J. C. Trichoptera World Checklist 2021 [Online]. http://entweb.clemson.edu/database/trichopt/index.htm (accessed Sept 25, 2021).

Morse, J.; Frandsen, P.; Graf, W.; Thomas, J. Diversity and Ecosystem Services of Trichoptera. *Insects* **2019,** *10 (5)*, 125.

Morton, K. J. Descriptions of New Species of Oriental Rhyacophilidae. *Trans. R. Entomol. Soc. Lond.* **1900,** *1900*, 1–7.

Morton, K. J. A New Indian Micro-Trichopteron. *Entomol. Mon. Mag.* **1902**, *13(2)*, 283.

Mosely, M. E. The Indian Caddis-Flies (Trichoptera) III. Inaequipalpia. *J. Bombay Nat. Hist. Soc.* **1935,** *38*, 123–133.

Pandher, M. S. Addition of a New Species to the Genus *Plectrocnemia* Stephens, 1836 (Trichoptera: Polycentropodidae) from Indian Himalaya. *Rec. Zool. Surv. India* **2018,** *118* (2), 174–177.

Pandher, M. S. New Species and a New Record of the Genus *Chimarra* Stephens, 1836 (Insecta: Trichoptera) from India. *Braueria (Lunz am See, Austria)* **2019a**, *46*, 11–12.

Pandher, M. S. A New Species to the Genus *Nyctiophylax* Brauer (1865) (Trichoptera: Polycentropodidae) from India. *Rec. Zool. Surv. India* **2019b,** *119* (3), 234–237.

Pandher, M. S. Genus *Chimarra* Stephens, 1829 (Insecta: Trichoptera: Philopotamidae) from Suntalekhola Biodiversity Camp, Neora Valley, National Park, West Bengal. *Rec. Zool. Surv. India* **2020,** *120* (2), 113–122.

Pandher, M. S.; Chandra, K. An Updated Checklist of Family Hydropsychidae Curtis, 1835 (Insecta: Trichoptera) from India. *Rec. Zool. Surv. India* **2019,** *119* (1), 64–68.

Pandher, M. S.; Kaur, S.; Garima, D. New Records of *Glossosomaatitto* Malicky & Chantaramongkol, 1992 and *Ecnomusareion* Malicky, 1999 from India. *Braueria (Lunz am See, Austria)* **2020a,** *47*, 39–40.

Pandher, M. S.; Kaur, S.; Garima, D. New Species, A New Name, and New Synonyms for *Chimarra* spp. (Insecta: Trichoptera: Philopotamidae) of India. *Zootaxa* **2020d,** *4790* (3), 577–585.

Pandher, M. S.; Kaur, S.; Parey, S. H.; Saini, M. S. Review of Genus *Hydropsyche* Pictet (Insecta: Trichoptera: Hydropsychidae) from India. *Zootaxa* **2017,** *4365* (3), 331–360.

Pandher, M. S.; Kaur, S.; Parey, S. H. Three New Species of the Genus *Kisaura* Ross (1956) (Trichoptera: Philopotamidae) from Arunachal Pradesh, India. *Zootaxa* **2018b,** *4403* (3), 586–593.

Pandher, M. S.; Kaur, S.; Parey, S. H. New Species of Genus *Hydromanicus* Brauer 1865 (Trichoptera: Hydropsychidae) from India. *Zootaxa* **2020c,** *4742* (2), 343–350.

Pandher, M. S.; Kaur, S.; Parey, S. H. Review of the Genus *Kisaura* Ross 1956 (Trichoptera: Philopotamidae) from India. *Zootaxa* **2020e,** *4845* (2), 225–238.

Pandher, M. S.; Malicky, H.; Parey, S. H. New Species of the Genus *Cheumatopsyche* Wallengren, 1891 (Insecta: Trichoptera: Hydropsychidae) from India Himalaya. *Zootaxa* **2018a,** *4379* (3), 407–420.

Pandher, M. S.; Parey, S. H. New Species of the Genus *Polyplectropus* Ulmer 1905 (Insecta: Trichoptera: Polycentropodidae) from India Himalaya. *Zootaxa* **2018,** *4504* (3), 431–438.

Pandher, M. S.; Parey, S. H. Three New Species of Genus *Chimarra* Stephens, 1829 (Trichoptera: Philopotamidae) from Indian Himalaya. *Zoosymposia* **2019,** *14*, 250–256.

Pandher, M. S.; Parey, S. H.; Kaur, S. Two New Species of Genus *Polyplectropus* Ulmer 1905 (Insecta: Trichoptera: Polycentropodidae) from the Indian Himalaya. *Zoosymposia* **2020b,** *18*, 112–117.

Pandher, M. S.; Parey, S. H.; Saini, M. S. Review of the Genus *Pseudoneureclipsis* Ulmer (Trichoptera: Pseudoneureclipsidae) from India with Description of *P. ramosa* from India. *Braueria* **2018c,** *45*, 19–22.

Pandher, M. S.; Saini, M. S. First Report of the Genus *Kisaura* Ross (Trichoptera, Philopotamidae) from India with the Description of Six New Species. *Zookeys* **2011,** *152*, 71–86.

Pandher, M. S.; Saini, M. S. Three New Species of Genus *Chimarra* Stephens, 1829 (Trichoptera: Philopotamidae) from the Indian Himalayas. *Pol. J. Entomol.* **2012,** *81*, 63–72.

Pandher, M. S.; Saini, M. S. Three New Species of Genus *Chimarra* Stephens (Insecta: Trichoptera) from Indian Himalaya. *Acta Zool. Acad. Sci. Hungaricae* **2013a,** *59* (3), 267–277.

Pandher, M. S.; Saini, M. S. Addition of a New Species to *Ecnomus* McLachlan, 1864 (Trichoptera: Ecnomidae) Along with Key to its Indian Fauna. *Halteres* **2013b**, *4*, 19–24.

Pandher, M. S.; Saini, M. S. New Additions to the Genus *Kisaura* Ross (Trichoptera: Philopotamidae) from the Indian Himalaya. *Zootaxa* **2014,** *3793* (5), 538–540.

Pandher, M. S.; Saini, M. S. Five New Species of Genus *Kisaura* Ross (Trichoptera: Philopotamidae) from Himachal Pradesh (India). *Zootaxa* **2015,** *4021* (2), 377–386.

Pandher, M. S.; Saini, M. S.; Parey, S. H. Review of the Genus *Stenopsyche* McLachlan (Trichoptera: Stenopsychidae) in India. *Zootaxa* **2013,** *3717* (1), 065–075.

Pandher, M. S.; Saini, M. S.; Parey, S. H. Four New Species of *Chimarra* Stephens (Trichoptera: Philopotamidoea: Philopotamidae) from India Himalaya. *J. Asia Pac. Entomol.* **2014,** *17*, 183–189.

Pandher, M. S.; Saini, M. S.; Ramamurthy, V. V. Addition of Four New Species to the Genus *Kisaura* Ross, 1956 (Trichoptera: Philopotamidae) from the Indian Himalaya. *Pol. J. Entomol.* **2012,** *81*, 185–194.

Parey, S. H. An Updated Checklist and Distribution of Plenitentoria Group of Caddisflies (Trichoptera: Integripalpia) from India. *Indian J. Appl. Res.* **2015,** *5 (4),* 6–14.

Parey, S. H.; Morse, J. C.; Pandher, M. S. Three New Species of the Genus *Lepidostoma*Rambur (Lepidostomatidae: Trichoptera) from India. *Zootaxa* **2016,** *4136* (1), 181–187.

Parey, S. H.; Pandher, M. S. Two New Species of *Apsilochorema* Ulmer, 907 (Trichoptera: Hydrobiosidae) from India. *Aquat. Insects* **2016,** *37* (4), 267–272.

Parey, S. H.; Pandher, M. S. A New Species of Genus *Lepidostoma* Rambur (Trichoptera: Lepidostomatidae) from India. *Zoosymposia* **2019,** *14*, 257–260.

Parey, S. H.; Saini, M. S.; Pandher, M. S. Two New Species of the Genus *Pseudostenophylax* Martynov (Trichoptera: Limnephilidae) from the Indian Himalaya. *Zootaxa* **2013,** *3616* (3), 287–290

Prommi, T. Ecological and Economic Importance of Trichoptera (Aquatic Insect). *J. Food Health Bioenviron. Sci.* **2018,** *11* (1).

Rodgers W. A.; Panwar H. S.; Mathur V. B. In *Wildlife Protected Area Network in India: A Review, Executive Summary*; Wildlife Institute of India: Dehradun, 2002; pp 1–51.

Saini, M. S.; Kaur, L.; Pandher, M. S.; Parey, S. H. Two New Species of Genus *Apsilochorema* Ulmer (Trichoptera: Hydrobiosidae) from India. *Zootaxa* **2013**, *3664* (2), 226–232.

Saini, M. S.; Pandher, M. S. New Species and Records of the Genus *Dolophilodes* Ulmer (Trichoptera: Philopotamidae) from India. *Zootaxa* **2011,** *3137*, 46–55.

Saini, M. S.; Parey, S. H. Four New Species of the Genus *Lepidostoma* Rambur (Trichoptera: Lepidostomatidae) from the Indian Himalayas, With a Checklist to its Indian Species. *Zootaxa* **2011,** *3062*, 25–36.

Saini, M. S.; Parey, S. H. *Limnephilusmorsei* sp. n., A New Species of the Genus *Limnephilus* Leach (Trichoptera: Limnephilidae) from India. *Polish J. Entomol.* **2012,** *81,* 57–61.

Saini, M. S.; Parey, S. H.; Pandher, M. S. Three New Species of Genus *Chimarra* Stephens (Trichoptera: Philopotamidae) from the Indian Himalayas. *Biosystematica* **2011,** *5,* 17–24.

Saini, M. S.; Parey, S. H.; Pandher, M. S.; Bajwa, P. Three New Species of Caddisfly Genus *Chimarra* Stephens from Indian Himalaya (Trichoptera: Philopotamidae). *Bionotes* **2010,** *12* (3), 86–88.

Saini, M. S.; Pandher, M. S.; Ramamurthy, V. V. Three New Species of the Genus *Kisaura* (Trichoptera, Philopotamidae) from Indian Himalaya. *Vestn. Zool.* **2012,** *46* (6), 25–30.

Schmid, F. Quelques*Himalopsyche*indiènnes (Trichopteres, Rhyacophilidae). *Bonn. Zool. Beitr.* **1963,** *14*, 206–223.

Schmid, F. Le Genre *Gunungiella* Ulmer (Trichoptéres: Philopotamides). *Can. Entomol.* **1968a,** *100*, 897–957.

Schmid, F. Le Genere*Poecilopsyche* N. Gen. (Trichoptera, Leptoceridae). *Annales de la Société Entomologique* **1968b,** *13 (1),* 3–31.

Schmid, F. Le Genre *Rhyacophila* et la famille des Rhyacophilidae (Trichoptera). *Memoires de la societeentomologique du Canada* **1970,** *66*, 5–224.

Schmid, F. Quelques Nouveaux *Glossosoma* Orientaux (Trichoptera: Glossosomatidae). *Nat. Can.* **1971,** *98*, 607–631.

Schmid, F. Sur Quelquesnouvelles Psychomyiines Tropicales (Trichoptera: Psychomyiidae). *Nat. Can.* **1972,** *99*, 143–172.

Schmid, F. La famille des Xiphocentronides (Trichoptera: Annulipalpia). *Memoires de la societeentomologique du Canada* **1982,** *114* (121), 3–127.

Schmid, F. Encore quelques *Stactobia* McLachlan (Trichoptera, Hydroptilidae). *Nat. Can.* **1983**, *110*, 239–283.

Schmid, F. In *Essaid'evaluation de la faunemondiale des trichoptères* (Abstract), Proceedings of the 4th International Symposium on Trichoptera; Morse, J. C., Ed., 1984; p 337.

Schmid, F. Considérationsdiverses sur quelques genres leptocérins (Trichoptera, Leptoceridae). *Bulletin de l'Institut Royal des Sciences Naturelles de Belgique, Entomologie* **1987,** *57*, 1–147

Schmid, F. Quelques Philopotamides Orientaux Nouveaux OuPeuConnus (Trichoptera, Annulipalpia). *Beaufortia* **1991,** *42 (4)*, 89–107.

Schmid, F. Les*Adicella*dugroupe de Pulcherrima (Trichoptera, Integripalpia, Leptoceridae). *Fabreries* **1994a,** *19 (2)*, 37–44.

Schmid, F. Quelques*Adicella* Indiennes (Trichoptera, Leptoceridae). *Fabreries* **1994b,** *19 (4)*, 85–127.

Schmid, F. Les *Oecetis* du grouped'eburneaenInde (Trichoptera: Leptoceridae). *Fabreries* **1995a,** *20* (2), 41–56.

Schmid, F. Les *Oecetis* du Groupe de TestaceaenInde (Trichoptera: Leptoceridae). *Fabreries* **1995b,** *20* (2), 57–78.

Schmid, F. Les Genre *Oecetis*enInde (Trichoptera: Leptoceridae). *Fabreries* **1995c,** *20* (4), 113–151.

Vannote, R. L.; Minshall, G. W.; Cummins, K. W.; Sedell, J. R.; Cushing, C. E. The River Continuum Concept. *Can. J. Fish. Aquat. Sci.* **1980,** *37*, 130–137.

Walker, F. In *Cataloque of the Specimens of Neuropterous Insects in the Collection of the British Museum*; Part I, Phryganides-Perlides; British Museum: London, 1852; p 135.

CHAPTER 7

Trichoptera of Europe: Diversity, Importance, and Conservation

HALIL IBRAHIMI

Department of Biology, Faculty of Mathematics and Natural Sciences, University of Prishtina "Hasan Prishtina," Prishtinë, Kosovo

ABSTRACT

Trichoptera is one of the most diverse orders of aquatic insects and important water quality indicators, due to their presence in almost all segments of freshwater ecosystems. Caddisflies of Europe and adjacent areas are represented by 132 genera and 1855 species up to date. The most species-rich genera in Europe are *Drusus, Rhyacophila, Hydropsyche*, and *Tinodes*, each with more than 100 species, constituting thus slightly more than 50% of the total caddisfly species. Europe and adjacent areas have a high number of endemic species, mostly belonging to the families Limnephilidae, Beraeidae, Glossosomatidae, Hydroptilidae, Philopotamidae, Psychomyiidae, and Rhyacophilidae.

Despite their importance and routine usage in biomonitoring, caddisflies are still not given proper conservation importance. International Union for Conservation of Nature (IUCN) still does not have on European or global level a list of assessed caddisfly species. However, on a regional scale, several national and regional Red Lists of caddisflies exist, or there are lists of caddisfly species included in national or regional Red Lists or Books of Fauna. While in most of the European countries which are part of European Union, threatening activities in freshwater ecosystems affecting caddisflies are addressed seriously, legally, and practically; in Southeastern

Insect Diversity and Ecosystem Services, Volume 1: Importance, Threats, Conservation and Economic Perspectives. Younis Ahmad Hajam, Sajad Hussain Parey, and Rouf Ahmad Bhat (Eds)

European countries there is no major action toward the prevention of these activities, or restoration of damages caused by their long impact.

7.1 WORLDWIDE DIVERSITY AND DISTRIBUTION OF CADDISFLIES

Trichoptera is the seventh most species-rich order of insects and, after Diptera the most species-rich order of aquatic insects (Morse et al., 2019), abundant primarily in almost all types of natural aquatic habitats, with some species able to adapt in man-made habitats as well (e.g., Previšić et al., 2007). Except Antarctica, caddisflies are present in all continents and are crucial for functioning of freshwater ecosystems. Larvae are exclusively aquatic, occurring mainly in cool and lotic habitats, although many species can be found in higher temperatures and standing waters as well. The aquatic larvae of caddisflies (Fig. 7.1) have a soft body and can live free life, while some species construct cases from different material, such as sand, plant debris, and other material, which serve primarily as a protection. Larvae develop through up to seven instars, during a year or two, after which they undergo a pupation under the water. Larvae can be herbivorous or predators and feed on algae, different material of plant origin or organic material, while predators feed on a variety of organisms living in the water. Adults (Fig. 7.2) live only a

FIGURE 7.1 A Trichoptera larvae. East Azerbaijan, Iran. Photo by Halil Ibrahimi.

few days up to a few weeks, during which period they do not feed. After the copulation, females lay hundreds of eggs on the water or in the nearby vegetation. Adults of some of the species undergo a diapause in cooler places, mainly during the hot summer months.

FIGURE 7.2 Male and female adults of *Potamophylax coronavirus* Ibrahimi, Bilalli, and Vitecek, 2021. Bjeshkët e Nemuna, Kosovo. Photo by Halil Ibrahimi.

As of today, 16,267 extant species of caddisflies are known, according to the Trichoptera World Checklist (Morse, 2021), and the number is constantly increasing with the description of new species from the least investigated areas of the world. Less than 13,000 species were known 5 years ago, and in such a short period there is almost 30% increase in the number of known caddisfly species. It is expected that the number of caddisfly species may be as high as 50,000, meaning that currently we know only a small portion of the caddisfly diversity (de Moor and Ivanov, 2008). The two most species diverse regions are Oriental and Neotropical, containing more than half of the known species of caddisflies (de Moor and Ivanov, 2008; Morse et al., 2019). Most of the new species are being described from the above-mentioned two regions, although constantly new species are being described even from the best investigated areas, such as Europe or North America, for example.

The most species-rich families worldwide are: Limnephilidae, Hydroptilidae, Hydrobiosidae, Leptoceridae, Rhyacophilidae, and Hydropsychidae.

The highest number of endemic genera/subgenera is present in the Australian Region, followed by almost the same level in the Neotropical Region (73% and 69% accordingly), and followed by the Afrotropical Region, West Palearctic Region, and Oriental Region (43%, 41%, and 33% accordingly) (de Moor and Ivanov, 2008; Morse et al., 2019).

7.2 IMPORTANCE OF CADDISFLIES

Morse et al. (2019) listed the following ecosystem services provided by Trichoptera: crucial role in food webs, usage in biological monitoring of water quality, food for fish and other predators of human concern, important role in engineering the freshwater habitats, usage in solving forensic issues and usage of silk as a potential applying model in medicine and textile.

Caddisflies are an essential link in food webs due to their role in the transfer of matter and energy from the organic resources up to the top levels. In this regard, caddisflies are the most important constituents of biomass in freshwater ecosystems, serving as a food for many organisms and by their diversity of food habits maintain this flow of energy. In addition, caddisflies are important part of terrestrial food webs as well, since the adult stage most often becomes food for many bats, bird species, or even other insects. In this regard, adults serve as a linkage between freshwater ecosystems and adjacent terrestrial ones.

Trichoptera are one of the most important water quality indicator groups due to the high number of species adaptable to different levels of pollution and their presence in all segments of freshwater ecosystems. Since several decades ago, Trichoptera have been used as indicators of organic pollution, habitat deterioration and disturbance from agriculture, urban pollutants, some industries, logging, and deteriorations of nearby terrestrial ecosystems. Trichoptera constitute an integral and crucial part of biomonitoring as part of the European Water Framework Directive, aiming to prevent deterioration and enhance status of aquatic ecosystems, promote sustainable water use, and reduce pollution. All major water quality indices use Trichoptera for quality assessment up to the family, generic, or species level. Usage of species level indices in many parts of Europe was shown to be problematic (e.g., Ibrahimi et al., 2014) due to the fact that many species of caddisflies still do not have their larvae described. Several studies have shown that in addition to larvae, adults of the light-trapped caddisflies can be used as indicators of the ecological status of water bodies, as well (e.g., Stanić-Koštroman et al., 2012).

7.3 DIVERSITY OF EUROPEAN CADDISFLIES

Caddisflies of Europe and adjacent areas are represented by 132 genera and 1855 species up to date (e.g., Neu et al., 2018; Olah et al., 2019, 2021; Ibrahimi et al., 2021a, 2021b). Here are included partly caddisflies of Asia Minor and some North African countries which are continuation of European fauna. Many of the species from these adjacent areas of Europe occur in the continent itself. This number is slightly a bit more than 10% of the worldwide known number of caddisflies today. The most species-rich genera in Europe are *Drusus, Rhyacophila, Hydropsyche, and Tinodes*, each with more than 100 species, constituting thus slightly more than 50% of the total caddisfly species (Fig. 7.3). Thirty-one genera are represented by a single species each. The most species-rich areas are the Iberian Peninsula, Italy and Corsica, the Alps and the Western Balkan (Kobler et al., 2017). A dozen of new species of the genus *Drusus* have been described during the past decade, mostly from the Balkan Peninsula, but Ibero-Macaronesian Region and other areas of Europe as well.

Western Europe is relatively well researched in terms of Trichoptera diversity, although there are still areas from which new species have been described in recent years and will likely still be described. Countries like France, or regions like the Alps and Iberian Peninsula can still carry unknown species of Trichoptera. Currently, the most unexplored area of Europe is the Balkan Peninsula. More than a hundred species of Trichoptera have been described in recent years from this region, from countries such as Albania, Kosovo, Bosnia and Herzegovina, and Montenegro. Until the beginning of this century, only a few species were known by many Western Balkan countries, but over the past decade this has changed drastically. Currently from this area of Europe, countries like Bulgaria, Croatia, Kosovo, Slovenia, and Bosnia and Herzegovina are the best researched. Macedonia and Serbia are relatively well researched, while Montenegro and Albania are the least investigated countries. Most of the species described in recent years from this region belong to the genus *Drusus,* where only from Albania were described dozens of them, but also from Kosovo, Bosnia and Herzegovina, and Montenegro. Also in recent years the *Potamophylax winneguthi* and *Potamophylax nigricornis* species groups have been revised, from which several new species have been described, which are microscale endemics of certain mountains. In addition to this, one of the most neglected habitats in terms of caddisfly investigations are midstream and downstream large rivers of the South-eastern Europe. A brief survey of such habitats in Kosovo, Albania, and Montenegro during the past years (unpublished results) showed that these segments of the rivers harbor a very interesting composition of species of the Hydroptilidae family. Many species of this family, previously recorded only from certain countries of Western and Central Europe, were recorded during this rapid survey.

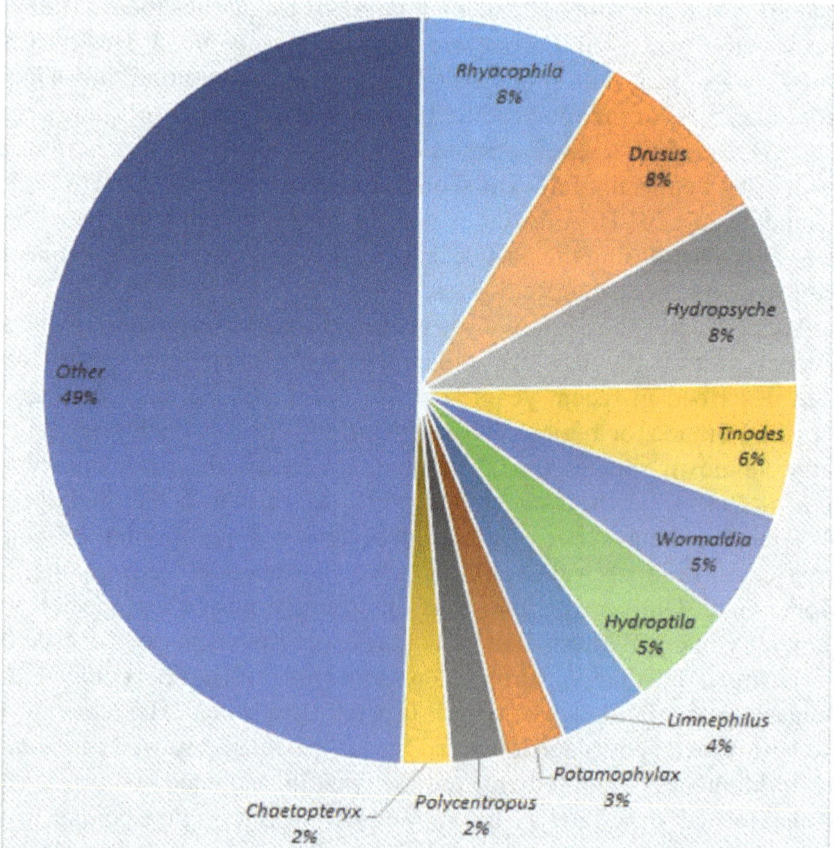

FIGURE 7.3 Percentage of the 10 most species-rich genera of Trichoptera in Europe.

Europe and adjacent areas have a high number of endemic species, mostly belonging to the families Limnephilidae, Beraeidae, Glossosomatidae, Hydroptilidae, Philopotamidae, Psychomyiidae, and Rhyacophilidae. The mountain barrier together with the limited dispersal capacities have acted as a favorable driver in the speciation process of these families. A recent study (unpublished results) on the distribution of endemic species in the Western Balkan showed that there are 69 caddisfly species endemics for this area. Out of them, almost 80% belong to the family Limnephilidae, 10% to the family Rhyacophilidae, while the rest belong to families Glossosomatidae, Beraeidae, Lepidostomatidae, Polycentropodidae, and Psychomiidae. The fact that the largest portion of the total number of endemic caddisflies distributed in the Western Balkan countries associates with the family Limnephilidae is due to the favorable habitats for the species of this family. Besides family Limnephilidae, the extremely rare *Rhyacophila* species within the family Rhyacophilidae are also endemics of certain mountains in this area. Although most of the Balkan endemics within this family are widely distributed over the Peninsula, some species are currently known only from isolated habitats.

Besides biogeographic, ecologic, and morphotaxonomic investigations and studies of caddisflies in Europe, during the past decade we have seen enormous increase of studies involving molecular analysis, which in addition have refined and changed many of the previously well-established lineages, classifications, and even synonymized many genera or revealed new ones (e.g., Kučinić et al., 2020; Pauls et al., 2009; Previšić et al., 2014; Valladolid et al., 2020, 2021). Molecular analysis of caddisflies also aided the understanding the past processes and timings of speciation events, main triggers of unevenly distributed high diversity of caddisflies in Europe and importance of refugia in distribution of high number of endemics, sometimes in very narrow areas.

7.4 CONSERVAITON OF CADDISFLIES IN EUROPE

Despite their importance and routine usage in biomonitoring, caddisflies are still not given proper conservation importance. International Union for Conservation of Nature (IUCN) still does not have on European or global level a list of assessed caddisfly species. IUCN has however created a Working Group of Mayfly, Stonefly, and Caddisfly species, with the goal of raising awareness of these three important small insect orders, and undertaken Red List Assessment, in the function of conservation activities. Among the other targets of the Group was that by 2020 select randomly 25 species of European Trichoptera for assessment. However, it seems that currently due to the lack of funds, the activities of this group are suspended. Financed by BioFresh Project, a major project Distributional Atlas of European Trichoptera (DAET) has been carried out a few years ago where more than 600,000 geo-referenced occurrence records of caddisflies have been generated (Neu et al., 2018). This is a great step in the conservation of European caddisflies and could serve as a basis for a European Red List of threatened caddisfly species (Kobler et al., 2017). However, this project did not evaluate species as per the IUCN criteria or any other one, but only provided the distribution of species.

On a regional scale, several national and regional Red Lists of caddisflies exist, or there are lists of caddisfly species included in national or regional Red Lists or Books of Fauna. In this regard, Western European countries have relatively long tradition of evaluating caddisfly species and including them in national protection legislation. Switzerland has a Red List of Swiss mayflies, stoneflies, and caddisflies (Federal Office for the Environment FOEN, Swiss Biological Records Center, CSCF, 2012), where 308 species are evaluated as per the IUCN criteria, and 51% of caddisflies found in Switzerland are included in the Red Lists of threatened species based on these criteria. Great Britain has a Red List of Trichoptera (Wallace, 2016), where 195 species known from the country were evaluated based on the IUCN criteria, and was updated for the last time during 2016. Sweden has published in 2021 the newest version of the Red List, where 209 caddisfly species in the country have been evaluated as per the IUCN criteria (Swedish Red List, 2021). In Finland during 2019 (Hyvärinen et al., 2019), 2014 species have been evaluated as per the IUCN criteria and 30 of them are red listed. In Austria for the last time during 2009 (Malicky, 2009), 308 species have been evaluated and included in the Red list of Trichoptera. Germany has published for the last time during 2016 its Red List of Trichoptera (Gruttke, 2016). The German national Red List threat categories do not correspond to the international IUCN categories. However, caddisflies do not have similar attention in most of the South East European countries, where usually caddisflies are not part of the Red Books, Red Lists or protected under national legislations. Kosovo is an exception, where during 2019 (Ibrahimi et al., 2019),

46 species were evaluated as per the IUCN criteria and included in the Red Book of Fauna.

TABLE 7.1 Distribution of Caddisfly Species in the Endangerment Categories as per IUCN in Some European Countries.

Country	RE	CR	EN	VU	NT	LC	
Austria	1	9	50	96	32	85	35
Finland	1	0	2	6	21	185	0
Great Britain		6	3	8	6	154	18
Kosovo		15	1	7	20	1	2
Sweden				5	5	206	6
Switzerland	17	29	47	54	43	101	11

CR, critically endangered; DD, data deficient; EN, endangered; LC, least concern; NT, near threatened; RE, regionally extinct; VU, vulnerable.

An overview of the categorization of caddiflies in the endangerment classes as per the IUCN in some European countries where Red Lists and Books exist, shows that there is a high number of species which are threatened for extinction (Table 7.1). Seventeen species in Switzerland, 8 in Germany, 1 in Finland, and 1 in Austria are already assessed as Regionally Extinct, 29 species in Switzerland and 15 in Kosovo are evaluated as Critically Endangered. Seventeen species in Germany have been assessed as Threatened with Extinction and 28 as Highly Threatened, as per the local evaluation criteria. Inclusion of more species in the evaluation will most certainly show that many caddisfly species are seriously threatened during the past years by anthropogenic activities, and especially in Southeastern Europe.

7.5 MAIN THREATS TO CADDISFLIES IN EUROPE

Several investigations have suggested and/or proved that caddisfly communities are declining due to the habitat degradation, pollution, and climate changes. Several studies have suggested that globally caddisfly communities are changing due to the anthropogenic activities. Several species may have gone extinct on different scales due to this. Studies suggest that shredder and predatory species are the most vulnerable ones and mostly affected by these activities, shifting toward the domination of filtering collectors, where most of the pollution tolerant species belong. In many large river segments these tolerant species communities may account up to the 95% of the original prairie habitats (Sanchez-Bayo and Wyckhuys, 2019). Vulnerability of endemic freshwater biota to climate change, pollution, and habitat degradation is very high (Hering et al., 2009; Conti et al., 2014) and due to their high rate of microscale endemic species, caddisflies are suffering greatly.

In many parts of Europe, outside European Union, but in some cases even within, one major threat to caddisflies and freshwater biota is discharge of untreated sewage and industrial effluents, runoff from agricultural areas with pesticides, etc. In many parts of Europe and especially in Southeastern Europe, midstream and downstream segments of rivers are heavily impacted by sewage pollution. These segments or the rivers at many places otherwise offer a very favorable condition for many interesting species of Hydroptilidae and Hydropsychidae, which are greatly impacted by these activities.

Construction of hydropower plants was demonstrated to be one of the major threats to freshwater biodiversity in Southeastern Europe (Zarfl et al., 2014). Consequences caused by these small hydropower plants and associated damming of streamlets, many times nearby spring areas and type localities of many endemic caddisfly species, are one of the main drivers of the freshwater habitat degradation in the Balkans, but also other parts of Europe (e.g., Freyhof, 2012; Schwarz, 2012). It has been observed that hydropower plants affect the river's physico-chemical characteristics and change the abundance and composition of aquatic organisms (Cesoniene et al., 2021). These studies suggest that trichopteran communities are greatly influenced by disturbances in physic-chemical parameters in streams and rivers caused by hydropower plants, such as nitrogen and suspended matter concentration, as well as vegetation cover on the riverbed, and water velocity. Nearly 9000 new hydropower plants are planned to be built all over Europe, in addition to the existing thousands ones, impacting greatly the freshwater biota (European Commission, 2011). Many of these hydropower plants operate, or are planned to be built inside the protected areas, threatening greatly microscale endemic caddisfly species. The EU Water Framework Directive (WFD) suggests that by 2027 Member States must ensure that all rivers are healthy; however, this figure seems to be hard to reach, in many cases mostly due to hydropower dams.

Urban developments are also listed as a considerable threat to the decrease of natural freshwater habitats, impacting greatly caddisflies. Spread of invasive alien species is also one of the threats to caddisflies due to the proven fact that they cause replacement and/or reduction in species richness and abundance of native organisms.

Deforestation is one of the major threats to caddisflies communities in Southeastern Europe and especially in Balkans. During the past two decades deforestation has been occurring in this area at a rapid scale. For example, it has been reported that woodlands in Albania are not half of what they used to be prior to 1990s. Similar activities have been reported from most of the Southeastern countries, including those that are part of European Union such as Romania, Bulgaria, and Hungary. Several studies have indicated that riparian deforestation affects functioning of freshwater ecosystems, including diversity and abundance of caddisflies (Silva-Araujo et al., 2020). Deforestation most probably affects many rare and microscale endemic species of caddisflies, due to the fact that forests provide specific habitats which enabled speciation process and its maintenance throughout the ages.

While in most of the European countries which are part of European Union, these threatening activities in freshwater ecosystems are addressed seriously, legally, and practically; in Southeastern European countries there is no major action toward the prevention of these activities or restoration of damages caused by their long impact.

7.6 CONCLUSIONS

Caddisflies Are an important component in functioning of ecosystems which are crucial for humankind. Their inclusion in conservation and mitigation activities for nature protection requires increased efforts in evaluating species as per the international conservation standards and increase in taxonomic and biogeographic studies. This will consequently result in the protection of crucial species and their associated habitats, as important indication of environmental and human health.

KEYWORDS

- **caddisflies**
- **Europe**
- **biodiversity conservation**
- **freshwater ecosystems**
- **water pollution**

REFERENCES

Cesoniene, L.; Dapkiene, M.; Punys P. Assessment of the Impact of Small Hydropower Plants on the Ecological Status Indicators of Water Bodies: A Case Study in Lithuania. *Water* **2021,** *13, 433.* doi.org/10.3390/ w13040433.

Conti, L.; Schmidt-Kloiber, A.; Grenouillet, G.; Graf, W. A Trait-Based Approach to Assess the Vulnerability of European Aquatic Insects to Climate Change. *Hydrobiologia* **2014,** *7211*, 297–315.

de Moor, F. C.; Ivanov, V. D. Global Diversity of Caddisflies (Trichoptera: Insecta) in Freshwater. *Hydrobiologia* **2008,** *595*, 393–407.

Federal Office for the Environment FOEN, Swiss Biological Records Center, CSCF Red List Mayflies, Stoneflies and Caddisflies, 2012; p 111.

Freyhof, J. Threatened Freshwater Fishes and Molluscs of the Balkan, Potential Impact of Hydropower Projects. *Unpublished Report, ECA Watch Austria&EuroNatur.*, 2012, p 81.

Hering, D.; Schmidt-Kloiber, A.; Murphy, J.; Lucke, S.; Zamora-Munoz, C.; Lopez-Rodrıguez, M. J.; Huber, T.; Graf, W. Potential Impact of Climate Change on Aquatic Insects: A Sensitivity Analysis for European Caddisflies (Trichoptera) Based on Distribution Patterns and Ecological Preferences. *Aquat. Sci.* **2009,** *71*, 3–14.

Hyvärinen, E.; Juslén, A.; Kemppainen, E.; Uddström, A.; Liukko, U. M. Eds.; The 2019 Red List of Finnish Species. Ympäristöministeriö & Suomen ympäristökeskus: Helsinki, 2019; p 704.

Ibrahimi, H.; Bilalli, A. A New Species of the Genus *Potamophylax* Wallengren, 1891 (Trichoptera: Limnephilidae) from Qafështamë National Park in Albania. *Ecol. Mont.* **2021,** *42*, 85–95.

Ibrahimi, H.; Bilalli, A.; Vitecek, S.; Pauls, S.; Erzinger, F.; Gashi, A.; Grapci-Kotori, L.; Geci, D.; Musliu, M.; Kasumaj, E. *Potamophylax coronavirus* sp. nov. (Trichoptera: Limnephilidae), A New Species from Bjeshkët e Nemuna National Park in the Republic of Kosovo, With Molecular and Ecological Notes. *Biodivers. Data J.* **2021,** *9*, e64486.

Ibrahimi, H.; Gashi, A.; Regjepaj, D.; Zhushi, F.; Grapci-Kotori, L.; Feher, Z.; Bino, T.; Jelaska, L. S.; Theou, Ph.; Mesaros, G. In *The Red Book of Fauna of the Republic of Kosovo*, 2019; p 410.

Ibrahimi, H.; Kučinić, M.; Gashi, A.; Grapci-Kotori, L. Trichoptera Biodiversity of the Aegean and Adriatic Sea basins in Kosovo. *J. Insect. Sci.* **2014,** *14* (209), 1–8. DOI: 10.1093/jisesa/ieu071

Kučinić, M.; Ćukušić, A.; Ibrahimi, H.; Plantak, M.; Vuković, M.; Bukvić, V.; Vučković, I. DNA Barcoding and a New Taxonomic Status of the *Triaenodes ochreellus lefkas* Malicky, 1974 (Insecta, Trichoptera) With New Distribution Data. *Ecol. Mont.* **2020,** *38*, 172–185.

Malicky, H. Rote Liste der Köcherfliegen Österreichs (Insecta, Trichoptera). In *Rote Liste gefährdeter Tiere Österreichs*; Zulka, K. P. (Red.); Checklisten, Gefährdungsanalysen, Handlungsbedarf; Teil 3: Flusskrebse, Köcherfliegen, Skorpione, Weberknechte, Zikaden; Grüne Reihe des Lebensministeriums (Gesamtherausgeberin Ruth Wallner) Band 14/3; Wien, Böhlau, 2009; pp 319–358.

Morse, J. C. Ed.; Trichoptera World Checklist 2021 [Online]. http://entweb.clemson.edu/database/trichopt/index.htm. (Accessed Aug 24, 2021).

Morse, J. C.; Frandsen, P. B.; Graf, W.; Thomas, J. A. Diversity and Ecosystem Services of Trichoptera. *Insects* **2019,** *10*, 125. https://doi.org/10.3390/insects10050125

Neu, P.; Malicky, H.; Graf, W.; Schmidt-Kloiber, A. In *Distribution Atlas of European Trichoptera*; ConchBooks, 2018; p 899.

Oláh, J.; Beshkov, S.; Chvojka, P.; Ciubuc, C.; Coppa, G.; Ibrahimi, H.; Kovacs, T.; Mey, W.; Oláh, J. Jr. Revision of *Drusinae* subfamily (Trichoptera, Limnephilidae): Divergence by Paraproct and Paramere: Speciation in Isolation by Integration. *Opusc. Zool. Bud.* **2017,** *48* (1), 3–228. https://doi.org/10.18348/opzool.2017.S1.3

Olah, J.; Vinçon, G.; Kerimova, I.; Kovacs, T.; Manko, P. On the Trichoptera of the Caucasus with Western and Eastern Relatives. *Opusc. Zool. Bud.* **2021,** *51* (3), 3–174. http://dx.doi.org/10.18348/opzool.2020.S3.3

Pauls, S. U.; Theißinger, K.; Ujvarosi, L.; Balint, M.; Haase, P. Population Structure in Two Closely Related, Partially Sympatric Caddisflies in Eastern Europe: Historic Introgression, Limited Dispersal and Cryptic Diversity. *J. N. Am. Benthol. Soc.* **2009,** *28*, 517–536.

Previšić, A.; Mihaljević, Z.; Kerovec, M. Caddisfly (Insecta: Trichoptera) Fauna of Altered and Man-Made Habitats in the Drava River, NW Croatia. *Nat. Croat.* **2007,** *16* (3), 181–187.

Previšić, A.; Schnitzler, J.; Kučinić, M.; Graf, W.; Ibrahimi, H.; Kerovec, M.; Pauls, S. U. Micro-Scale Vicariance and Diversification of Western Balkan Caddisflies Linked to Karstification. *Fresh. Sci.* **2014,** *33*, 250–262.

Robert, B. Rote Liste und Gesamtartenliste der Köcherfliegen (Trichoptera) Deutschlands. In *(Bearb.): Rote Liste der gefährdeten Tiere, Pflanzen und Pilze Deutschlands*; Gruttke, H., Balzer, S., Binot-Hafke, M., Haupt, H., Hofbauer, N., Ludwig, G., Matzke-Hajek, G., Ries, M; Band 4: Wirbellose Tiere (Teil 2); Bonn (Bundesamt für Naturschutz). *Naturschutz und Biologische Vielfalt* **2016,** *70* (4), 101–135.

Sanchez-Bayo, F.; Wyckhuys, K. A. G. Worldwide Decline of the Entomofauna: A Review of its Drivers. *Biol. Conserv*. **2019,** *232*, 8–27.

Schmidt-Kloiber, A.; Neu, P. J.; Malicky, M.; Pletterbauer, F.; Malicky, H.; Graf, W. Aquatic Biodiversity in Europe: A Unique Dataset on the Distribution of Trichoptera Species with Important Implications for Conservation. *Hydrobiologia* **2017,** *797* (1), 11–27. DOI: 10.1007/s10750-017-3116-4

Schwarz, U. Balkan Rivers—The Blue Heart of Europe. Hydromorphological Status and Dam Projects Report; Ulrich Schwarz, FLUVIUS; Floodplain Ecology and River Basin Management, Vienna, 2011; p 151.

Silva-Araujo, M.; Silva-Junior, E. F.; Neres-Lima, V.; Feijo-Lima, R.; Tromboni, F.; Lourenco-Amorim, C.; Thomas, S. A.; Moulton, T. P. Effects of Riparian Deforestation on Benthic Invertebrate Communityand Leaf Processing in Atlantic Forest Streams. *Perspect. Ecol. Conserv*. **2021,** *18*, 277–282.

Stanić - Koštroman, S.; Kučinić, M.; Kolobara, A.; Škobić , D.; Knezović, L.; Durbešić , P. Light-Trapped Caddisflies (Insecta: Trichoptera) as Indicators of the Ecological Integrity of the Lištica River, Bosnia and Herzegovina. *Entomol. Croat.* **2012,** *16* (1-4), 21–36.

Swedish Red List, 2021 [Online]. https://www.artdatabanken.se/en/the-red-list/ (accessed Sept 17, 2021).

Valladolid, M.; Arauzo, M.; Chertoprud, M. V.; Chvojka, P.; Czachorowski, S.; Dorda, B. A. Hinic, J.; Ibrahimi, H.; Karaouzas, I.; Krpac, V.; Kucinic, M.; Salokannel, J.; Slavevska-Stamenkovic, V.; Stojanovic, K.; Wallace, I. The Rhyacophila Fasciata Group in Europe: Rhyacophila Fasciata Hagen 1859 and Formerly Synonymized Species (Trichoptera: Rhyacophilidae), With New Description of Rhyacophila Fasciata and Rhyacophila Septentrionis McLachlan 1865 (stat. prom.). *Zootaxa* **2021,** *4975* (1), 1–57.

Valladolid, M.; Kučinić, M.; Arauzo, M., Cerjanec, D.; Ćuk, R.; Dorda, B. A.; Lodovici, O.; Stanić-Koštroman, S. The *Rhyacophila fasciata* Group in Croatia and Bosnia and Herzegovina: *Rhyacophila f. fasciata* Hagen 1859 and the Description of Two New Subspecies, *Rhyacophila fasciata delici* Kučinić & Valladolid (ssp. nov.) from Croatia and Bosnia and Herzegovina and *Rhyacophila fasciata viteceki* Valladolid & Kučinić (ssp. nov.) from Bosnia and Herzegovina (Trichoptera: Rhyacophilidae). *Zootaxa* **2020,** *4885* (1), 51–075. https://doi.org/10.11646/zootaxa.4885.1.3

Wallace, I. A Review of the Status of the Caddis Flies (Trichoptera) of Great Britain; *Natural Englan*d, 2016; p 132.

Zarfl, C.; Lumsdon, A. E.; Berlekamp, J.; Tydecks, L.; Tockner, K. A Global Boom in Hydropower Dam Construction. *Aquat. Sci.* **2014,** *77* (1), 161–170. 10.1007/s00027-014-0377-0

CHAPTER 8

Current Status and Future Prospects of the Aphidophagous Hoverflies (Diptera: Syrphidae) of Pakistan

MUHAMMAD ASGHAR HASSAN[1], IMRAN BODLAH[1], RIAZ HUSSAIN[1], ZERSHINA MARYAM[2], SAJJAD HUSSAIN[2], MUHAMMAD QASIM[3], and TAYYABA NASEEM[4]

[1]*Department of Entomology, Pir Mehr Ali Shah Arid Agricultural University, Rawalpindi, Pakistan*

[2]*Department of Biological Sciences, University of Baltistan, Skardu, Gilgit-Baltistan, Pakistan*

[3]*Department of Agriculture and Forestry, Kohsar University Murree, Punjab, Pakistan*

[4]*Department of Entomology, The University of Poonch, Rawalakot, Azad Kashmir, Pakistan*

ABSTRACT

The family Syrphidae (Insecta: Diptera), being dominant among the other dipterous families in Pakistan, currently includes four subfamilies, 83 species under 43 genera, and is widely distributed throughout the country. Among the four subfamilies, the immatures of subfamily Syrphinae are as important as the Coleopteran (Coccinellidae) biocontrol agents against the soft sucking insect-pests, particularly against aphids: the notorious pests of fruits, vegetables, cultivated crops, and ornamental plants in Pakistan. Aphids not only serve as vectors of

Insect Diversity and Ecosystem Services, Volume 1: Importance, Threats, Conservation and Economic Perspectives. Younis Ahmad Hajam, Sajad Hussain Parey, and Rouf Ahmad Bhat (Eds)

many diseases in plants but also cause direct and indirect damages by sucking phloem sap and secretion of honeydews resulting in yield losses. In order to understand the current situation of known and new records of the recently collected potential aphidophagous hoverfly species in Pakistan, we have enumerated the list of all known and newly recorded species along with their associated prey in Pakistan. In addition, we have studied the tritrophic association of aphidophagous hoverfly species with aphids and their preferable host plants collected during 2016–2018 from the Pothwar region of Punjab, Pakistan are discussed. As the majority of immature stages in subfamily Syrphinae are zoophagous and their preferable hosts are hemipterans (aphis, psyllids, adelgids, and thrips), except some species (*Toxomerus*) are pollen feeders. So, the subfamily Syrphinae are particularly interesting due to their aphidophagy; it is used as a biological control agent against various aphid species in the field or laboratory experiment. The immature stages of hoverflies are confined to the breeding sites; thus for the effective management of soft sucking insectpests, the different ecological environment of the pest and its natural enemies and their trophic association should be investigated in future studies. Hence, the present review aims to summarize our knowledge based on published literature and the current research from Pakistan. Also we have provided detailed photographs of the key diagnosis for the two common widespread *Episyrphus* species in Pakistan: *Episyrphus balteatus* (De Geer, 1776) and *E. viridaureus* (Wiedemann, 1824). However, *Episyrphus balteatus* is a well-known potential aphidophagous species in Pakistan; during our present study the following three species: *Episyrphus viridaureus*, *Ischiodon scutellaris* (Fabricius, 1805), and *Scaeva pyrastri* (Linnaeus, 1758) exhibit similar ecological requirements, feeding potential, wide distribution, and wide host range against aphids in Pakistan. The tritrophic association shows a highly significant positive relationship among the predators, pests, and their host plants. For example, eight aphidophagous species were associated with 11 host plants that were infested by 13 aphid species. Among those, *Episyrphus balteatus* was noticed while feeding on 11 aphid species and is the most important species in comparison with *E. viridaureus*, *Ischiodon scutellaris*, and *Scaeva pyrastri* that feed on nine, eight, and four species, respectively.

8.1 INTRODUCTION

Syrphid flies (Diptera: Syrphidae) are one of the largest, most diverse, and worldwide distributed groups of true flies, comprising more than 6000 described species (Pape and Evenhuis, 2013; Miranda et al., 2013). Currently, four subfamilies including Syrphinae, Eristalinae, Pipizinae, and Microdontinae, and 14 tribes are recognized in this family (Thompson and Routheray, 1998; Mengual et al., 2015).

In contrast to the uniform flower-feeding habitats of adult hoverflies, the larval habitats are diverse and prefer a variety of foods. The immature stages of the subfamily Eristalinae are saprophagous in deadwood, coprophagous, phytophagous, aquatic filter feeders or inquilines in social insects' nests and Microdontinae are myrmecophiles (Ghorpadé, 2015). The immature stages of most species of subfamily Syrphinae are zoophagous and their preferable hosts are hemipterans (aphids, psyllids, adelgids, and thrips), except that some species of the genus *Toxomerus* are pollen feeders (Reemer and Rotheray, 2009). Among these subfamilies, the immatures of Syrphinae are particularly interesting due to its aphidophagous nature; it is used as a biological control agent against various aphid species worldwide. Sucking pest like aphids (Hemiptera: Aphididae) are the serious pests of many agricultural crops. They are phloem-feeders that reduce plants performances, change plant architecture, suppress growth, and photosynthesis (Sudderth et al., 2005; Flynn et al., 2006).

Aphidophagous predators, like hoverflies, ladybirds, and lacewings, are commonly present in agricultural crops, herbaceous plants, and on trees. The adult aphid predators are winged, can easily move between patches of prey, and therefore can find suitable patches of prey (Kindlmann et al., 2010). The immature stages of hoverflies are confined to the breeding sites; thus, the investigation of host plants and associated aphid prey can be easily well-studied.

Unfortunately, studies on the aphidophagous hoverflies from the Indo-Pak subcontinent are scattered in published literature (Alam and Hafiz, 1963; Alam et al., 1969; Ghorpadé, 1981; Agarwala et al., 1984; Irshad, 2001, 2014; Ghorpadé, 2015; Jamali et al., 2018). Among these, the comprehensive catalogue of the Indian and Pakistani aphidophagous hoverflies was presented by Ghorpadé (1981), who provided the checklist of 47 aphidophagous hoverflies against 77 prey species from India and neighboring countries including Pakistan. Later, Agarwala et al. (1984)

presented the aphid prey catalog of India and listed 23 known and 4 unidentified aphidophagous hoverflies. From the territory of Pakistan, Alam and Hafiz (1963) and Alam et al. (1969) presented some aphidophagous hoverflies, which were later updated by Ghorpadé (2015). Irshad (2001) provided the detailed list of host and natural enemies of aphid pests of different crops, vegetables, and fruit trees from Pakistan, and in 2014, he discussed the role of biotic agents and pollinators aspects and provided the local distribution of all known species of aphidophagous hoverflies in Pakistan. Recently, Jamali et al. (2018) investigated the great diversity of *Episyrphus balteatus*, followed by *Ischiodon scutellaris* and *Eupeodes luniger* in the brassica field (*Brassica rapaoleifera*) against *Myzus persicae* in Sindh province of Pakistan. *Eupeodes luniger*is known from Afghanistan and India and not formerly reported from Pakistan, but the photographs provided in this paper are almost identical to *Scaeva latimaculata* (see for the comparison; Hassan et al., 2018: fig 3J).

The taxonomy of adult hoverflies from Pakistan was initiated by Khan et al. in 1997, after the independence in 1947, they reported 17 species and provided a detailed list of previously known 22 species along with 17 new records for the country. Later, Arif (2001) presented the first comprehensive work on the syrphid fauna of Pakistan and reported 37 species of 20 genera in 3 subfamilies. After that, the following authors worked from the different localities in Pakistan (Arif et al., 2001, 2002, 2004; Saleem et al., 2001; Nielsen, 2003a, 2003b; Sajjad and Saeed, 2009, 2010; Turk et al., 2015a, 2015b). Ghorpadé and Shehzad (2013) presented an annotated checklist of Syrphidae of Pakistan, the adjacent areas of India in Jammu and Kashmir and in adjoining areas of Afghanistan, and reported some 84 species of 42 genera from Pakistan and listed more than 153 species of 58 genera from areas adjacent to Pakistan. Recently, Shehzad et al. (2017) and Hassan et al. (2017, 2018a, 2018b, 2019a, 2019b, 2020a, 2020b, 2021) updated the list of all known 83 species under 43 genera from Pakistan.

Still, there was no such complete list about the tritrophic association of aphidophagous hover flies with aphids and other soft-sucking insect pests in Pakistan. So, the current study was aimed to provide the association of aphidophagous hoverflies with aphids and to find the potential aphidophagous hoverflies in Pakistan. We hope this study will be helpful for future researches in exploring the important biocontrol agent of this group in Pakistan.

8.2 MATERIALS AND METHODS

The adults and immatures of hoverflies, along with their associated host plants and aphid species, were collected by the sweeping aerial net and hand pickling from the various localities in the Pothwar region of Punjab, Pakistan viz; Rawalpindi: Ayub National Park, Kahuta, PMAS-Arid Agriculture University; Murree: Chara Pani, Guldana, Ghora Gali, Numb Behra Mall; Islamabad Capital Territory: Daman-e-Koh, and Kachnar Park. The immature aphidophagous hoverflies were reared in the Laboratory of Biosystematics, Department of Entomology, Pir Mehr Ali Shah Arid Agriculture University Rawalpindi, Pakistan, and fed upon their specific aphid species till the emergence of adults. The emerged adults were fed upon sugary or honey solutions for at least 2 or 3 days to develop the morphological characteristics which aid in the identification of adult hoverflies. The adult species were identified by using the following available literature viz. Brunetti (1923), Ghorpadé (1994), Thompson and Ghorpadé (1992), Stubbs and Falk (2002), Miranda et al. (2013), and Van-Veen (2010). For the identification of aphid species by Blackman and Eastop (2006), CorelDraw version 12 software was used to check the association among the immature stages of aphidophagous hoverflies, their associated aphid species, and host plants.

8.3 RESULTS AND DISCUSSION

The immature stages of aphidophagous hoverflies are well known due to their greater economic importance, as a predator of aphids and other soft sucking insect pests of ornamental and field crops. During the current study, the tritrophic association of the aphidophagous hoverflies with aphids and their host plants in Pakistan are presented: we found 8 aphidophagous hoverflies species from 11 host plants that were infested by 13 aphid species in the Pothwar region of Punjab, Pakistan. *Episyrphus balteatus* (De Geer, 1776) was noticed while feeding on 11 aphid species and is the most dominant in comparison with *E. viridaureus* (Wiedemann, 1824), *Ischiodon scutellaris* (Fabricius, 1805), and *Scaeva pyrastri* (Linnaeus, 1758) on 9, 8, and 4, respectively.

Episyrphus balteatus can be confused with *E. viridaureus* in the field due to their similar body colorations. Kapoor et al. (1985) reported six separate color variants of *E. balteatus* from India. Although *Episyrphus balteatus* is widespread in Pakistan: Azad Kashmir, Gilgit-Baltistan, Khyber Pakhtunkhwa, Punjab, and Sindh (Saleem et al., 2001; Arif, 2001; Haq et al., 2014; Arif et al., 2014; Shehzad et al., 2017; Hassan et al., 2018), but *E. viridaureus* is only known from a few localities in Pakistan (Azad Kashmir, Khyber Pakhtunkhwa, and Punjab) (Ghorpa-déand Shehzad, 2013; Hassan et al., 2018). Paray et al. (2014) studied the pollinators of an apple orchard of Kashmir valley (India) and reported *E. balteatus* with high diversity in district Baramulla. Sajjad and Saeed (2010) studied the floral host of Syrphidae and reported *E. balteatus* as a frequent visitor of maximum plant species. Irshad (2014) studied the role of Syrphidae as biotic agents and pollinators in Pakistan and reported *E. balteatus* to have a great ability of biological control in IPM and high pollinating ability due to its diverse nature. Ghorpadé (2009) examined more than 1000 specimens of *E. viridaureus* from different locations of the Oriental Region and concluded that this is only a single variable species of the genus *Episyrphus* and has the diagnostic character of sternal markings. Ghorpadé (1994) presented the diagnostic character for both male and female *E. balteatus* and *E. viridaureus*. Ghorpadé and Shehzad (2013) listed both E. *balteatus* and *E. viridaureus* from Pakistan. Ghorpadé (2014) reported both *E. balteatus* and *E. viridaureus* in the checklist of Syrphidae of the Indian sub-continent. Mitra et al. (2015) also listed both *E. balteatus* and *E. viridaureus* from India.

Based on high feeding potency, wide distribution, and wide host range, we recommend other than *E. balteatus* (De Geer); *E. viridaureus*, *Ischiodon scutellaris*, and *Scaeva pyrastri* can also be used as potential biocontrol agents in Pakistan (for the detailed list of all known and new records of aphidophagous hoverflies and its prey in Pakistan, see Table 8.1).

No doubt *Episyrphus balteatus* is widely known and the potential aphidophagous hoverflies in Pakistan, now the special focus needs to be done on other aphidophagous hoverflies species like *Episyrphus viridaureus*, *Ischiodon scutellaris*, and *Scaeva pyrastri* as these species were noticed throughout the year in Pakistan. The detailed list of aphidophagous hoverflies species and their trophic association with host plants and aphid prey are enlisted in Figures 8.1 and 8.2. For the purpose of identification, the

TABLE 8.1 List of Known and New Records of Aphidophagous Hoverflies and its Associated Prey in Pakistan.

S.No.	Hoverfly species	Aphid species	References
1	*Episyrphus balteatus*	*Myzus persicae*	Saljoqi et al. (2009)
		Erisoma lanigerum	Irshad (2014)
		Sitobion avenae	Irshad (2014)
		Myzus obtusirostris	Irshad (2014)
		Schizaphis graminum	Irshad (2014); Faheem et al. (2019); Present study
		Rhopalosiphum maidis	Irshad (2014)
		Brevicoryne brassicae	Habib (1973); Present study
		Acyrthosiphon pisum	Hamid et al. (1977)
		Tuberolachnus salignus	Present study
		Macrosiphum rosae	Present study
		Hyalopterus pruni	Present study
		Aphis gossypii	Present study
		Aphis eugeniae	Present study
		Aphis fabae	Present study
		Lipaphis erysimi	Present study
		Cinara tujafilina	Present study
2	*Ischiodon scutellaris*	*Acyrthosiphon pisum*	Hamid et al. (1977)
		Macrosiphum rosae	Present study
		Hyalopterus pruni	Present study
		Aphis gossypii	Present study
		Aphis eugeniae	Present study
		Schizaphis graminum	Faheem et al. (2019); Present study
		Cinara tujafilina	Present study
		Brevicoryne brassicae	Habib (1973); Present study
		Tuberolachnus salignus	Present study
3	*Scaeva pyrastri*	*Aphis gossypii*	Irshad (2014)
		Aphis eugeniae	Present study
		Myzus persicae	Irshad (2014)
		Lipaphis pseudobrassicae	Irshad (2014)
		Cinara tujafilina	Present study
		Tuberolachnus salignus	Present study
		Cavariella theobaldi	Present study

TABLE 8.1 *(Continued)*

S.No.	Hoverfly species	Aphid species	References
4	*Episyrphus viridaureus*	*Macrosiphum rosae*	Present study
		Hyalopterus pruni	Present study
		Aphis eugeniae	Present study
		Aphis fabae	Present study
		Cinara tujafilina	Present study
		Schizaphis graminum	Present study
		Brevicoryne brassicae	Present study
		Tuberolachnus salignus	Present study
		Cavariella theobaldi	Present study
5	*Eupeodes corolla*	*Melanaphis sacchari*	Irshad (2014)
		Rhopalosiphum maidis	Irshad (2014)
		Schizaphis graminum	Irshad (2014)
		Cinara tujafilina	Present study
		Brevicoryne brassicae	Present study
		Myzus obtusirostris	Irshad (2014)
6	*Paragus indicus?*	*Schizaphis graminum*	Irshad (2014)
		Myzus obtusirostris	Irshad (2014)
7	*Serratoparagus serratus*	*Melanaphis sacchari*	Irshad (2014)
		Acyrthosiphon pisum	Hamid et al. (1977)
		Aphis fabae	Present study
8	*Syrphus ribesii*	*Lipaphis erysimi*	Present study
		Tuberolachnus salignus	Present study
9	*Melanostoma orientale*	*Cavariella theobaldi*	Present study
		Macrosiphum rosae	Present study
10	*Betasyrphus isaaci*	*Hyalopterus pruni*	Present study
		Macrosiphum rosae	Present study
11	*Sphaerophoria menthastri*	*Acyrthosiphon pisum*	Hamid et al. (1977)
12	*Sphaerophoria indiana*	*Brevicoryne brassicae*	Habib (1973)
13	*Eupeodes confrator*	*Eriosoma lanigerum*	Hamid et al. (1977)

distinguishing morphological characters of *E. viridaureus* and *E. balteatus* are also provided (Fig. 8.1).

FIGURE 8.1 *Episyrphus balteatus* (De Geer). (A) Living female adult; (B) Dorsal habitus; (C) Close view of the head in lateral view, showing the black hairs on frons; (D) Sternites of female, without black spots; (E) Sternites of male, with small circular black spots.

Based on our recent findings, we recommend *Episyrphus viridaureus*, *Ischiodon scutellaris*, and *Scaeva pyrastri* can also be utilized as potential biocontrol agents in Pakistan (for the detailed list of all known and new records of aphidophagous hoverflies and their prey in Pakistan, see Table 8.1). No doubt *Episyrphus balteatus* is a widely known and potential biocontrol agent in Pakistan, now the special focus needs to

FIGURE 8.2 *Episyrphus viridaureus* (Wiedemann). (A) Living male adult (on the left side); (B) Dorsal habitus; (C) Sternites of female, with a distinct complete blackstripe on the posterior margin; (D) Close view of the head in frontal view, showing the pale-yellow hairs on frons.

be done on other aphidophagous hoverflies species like *E. viridaureus*, *Ischiodon scutellaris*, and *Scaeva pyrastri* as these species were noticed throughout the year in Pakistan. The detailed list of their host plants and aphid prey are enlisted in Figure 8.3.

ACKNOWLEDGMENTS

The first author is grateful to Mr. Muhammad Tariq Rasheed, Mrs. Amara Gull e Fareen, and Dr. Zafar Iqbalfor their support in the collection of hoverflies from the Pothwar region of Punjab, Pakistan.

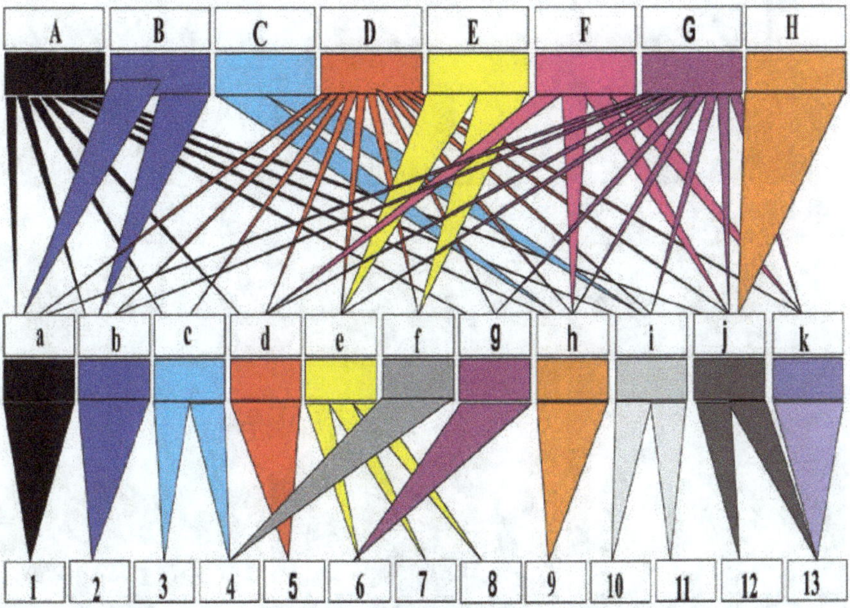

FIGURE 8.3 Tritrophic studies of aphidophagous hoverflies (Diptera: Syrphidae) with aphids and their host plants by using Corel Draw. (A–H) (Hoverfly species): (A) *Ischiodon scutellaris*, (B) *Betasyrphus isaaci*, (C) *Eupeodes corollae*, (D) *Episyrphus balteatus*, (E) *Syrphus ribesii*, (F) *Scaeva pyrastri*, (G) *E. viridaureus*, (H) *Melanostoma orientale*; (a–k) (Aphid species): (a) *Macrosiphum rosae*, (b) *Hyalopterus pruni*, (c) *Aphis gossypii*, (d) *A. eugeniae*, (e) *A. fabae*, (f) *Lipaphis erysimi*, (g) *Schizaphis graminum*, (h) *Cinara tujafilina*, (i) *Brevicoryne brassicae*, (j) *Tuberolachnus salignus*, (k) *Cavariella theobaldi*; (1–13) (Host plants): (1) *Rosa indica*, (2) *Prunus persica*, (3) *Hibiscusrosa-sinensis*, (4) *Parthenium hysterophorus*, (5) *Eriobotrya japonica*, (6) *Triticum aestivum*, (7) *Rumex dentatus*, (8) *Solanum nigrum*, (9) *Thuja occidentalis*, (10) *Brassica campestris*, (11) *Brassica napus*, (12) *Prunus persica*, (13) *Salix acmophylla.*

FIGURE 8.4 (A–M) (List of host plants): (A) *Rosa indica*; (B) *Prunus persica*; (C) *Hibiscus rosa-sinensis*; (D) *Parthenium hysterophorus*; (E) *Solanum nigrum*; (F) *Prunus persica*; (G) *Brassica campestris*; (H) *Rumex dentatus*; (I) *Brassica napus*; (J) *Eriobotrya japonica*; (K) *Triticum aestivum*; (L) *Salix acmophylla*; (M) *Thuja occidentalis*.

KEYWORDS

- **tritrophic association**
- **aphidophagous**
- **syrphidae**
- **insect pest**

REFERENCES

Agarwala, B. K.; Laska, P.; Raychaudhuri, D. N. Prey Records of Aphidophagous Syrphid Flies from India (Diptera: Syrphidae). *Acta Entomolol. Bohemoslov.* **1984,** *31*, 15–21.

Alam, M. M.; Beg, M. N.; Syed, R. A.; Shah, S. *Survey of Parasites of Insect Pests of Cultivated and Useful Plants and Survey of Insects Destroying Weeds and their Parasites*, Final Report; Common Wealth Institute of BiologicalControl (CIBC): Pakistan Station: Rawalpindi, 1969; p 243.

Alam, M. M.; Hafiz, I. A. Some Natural Enemies of Aphids in Pakistan. *Tech. Bull. Commonw. Inst. Biol. Contr*. **1963,** *3*, 41–44.

Ali, S.; Shehzad, A.; Rafi, M. A.; Zia, A. Insect Pollinators of Litchi from District Haripur, Pakistan. *Pak. J. Agricul. Res*. **2013,** *26*, 220–229.

Arif, M. J. Taxonomic Study of Syrphidae (Diptera) of Pakistan. Unpublished Ph.D. Thesis Submitted to the Agricultural University Faisalabad, Pakistan, 2001; p 154.

Arif, S.; Khan, M. R.; Shehzad, A.; Nazir, N.; Rahim, J.; Naz, F.; Rafi, R. Biodiversity of Syrphid Flies (Syrphidae: Diptera) from Poonch District of Azad Kashmir, Pakistan. *J. Agril. Technol.* **2014,** *10*, 1465–1472.

Arif, M. J.; Suhail, A.; Yousuf, M. Genus *Sphaerophoria* St. Farg. *Et* Serv. from Pakistan. *Int. J. Agric. Biol.* **2001,** *3*, 260–261.

Arif, M. J.; Suhail, A.; Yousuf, M. Taxonomic Studies of Saprophagous Syrphids fromPakistan. *Int. J. Agric. Biol*. **2002,** *4*, 76–77.

Aslamkhan, M.; Safdar, S.; Azizullah. Biodiversity of Syrphidae of Pakistan. *Biologia* **1997,** *43*, 19–25.

Blackman, R. L.; Eastop, V. F. Aphids on the World's Herbaceous Plants and Shrubs. The Natural History Museum; John Wiley & Sons: Chichester, Great Britain, 2006; p 1439.

Brunetti, E. In *The Fauna of British India, Including Ceylon and Burma. Diptera. Pipunculidae, Syrphidae, Conopidae, Oestridae*; Taylor and Francis, 1923; vol *3*, pp 23–339.

Faheem, M.; Saeed, S.; Sajjad, A.; Razaq, M.; Ahmad, F. Biological Parameters of Two Syrphid Fly Species *Ischiodon scutellaris* (Fabricius) and *Episyrphus balteatus* (DeGeer) and their Predatory Potential on Wheat Aphid *Schizaphis graminum* (Rondani) at Different Temperatures. *Egypt. J. Biol. Pest Control* **2019,** *29*, 1–8.

Ghorpadé, K. Insect Prey of Syrphidae (Diptera) from India and Neighboring Countries: Are View and Bibliography. *Trop. Pest Manag*. **1981,** *27*, 62–82.

Ghorpadé, K. Diagnostic Keys to New and Known Genera and Species of Indian Subcontinent Syrphini (Diptera: Syrphidae). *Colemania* **1994,** *3*, 1–15.

Ghorpadé, K. Hoverflies (Diptera—Syrphidae) Documented from the Northwest Frontier of the Indian Sub-continent: A Circumstantial History and Inclusive Bibliography. *Colemania* **2015,** *50*, 1–151.

Ghorpadé, K.; Shehzad, A. An Annotated Checklist and Select Bibliography of the Hover-Flies (Diptera—Syrphidae) of Pakistan, Indian Subcontinent. *Colemania* **2013,** *37*, 1–26.

Habib, R. Natural Enemies and their Role in Limitation and Control of *Brevicoryne brassicae* L. in Pakistan. *Tech. Bull. Common. Inst. Biol. Control* **1973,** *16*, 69–78.

Hamid, S.; Shah, M. A.; Anwar, A. M. Investigations on *Acyrthosiphon pisum* Harris with Special Reference to its Parasite *Aphidius smithi* Sharma and Subba Rao. *Tech. Bull. Common. Inst. Biol. Control* **1974,** *17*, 69–85.

Hamid, S.; Shah, M. A.; Anwar, A. M. Some Ecological and Behavioral Studies on *Aphiscraccivora* Koch (Hem.: Aphididae). *Tech. Bull. Common Inst. Biol. Control* **1977,** *18*, 99–111.

Hassan, M. A.; Bodlah, I.; Ahmad, M.; Kayani, A. R.; Mahmood, K. First Record of Thegenus *Graptomyza* Wiedemann, 1830 (Diptera: Syrphidae) from Pakistan. *J. Anim. Plant Sci.* **2020a,** *30*, 512–516.

Hassan, M. A.; Bodlah, I.; Aihetasham, A. First Record of *Baccha maculata* Walker,1852 (Diptera: Syrphidae) from the Pothwar Punjab, Pakistan. *Punjab Uni. J. Zool.* **2019a,** *32*, 133–135.

Hassan, M. A.; Bodlah, I.; Bodlah, M. A.; Hussain, R. New Records of the Genus *Ceriana* Rafinesque, 1815 (Diptera: Syrphidae) from Pakistan. *Mun. Ent. Zool.* **2019b,** *14*, 185–187.

Hassan, M. A.; Ghorpadé, K.; Bodlah, I.; Mahmood, K.; Iqbal, Z. Additional Notes on the Genus *Paragus* Latreille (Diptera: Syrphidae) from Pakistan with a New Country Record. *J. Anim. Plant Sci.* **2018a,** *28*, 708–714.

Hassan, M. A.; Ghorpadé, K.; Mahmood, K.; Shehzad, A.; Nazir, N.; Fatima, N. Preliminary Studies on the Syrphidae (Diptera) of Poonch District, Azad Kashmir, Pakistan. *Orient. Insects* **2018b,** *52*, 190–209.

Hassan, M. A.; Mahmood, K.; Nazir, K.; Fatima, N.; Aslam, M. A. Faunistic Work on the Hover Flies (Diptera: Syrphidae) of District Narowal, Pakistan. *J. Entomol. Zool. Stud.* **2017,** *5*, 626–630.

Hassan, M. A.; Bodlah, I.; Shehzad, A.; Fatima, N. New Records of the Genus *Volucella* Geoffroy (Diptera: Syrphidae) from Pakistan Along with a Checklist of Known Species. *J. Threat. Taxa* **2020b,** *12*, 16632–16635.

Hassan, M. A.; Bodlah, I.; Shehzad, A.; Fazal, F.; Fatima, N. First Record of Genus *Myolepta* Newman, 1838 (Diptera: Syrphidae) from Pakistan, With Description of a Newspecies. *Orient. Insects* **2021,** *55*, 564–573.

Irshad, M. Aphids and their Biological Control in Pakistan. *Pak. J. Biol. Sci.* **2001,** *4*, 537–541.

Irshad, M. Role of Syrphids (Diptera: Syrphidae) as Biotic Agents and Pollinators in Pakistan. *J. Biores. Manag.* **2014,** *1*, 1–9.

Jamali, R. A.; Memom, N.; Shah, M. A.; Khan, K.; Ansari, A. Prevalence of Aphidophagous Hoverflies (Syrphidae: Syrphinae) in Relation to their Prey, Green Aphids (*Myzus persicae*) on Brassica (*Brassica rapa oliefera*) in Dadu. *J. Anim. Plant Sci.* **2018,** *28*, 1447–1456.

Kindlmann, P.; Dixon, A. F. G.; Michaud, J. P. In *Aphid Biodiversity under Environmental Change: Patterns and Processes*; Springer, 2010; pp 1–199.

Mengual, Z.; Stahls, G.; Rojo, S. Phylogenetic Relationships and Taxonomic Ranking of Pipizine Flower Flies (Diptera: Syrphidae) with Implications for the Evolution of Aphidophagy. *Cladistics* **2015,** 1–18.

Miranda, G. F. G.; Young, A. D.; Locke, M. M.; Marshall, S. A.; Skevington, J. H.; Thompson, F. C. Key to the Genera of Nearctic Syrphidae. *Can. J. Arthropod Identif.* **2013,** *23*, 1–351.

Mitra, B.; Roy, S.; Imam, I.; Ghosh, M. A Review of the Hover Flies (Syrphidae:Diptera) from India. *Int. J. Fauna Biol. Stud.* **2015,** *2*, 61–73.

Nielsen, T. R. A New *Platycheirus* of the *ambiguous* Group (Diptera: Syrphidae) from Kashmir. *Volucella* **2003a,** *7*, 31–34.

Nielsen, T. R. *Platycheirus cryophilus* spec. nov. (Diptera, Syrphidae) from Kashmir. *Volucella* **2003b,** *8*, 95–97.

Pape, T.; Evenhuis, N. L. Systema Dipterorum, Version [1.5] 2015 [Online]. http://www.diptera.org/ (accessed Jan 30 and Feb 22, 2015).

Paray, M. A.; Parey, S. H.; Munazah, Y.; Rizwana, K.; Barkat, H.; Bhat, S. G.; Rather, A. Z. The Pollinators of Apple Orchards of Kashmir Valley (India). *Ecol. Env. Cons.* **2014,** *20*, S471–S477.

Reemer, M; Rotheray, G. E. Pollen Feeding Larvae in the Presumed Predatory Syrphinegenus *Toxomerus* Macquart (Diptera, Syrphidae). *J. Nat. Hist.* **2009,** *43*, 939–949.

Sajjad, A.; Saeed, S. New Records of *Eristalinus* (Syrphidae: Diptera) from Multan. *Pak. J. Zool.* **2009,** *41*, 238–239.

Sajjad, A.; Saeed, S. Floral Host Plant Range of Syrphid Flies (Syrphidae: Diptera) Under Natural Conditions in Southern Punjab, Pakistan. *Pak. J. Bot.* **2010,** *42*, 1187–1200.

Saleem, M.; Arif, M. J.; Suhail, A. Taxonomic Studies of Syrphidae of Peshawar, Pakistan. *Int. J. Agric. Biol.* **2001,** *3*, 533–534.

Saljoqi, A. U. R.; Khan, K.; Rahman, S. U. Integrated Management of Potato-Peach Aphid, *Myzus persicae* (Sulzer). *Sarhad J. Agric.* **2009,** *25*, 573–580.

Shehzad, A.; Ghorpadé, K.; Rafi, M. A.; Zia, A.; Bhatti, A. R.; Ilyas, M.; Shah, S. W. Faunistic Study of Hover Flies (Diptera: Syrphidae) of Pakistan. *Orient. Insects* **2017,** *51*, 197–220.

Stubbs, A. E.; Falk, S. J. In *British Hover Flies: An Illustrated Identification Guide*; The British Entomology and Natural History Society: Reading, U.K., 2002; p 469.

Thompson, F. C.; Ghorpadé, K. A New Coffee Aphid Predator, with Notes on Other Oriental Species of Paragus (Diptera: Syrphidae). *Colemania* **1992,** *5*, 1–24.

Turk, J. K.; Memon, N.; Mal, B.; Memon, S. A.; Shah, M. A.; Solangi, D. A. Redescription of Two Species of Genus *Eupeodes* Osten Sacken from Quetta Balochistan, Pakistan. *J. Anim. Plant Sci.* **2015a,** *25*, 1329–1334.

Turk, J. K.; Memon, N.; Mal, B.; Shah, M. A.; Memon, S. A.; Shah, N. A.; Ansari, A. New Record and Description of Two Species of Genus *Sphaerophoria* St. Farg. et Serv (Diptera: Syrphidae) from Balochistan, Pakistan. *Pure Appl. Biol.* **2015b,** *4*, 110–117.

Van-Veen, M. P. In *Hoverflies of Northwest Europe. Identification Keys to theSyrphidae*; KNNV Publishing: Utrecht, 2010; p 248.

CHAPTER 9

Bumblebees: Diversity, Role in Pollination Ecology, Threats, and Conservation

RIFAT H. RAINA[1], KESHAV KUMAR[1], PURNIMA PATHAK[1], TRILOK JANGID[1], PREETI CHOUDHARY[1], AEJAZ HUSSAIN PARREY[2], and INDU SHARMA[1]

[1]*Desert Regional Centre, Zoological Survey of India, Jodhpur, Rajasthan, India*

[2]*Department of Zoology, School of Biosciences and Biotechnology, Baba Ghulam Shah Badshah University, Rajouri, Jammu and Kashmir, India*

ABSTRACT

Pollination is indeed the most significant ecological process. Pollination is mainly done with the help of biotic and abiotic factors such as the wind (anemophily). Different plants show different responses related to pollination. They show a mutual relationship between the plant and the pollinators. Bees are small pollinators included in the group hymenoptera. Bumblebees are the key pollinator species in the Indian Himalayan Region (IHR). In India, bumblebees are mainly distributed in the IHR. Bumblebee shows their diversity ranging from Eastern Himalaya to the northwest Himalaya. Saini et al. (2015) revised and included some parts of the bumblebee and have described 47 species of bumblebee out of which 37 species are from the northeast Himalaya and 38 species from the northwest Himalaya while 27 species are common in both these regions. Bumblebees were different from other species of bees in their habitat as well as their seasonal activity.

Insect Diversity and Ecosystem Services, Volume 1: Importance, Threats, Conservation and Economic Perspectives. Younis Ahmad Hajam, Sajad Hussain Parey, and Rouf Ahmad Bhat (Eds)

Therefore, the present review aims to discuss the diversity of bumblebees and their role in pollination.

9.1 INTRODUCTION

Pollination is indeed the most significant ecological process. Pollination is mainly done with the help of biotic and abiotic factors such as the wind (Anemophily). Different plants show different responses related to pollination. They show a mutual relationship between the Plant and the pollinators. Some pollinators visit plants for the mean of nectar, oils, and scents for their growth and their young ones. Flowers exhibit some special characteristics such as shining leaves and perfumes to attract pollinators. Each pollinator is attracted by the different special characteristics of plants. Ollerton et al. (2007) show the "generalist and the specialized" flower to be pollinated by one or many species. Not any particular species play a role in pollination for a specific plant. Waser et al. (1996) stated that the major insect-pollinated plant uses a different group of pollinators but most species of bumblebee are generalist pollinators. Most of the plant species show a generalized mode of pollination. Proctor et al. (1996) mentioned in their book that different groups of organisms take part in the pollination of flowering plants. Through birds (ornithophily), bats (chiropterophily), insects (entomophily) beetles (cantharophily), butterflies (psychophily), bees (melittophily), and also done by humans, reptiles, etc. Keeping the situation in view most of the insect's order plays a major role in pollination. These include Lepidoptera, Diptera, Coleoptera, Odonata, Orthoptera, Hymenoptera, Hemiptera, etc. In the order hymenopteran, these include mainly bees, wasps, ants, and sawflies. They show complete metamorphosis and differ from each other. Bees differ from wasps in terms of their body structure and body hairs. Most of the bees show a high rate of pollination because they feed on nectars from different flowers.

9.2 BUMBLEBEE AS POLLINATOR

Bees are small pollinators included in the group Hymenoptera. Bumblebees are the key pollinator species in the Indian Himalayan Region (IHR). The superfamily Apoidea includes a few other populations of bee species like Megachilidae (leafcutter bees) Andrenidae (mining bees) which also play a significant role in pollination. There are nearly 20,000 species of bees throughout the world, and 715 species are found in India (Gupta, 2003; Ascher and Pickering, 2019). There are approximately 265 species of Bumblebees reported globally (Williams et al., 2008). In (1758) Linnaeus was the first who initiate work on bumblebee India. In the Himalayan range, around 47 (Raina et al., 2020) species of bumblebee reported so far (Williams, 1991, 1998, 2004) out of which nine are endemic to the region: *Bicyclus. abnormis*, *Bombus mirus*, *Bombus parthenius*, *Brachyponera luteipus*, *Bombus tunicatus*, *Bombus jacobsoni*, *Bombus simillimus*, and *Bombus miniatus*. The other species of bumblebee need to access urgently. They are eusocial insects and wild pollinators found mostly in high-altitude regions ranging from 1000 to 5500 m above mean sea level (AMSL) (Kearns and Thomson, 2001) and play an important role in the pollination of high-land crops. The bumblebee species are not found in the plains of India. Not most of the bees but bumblebees forage in a very cold environment. They have an annual life cycle with different castes, namely, workers, drones, and queens. Bumblebees are more similar to honey bees in terms of foraging for nectars. Bumblebee colonies die after the end of each growing season with the queen undergoing diapause. In the next season, the queen searches a nesting site and tries to make their nest in the hole over or in the ground. The rate of oviposition and the nesting sites are affected by several different factors, namely, predators, hygienic conditions, food availability nesting site, many environmental factors, etc.

9.3 HABITAT AND FORAGING PLANTS

Most of the bumblebees are restricted to high-altitude regions and mostly in the mountains of the IHR. they are considered to be the most effective indigenous pollinators at high altitudes and forage on a snowy day, or rainy day (Saini et al., 2012). Bumblebee show hairs on its outer body. When they forage for nectar, the pollen on the flower is attached to these hairs and this may result in "buzz" pollination by vibrating their flight muscles. With the help of this buzz pollination, many families of plant species are included in pollination. (Van den Eijnde et al., 1991) describes that these bumblebees are the best pollinators of the family Solanaceae; for example, tomatoes (Solanum Lycopersicon). Some of the other families of the plant species like Rosaceae, Fabaceae, etc., are pollinated by bumblebees. Many medicinal plants, aromatic, and some wildflowers in the IHR are mainly pollinated by bumblebees. Most of the wild plants are mainly pollinated by different species of bumblebee (Goulson, 2003a). Westrich (1996) reported that the species of bumblebee are polylactic, which means they can forage on many floral species. Raina et al. (2019) observed that the upper region of the Himalayan region predominates the tree of less alpine meadow between 3000 and 5500 m asl and the vegetation that is predominantly by the species of bumblebee include a wide range of herbs, shrubs, grasses, and medicinal plants. The extreme environment of these snow mountains predominantly the Karakoram range of Leh–Ladakh, Apharwat range of Jammu and Kashmir, and Lahaul Spiti regions of Himachal Pradesh where only some preferred species of medicinal plants maintain their existence. Therefore, bumblebees play an incredibly positive and decisive role in conserving them. The common vegetation existing at these ranges and most of them are pollinated by bumblebees include *Hyoscyamus niger* Linn (*Solanaceae*), *Sedum ewersii* Ladeh (*Crassulaceae*), *Cynathuslobatus* Wall. (*Campanulaceae*), *Gypsophila cerastiodies* (*Charophyceae*), *Rhodiola crenulate* Linn, *Anaphalisbusua* DC., *Taraxacum officinale*, *Sparganothoides albescens*, *Circbitacineria* Linn., *Senna auriculata*, *Senecio chrysanthemoides* DC., *Saussurea lappa*, *Saussurea jacea* (Klotz) CL., *Saussurea costus*, *Lacutaorientallis*, *Scorzonera virgata*, *Semiarundinaria fastuosa*, *Lentopodium alpinum*, *S. chrysanthemoides*, *Potentilla atrosanguinea* Lodd. (*Rosaceae*), *Rudbeckia laciniata*, *Tanacetum dolicophyllum* (*Asteraceae*), *Myosotis sylvatica* Ehrh. *ex Hoffm* (*Boraginaceae*), *Nepeta laevigata* (D. Don) Hand-Mazz. (Lamiaceae), *Pedicularis pectinate* Wall., *Melilotus officinalis* (L.) *Thymus serpyllum* Linn., *Pall* (*Leguminosae*), *Mentha longifolia* Linn., *G. cerastiodies* D. Don (*Caryophyllaceae*) *Phlomisbracteosa* Royle, *Eupharasia officinalis* (*Scrophulariaceae*), *Stachys serecia* Wall. ex Benth., *Polygonum amplexicaule* D. Don. (*Polygonaceae*) *Hyssopus officinalis* Linn., *Nepeta discolor* Linn., *Dracocephalum heterophyllum* Benth. Hence, all aforementioned plants show that this high-altitude vegetation is mostly conserved by the bumblebees. Other crops like cucumber, seed crops, blueberries, cane berries, vegetables, squash, and melons are more convenient with bumblebee pollination.

9.4 DIVERSITY AND DISTRIBUTION OF BUMBLEBEE

In India, bumblebees are mainly distributed in the IHR. Bumblebee shows their diversity ranging from Eastern Himalaya to the northwest Himalaya. Saini et al. (2015) revised and included some parts of the bumblebee and have described 47 species of bumblebee out of which 37 species are from the northeast Himalaya and 38 species from the northwest Himalaya while 27 species are common in both these regions. The Himalayan region falling into these areas includes The Union territories of Jammu and Kashmir, The Union territories of Ladakh, Himachal Pradesh, Uttarakhand, Arunachal Pradesh, Sikkim, etc. The region of Kashmir Valley and the upper region of the northeast Himalaya show a high diversity of bumblebee species. However, the region of Himachal Pradesh and Uttarakhand also favors the high diversity of bumblebees. These need extensive surveys as they show a prediction of a greater number of species of bumblebee that may be documented in the future.

9.5 BUMBLEBEE SPECIES IN INDIA

In India, 47 species of bumblebee are described by (Raina et al., 2020) out of which nine species are endemic to the IHR. These include *B. jacobsoni* (Skorikov, 1912), *Bombus asiaticus* (Morawitz, 1875), *Bombus avinoviellus* (Skorikov, 1914), *Bombus sibiricus* (Fabricius, 1781), *B abnormis* (Tkalcu, 1968), *Bombus funerarius* (Smith, 1852), *Bombus festivus* (Smith, 1861), *Bombus flavescens* (Smith, 1852), *Bombus ferganicus* (Radoszkowski, 1893), *Bombus cornutus* (Frison, 1933), *Bombus eximius* (Smith, 1852), *Bombus breviceps* (Smith, 1852), *Bombus biroi* (Vogt, 1911), *Bombus branickii* (Radoszkowski, 1893), *Bombus waltoni* (Cockerell, 1910), *Bombus turneri* (Richards, 1929), *B. tunicatus* (Smith, 1852), *Bombus albopleuralis* (Smith, 1852), *Bombus tanguticus* (Morawitz, 1887), *Bombus subtypicus* (Skorikov, 1914), *Bombus skorikovi* (Popov, 1927), *B. simillimus* (Smith, 1852), *Bombus semenovianus* (Skorikov, 1914), *Bombus rufofasciatus* (Smith, 1852), *Bombus rotundiceps* (Friese, 1916), *Bombus pyrosoma* (Morawitz, 1890), *Bombus pressus* (Frison, 1935), *Bombus personatus* (Smith, 1879), *B. parthenius* (Richards, 1934), *Bombus oberti* (Morawitz, 1883), *Bombus novus* (Frison, 1933), *Bombus nobilis* (Friese, 1905), *Bombus genalis* (Friese, 1918), *Bombus graham*

(Frison, 1933), *B. mirus* (Tkalcu, 1968), *Bombus morawitzianus* (Popov, 1931), *B. miniatus* (Bingham, 1897), *Bombus melanurus* (Lepeleitier, 1836), *Bombus luteipes* (Richards, 1934), *Bombus lepidus* (Skorikov, 1912), *Bombus lemniscatus* (Skorikov, 1912), *Bombus ladakhensis* (Richards, 1928), *Bombus keriensis* (Morawitz, 1886), *Bombus kashmirensis* (Friese, 1909), *Bombus hypnorum* (Linnaeus, 1758), *Bombus himalayanus* (Skorikov, 1914), *Bombus haemorrhoidalis* (Smith, 1852) (Fig. 9.1).

FIGURE 9.1 Different species of bumblebees.

9.6 IDENTIFICATION AND MIMICRY

Bumblebees are not easily identified without the proper key of identification. Bees generally show mimicry about their coloration. This may be warning coloration for their protection from the predators or maybe a signal received from the flowers for foraging. Bumblebee mostly shows yellow and black color patterns but sometimes with red or white color also. Other insects that lack stingers or other characteristics have evolved and act like bumblebees. Those insects that closely resemble bumblebees are flies, beetles, sawflies, and other bees of different genera. Color variations and convergence make it possible to identify the bumblebee in the field.

Without a microscopic examination of their morphological character and a closer view, we cannot distinguish one bee from another. During collection try to increase the expertise of their identification, so that fewer bee species need to be killed and the remaining others may be released in the field.

9.7 NESTINGBEHAVIOUR OF BUMBLEBEES

Bumblebees require a specific habitat for nesting. After the growing season, the bumble undergoes hibernation. The queen of the bumblebee shows the diapause stage. In the early summer season, the bumblebee queen visits preexisting burrows. Birds' nests, rodent dens, or above the ground in herbs. When the queen found a suitable nesting place, the first clutch of eggs was laid down. She made a nectar cup in her nest. After the hatching of these eggs, these young larvae feed on this nectar. This larva feeds for about 2 weeks, it spins a silk cocoon and pupate for another 2 weeks. The adult bees that emerge for the first time are female. Later on, produce their offspring and living worker in their nest. Now, the queen stays at her nest and lays eggs, working forage for resources like nectar, regulating the nest temperature, and defending the host. In early springs the bumblebees show fast and active movement for nectar.

9.8 THREATS TO BUMBLEBEE

The bumblebee is also facing threats of extinction. Both natural and anthropogenic activities affect their life cycle. Over the last few decades, the rate of urbanization, habitat fragmentation, pollution, overgrazing, excessive use of agricultural pesticides, deforestation, overgrazing, climate change, and other anthropogenic activities have affected bumblebees' life history and range shift or decline in their number from their natural ecosystem.

Some natural enemies like wasps, flies, bugs, spiders, birds, etc., attack the foraging bumblebee. The larva of the bumblebee is used by some mammals, for example, raccoons. Spiders do not spin webs; they ambush bumblebee at a flower. Some wasps inject venom into the live bees and make them paralyzed and used to feed their developing larvae.

Some other natural enemies such as parasitoids, fungi (Nosema), mites, etc., also decline their number. Bumblebees are host to a variety of parasitoids. Most of the bumblebee species are affected by these enemies. Some nematode worms enter early summer and cause sterilization of the queen.

Mites are transferred within the colony. Some mites live inside the bumblebee and decrease their foraging efficiency. Some protozoans and fungi are also present in their digestive gut. Later on, these may cause the decline of the species.

Some anthropogenic activities such as excessive use of a pesticide may affect their foraging behavior in the high-altitude mountain ecosystem.

9.9 CONSERVATION MEASURES

Bumblebees were different from other species of bees in their habitat as well as their seasonal activity. India shows rich and unique diversity for pollinators in the high-land-mountain ecosystem. It seems that the rapid decline of bumblebees in recent times. Some conservation measure needs to minimize the threat of their declining population. To raise awareness among the population UN General Assembly designated 20 May as "World Bee Day."

- ➢ Habitat management and restoration is the main tool for their conservation. Farmer uses low-harming activity insecticide and manipulation of many agricultural practices.
- ➢ Support organic agriculture practices with gardening.
- ➢ Plantation of the various species of flowers and plants rich in nectar and pollen into the barren areas and try to restrict low moves into the wild habitat.
- ➢ Queen spends their hibernation in rotting logs, burrows, and grasses until the summer arrives. Try to provide an area with these substrates without disturbing them.
- ➢ Forests are the main source in the conservation of bumblebee's ecology. Wildflowers are a suitable site for nesting and hibernation. Try to control deforestation as a conservation measure.
- ➢ Capacity building of forest officials, school children, farmers, and other local stakeholders and aware them of the benefits of bumblebees.
- ➢ On the occasion of "World Bee Day," Colleges and schools organize different awareness programs such as quizzes, and poster making, and teach them the benefits of bumblebees.
- ➢ Promote Artificial nesting and try to encourage queens to establish a nest in the artificial nest box, so that it may lead to awareness of conservation through local people's efforts.

9.10 WHY DO WE CONSERVE BUMBLEBEE?

Bumblebee is a keystone pollinator species for pollination in the high mountain ecosystems. As compared to other bees, Bumblebee is the only key species to survive at a very low temp. Bumblebees forage up to approximately –37°C. Other bees do not survive at this temperature. Countless benefits of bumblebee conservation are as under:

- Bumblebees are used to pollinate a large number of agricultural and horticultural crops such as tomatoes, potatoes, peas apples, etc.
- Bumblebees are considered to be important pollinators of wildflowers and aromatic and medicinal plants in the highland ecosystem.
- They show buzz pollination with fast vibration of the wing's muscle per second.
- Bumblebee is a reliable worker and works from dusk to dawn in harsh conditions under snowy, rainy, and cloudy days.
- They are also economically important to farmers by providing bumblebee colonies on a rental basis for pollination during the blooming season as it provides high yield with better quality.
- The body size of a bumblebee can accommodate more load of pollen than other bees and better contact with flowers. Therefore, helps in more pollination of flowers.
- Bumblebees are not aggressive and therefore help in the pollination of the greenhouse growing vegetables.
- Bumblebees are useful in determining the health of the mountain ecosystem by assessing their diversity and distribution.

ACKNOWLEDGMENTS

We are grateful to Dr. Dhriti Banerjee, Director of Zoological Survey of India, Kolkata for providing the necessary facilities. Special thanks are due to the National Mission on Himalayan Studies, MOEF & CC, Government of India for providing financial support under the project "Documentation conservation and utilization of indigenous mountain pollinators-with special reference to Himalayan bumblebees."

KEYWORDS

- **bumblebee**
- **pollination**
- **conservation**
- **threats**

REFERENCES

An, J.; Huang, J.; Shao, Y.; Zhang, S.; Wang, B.; Liu, X.; Wu, J.; Williams, P. H. The Bumblebees of North China (Apidae, Bombus Latreille). *Zootaxa* **2014,** *3830* (1), 1–89.

Ascher, J. S.; Gupta, A. Bee Fauna of India, 2019.

Cockerell, T. D. A. Some Bees from High Altitudes in the Himalayan Mountains. *Entomologist* **1910,** *43*, 238.

Faegri, K.; Van der Pijl, L. In *The Principles of Pollination Ecology*, 3rd Ed.; Oxford, Pergamon Press: UK, 1979.

Goulson, D. Conserving Wild Bees for Crop Pollination. *J. Food Agric. Environ.* **2003,** *1*, 142–144.

Gupta, R. K. The Diversity of Bees (Hymenoptera, Apoidea) in India. In *Advancements in Insect Biodiversity*; Agrobios: India, 2003; pp 1–337.

Kearns, C. A.; Thomson, J. D. In *Natural History of Bumblebees*; University Press of Colorado, 2001.

Matheson, A.; Buchmann, S. I.; O'Toole, C.; Westrich, P.; Williams, I. H. In *The Conservation of Bees*; Academic, 1996.

Ollerton, J.; Killick, A.; Lamborn, E.; Watts, S; Whiston, M. Multiple Meanings and Modes: On the Many Ways to be a Generalist Flower. *Taxon* **2007,** *56* (3), 717–728.

Parrey, A. H.; Raina, R. H.; Saddam, B.; Pathak, P.; Kumar, S.; Uniyal, V. P.; Gupta, D.; Khan, S. A. Role of Bumblebees (Hymenoptera: Apidae) in Pollination of High Land Ecosystem: A Review. *Agricult. Rev.* **2021a**. https://doi. org/10.18805/ag

Paul, H. W. The Bumble Bees of the Kashmir Himalaya (Hymenoptera: Apidae, Bombini). *Bull. Br. Mus. Entomol.* **1991,** *60* (1), 1–204.

Proctor, M.; Yeo, P.; Lack, A. In *The Natural History of Pollination*; HarperCollins Publishers, 1996.

Raina, R. H.; Parrey, A. H.; Chandra, K.; Khan, S. A. New Distributional Records of Bumblebees (Hymenoptera: Apidae) from Bangus Valley in North Kashmir of Union Territory of Jammu and Kashmir, India. *Rec. Zool. Surv. India* **2021 b,** *121* (2), 261–267.

Raina, R. H.; Saini, M. S. Morphology and Food Plants of Cuckoo Bees (Apidae: Hymenoptera) From Indian Himalayas. *Ecol. Balk.* **2015,** *7* (1), 63–80.

Raina, R. H.; Saini, M. S.; Khan, Z. H. Taxonomic Revision of Subgenus Subterrane obombusvogt (Apida: Hymenoptera) from Indian Himalayas. *Ann. Entomol.* **2012,** *30* (2), 41–51.

Raina, R. H.; Saini, M. S.; Khan, Z. H.; *Bombus morawitzianus* (Popov) A New Addition to the Genus Bombus (Apidae: Hymenoptera) from Indian Himalaya. *Entomol. News* **2013a,** *123* (1), 26–32.

Raina, R. H.; Saini, M. S.; Khan, Z. H. Taxonomy and Ecology of *Bombus pyrosoma* MORAWITZ (Hymenoptera: Apidae) from the North-West Indian Himalaya. *Polish J. Entomol.* **2013b,** *82* (2), 95.

Raina, R. H.; Saini, M. S.; Khan, Z. H. Taxonomy and Food Plants of *Bombus simillimus* SMITH (Hymenoptera: Apidae) from the Indian Himalaya. *Polish J. Entomol.* **2013c,** *82* (1), 35.

Raina, R. H.; Saini, M. S.; Khan, Z. H. Population Variation in *Bombus asiaticus* (Hymenoptera: Apidae) from Northwest Indian Himalaya. *Entomol. News* **2014,** *123* (5), 321–332.

Raina, R. H.; Saini, M. S.; Khan, Z. H. Morphology and Food Plants of Subgenus Bombus Sensustricto (Hymenoptera: Apidae: Apinae) from Indian Himalaya. *Indian J. Agric. Sci.* **2014,** *84* (10), 1217–1224.

Raina, R. H.; Saini, M. S.; Khan, Z. H. Taxonomy and Ecology of Bombus trifasciatus Smith (Hymenoptera: Apidae: Megabombus) from Indian Himalaya with Description of New Variants, 2017.

Raina, R. H.; Saini, M. S.; Khan, Z. H. Altitudinal Food Preference of Bumblebee Species (Hymenoptera: Apidae) from Indian Himalaya. *J. Entomol. Zool. Stud.* **2019,** *7*, 234–237.

Richards, O. W. LXXIV. —A New Species of Indian Humble-Bee in the Collection of the British Museum (Hymenoptera, Bombidæ). *Ann. Mag. Nat. Hist.* **1931,** *8* (47), 529–533.

Saini, M. S.; Raina, R. H.; Khan, Z. H. A Check List of Bumblebees (Hymenoptera: Apidae) from Indian Himalaya. *J. Insect Sci.* **2011,** *24* (4), 326–352.

Saini, M. S.; Raina, R. H.; Khan, Z. H. Species Diversity of Bumblebees (Hymenoptera: Apidae) from Different Mountain Regions of Kashmir Himalayas. *J. Sci. Res.* **2012a,** *4* (1), 263–263.

Sharma, H. K.; Devi, D.; Thakur, M.; et al. "Bumble Bees" In Vitro Rearing Methods – A Review. *Int. J. Trop. Insect Sci.* **2021,** *42* (2).

Van den Eijnde, J.; De Ruijter, A.; Van der Steen, J. In *Method for Rearing Bombus Terrestris Continuously and the Production of Bumblebee Colonies for Pollination Purposes*, VI International Symposium on Pollination, 1990; vol *288*, pp 154–158.

Williams, P. H. An Annotated Checklist of Bumble Bees with an Analysis of Patterns of Description (Hymenoptera: Apidae, Bombini). *Bull. Nat. Hist. Mus. Entomol. Ser.* **1998,** *67*, 79–152.

Williams, P. H.; Cameron, S. A.; Hines, H. M.; Cederberg, B.; Rasmont, P. A Simplified Subgeneric Classification of the Bumblebees (genus Bombus). *Apidologie* **2008,** *39* (1), 46–74.

Williams, P. H.; Ito, M.; Matsumura, T; Kudo, I. The Bumblebees of the Nepal Himalaya (Hymenoptera: Apidae). *Insecta Matsumurana* **2010,** *66*, 115–151.

Zurbuchen, A.; Landert, L.; Klaiber, J.; Müller, A.; Hein, S.; Dorn, S. Maximum Foraging Ranges in Solitary Bees: Only Few Individuals have the Capability to Cover Long Foraging Distances. *Biol. Conserv.* **2010,** *143* (3), 669–676.

CHAPTER 10

Review on Antlions (Neuroptera: Myrmeleontidae) of India

T. B. SURYANARAYANAN and BIJOY C.

Shadpada Entomology Research Lab, Department of Zoology, Christ College (Autonomous), Irinjalakuda, Thrissur, Kerala, India

ABSTRACT

Antlions belong to the order, Neuroptera, and family, Myrmeleontidae. In this work, a detailed review of existing works of diversity, ecology, taxonomy, distribution, seasonality, flight period, and endemism of the antlions of India are compiled. A total of 124 species belonging to 35 genera of antlions are reported from India. The percentage of antlions in each state and the number of antlion species in each genus of India are also provided.

10.1 INTRODUCTION

Antlions are the predominant member of the order, Neuroptera (Family, Myrmeleontidae) with 2138 species belonging to 299 genera (Machado et al., 2019). According to the recent classification by Machado et al. (2019), Myrmeleontidae currently comprises four subfamilies; Ascalaphinae, Myrmeleontinae, Dendroleontinae, and Nemoleontinae. Except Ascalaphinae, which includes owlflies, Myrmeleontinae, Dendroleontinae, and Nemoleontinae comprises antlions. Antlion gained its name because of the pit-building habit of some of its larval genera (Stange, 2002). The larvae

Insect Diversity and Ecosystem Services, Volume 1: Importance, Threats, Conservation and Economic Perspectives. Younis Ahmad Hajam, Sajad Hussain Parey, and Rouf Ahmad Bhat (Eds)

of antlion are voracious predators which will hide in sandy surfaces. After the larval stage, they make globular cocoons with sand grains in the pupal stage. Cocoon emerges in adults which are sexually mature insects. After a short period, the wings will be fully active for flight. Many of the adult antlions look like damselflies and dragonflies (order, Odonata), except for their clavate antennae and resting posture. There is a definite paucity of studies on Indian antlions with the exception of studies conducted by S. K. Ghosh. Earlier during the preindependent period, Navas, Westwood, Walker, and Banks contributed through their works on Indian antlions. Stange (2004) compiled the systematic catalog of the world antlions. In this work, a detailed review of existing works of Indian antlions (owlflies are not included in this work), their diversity, distribution, seasonality, and taxonomy are compiled. The details are updated based on the Oswald (2022) Neuroptera database.

10.2 DIVERSITY AND DISTRIBUTION OF ANTLIONS IN INDIA

Major antlion collections in India are opportunistic collections which are collected by non-taxonomists. Most of the antlion works in India after the postindependence period are done by Ghosh and Sen (1977), Ghosh (1977, 1980, 1981, 1983, 1984, 1987, 1990, 1991, 1997, 1998, 2000), Chandra and Sharma (2009), Chandra and Thilak (2009), Chandra et al. (2011), Sharma and Chandra (2012, 2013), Bhagat (2013), Chandra et al. (2014), Halder et al. (2018), Kaur et al. (2019), Singh et al. (2020), etc. Ghosh and Sen (1977) compiled the first detailed checklist of the antlion of India. According to them, the antlion diversity of India was 104 species belonging to 38 genera. Chandra and Sharma (2009) updated the checklist of antlions of India and it includes 123 species belonging to 30 genera. Some antlion species in the checklist got synonymized and some of the species in the same genera are separated into different genera. After those taxonomic changes, the current antlion diversity of India is 124 species belonging to 35 genera. The validity and status of each species are noted based on the Neuroptera database by Oswald (2022). The distribution of antlions of India in each state is represented in Table 10.1. Maharashtra is the state with the highest number of antlion species (29 species, nearly 24%) reported from India. The second most abundant states are Odisha and West Bengal (25 species). Gujarat, Kerala, Tripura, and Goa have

TABLE 10.1 Distribution and Percentage of Antlions in Each State of India.

Sl. No	State	Number of antlion species	Percentage (%)	Reference
1	Odisha	25	20.16	Ghosh, 1987
2	Himachal Pradesh	12	9.67	Ghosh, 1991
3	Maharashtra	29	23.38	Ghosh and Sen, 1977; Sharma and Chandra, 2012
4	Karnataka	7	5.64	Sharma and Chandra, 2013
5	Madhya Pradesh	16	12.90	Ghosh and Sen, 1977; Chandra et al., 2011
6	Meghalaya	3	2.14	Ghosh, 2000
7	West Bengal	25	20.16	Ghosh, 1998
8	Jammu and Kashmir	2	1.61	Bhagat, 2013
9	Rajasthan	12	9.67	Ghosh, 1977; Ghosh, 1981
10	Uttar Pradesh	7	5.64	Ghosh, 1983; Ghosh and Sen, 1977; Ghosh, 1984; Ghosh, 1987
11	Punjab	3	2.41	Ghosh and Sen, 1977
12	Andhra Pradesh	2	1.61	Ghosh and Sen, 1977; Ghosh, 1990
13	Tamil Nadu	12	9.67	Ghosh and Sen, 1977
14	Haryana	0	0	NA
15	Assam	8	6.45	Ghosh, 1984, 2000
16	Chhattisgarh	6	4.83	Ghosh and Sen, 1977; Chandra et al., 2014
17	Bihar	19	15.32	Ghosh and Sen, 1977; Ghosh, 1984
18	Gujarat	1	0.80	Ghosh and Sen, 1977
19	Kerala	1	0.80	Mathew, 1993
20	Jharkhand	0	0	NA
21	Uttarakhand	0	0	NA
22	Tripura	1	0.80	Ghosh, 2000
23	Telangana	4	3.22	Ghosh and Sen, 1977; Ghosh, 1990
24	Manipur	0	0	NA
25	Arunachal Pradesh	4	3.22	Ghosh and Sen, 1977; Ghosh, 2000
26	Mizoram	3	2.41	Ghosh and Sen, 1977; Ghosh, 2000
27	Goa	1	0.80	Ghosh, 1984, 1987
28	Nagaland	0	0	NA
29	Sikkim	8	6.45	Ghosh and Sen, 1977; Ghosh, 2000

sparse reports of antlion species (1 species) from India. Not a single antlion species has been reported so far from the states like Haryana, Jharkhand, Uttarakhand, Manipur, and Nagaland. This is not because of the absence of antlions in that particular state; it is due to a lack of detailed taxonomic studies on antlions.

10.3 FLIGHT PERIOD AND SEASONALITY OF ANTLIONS IN INDIA

There are 35 genera of antlions reported from India based on the details of from the available literature and database. It is listed in Table 10.2.

The flight period and seasonality of antlion genera in India are represented in Figure 10.1. This is prepared based on the collection details of antlions from the available literature. The flight period and seasonality data of some of the genera such as *Banyutus*, *Dendroleon*, *Epacanthaclisis*, *Iranoleon*, *Layahima*, *Macronemurus*, *Myrmecaelurus*, *Negrokus*, *Nuglerus*,

Palparellus, and *Solter* are unavailable. So, these genera are not included in Figure 10.1. A few genera were collected only in one or two particular months, while others, particularly *Distoleon* Banks (March, April, June, July, August, September, October, and November) were active virtually in 8 months. Genera such as *Nohoveus* Navás (October) and *Syngenes* Kolbe (June) have a single month of flight activity period. From the available information, the adult antlion flight period in India is mainly from March to November. Flight period and seasonality data of antlions will be helpful for future researchers for their field studies.

10.4 TAXONOMIC REVIEW ON ANTLIONS

Taxonomic studies on antlions will give the basic knowledge for future studies. Based on the Neuroptera database by Oswald (2022), some species in Chandra and Sharma (2009) checklist became invalid. Experts identified *Palpares infimus* (Walker, 1853), *Pinna nobilis* (Navas, 1931a), *Myrmeleon contractus* (Walker, 1860), *Myrmeleon (Myrmeleon) montanus* (Navas, 1914a), *Distoleon cubitalis* (Navas, 1914a), *Neuroleon (Neuroleon) roscidus* (Navás, 1937) as the synonyms (*P. infimus* = *P. nobilis* =

TABLE 10.2 Antlion Genera Reported from India.

Sl. No.	Genus
1	*Acanthaclisis* (Rambur, 1842)
2	*Baliga* (Navás, 1912d)
3	*Banyutus* (Navás, 1912a)
4	*Centroclisis* (Navás, 1909b)
5	*Chaetoleon* (Banks, 1920)
6	*Creoleon* (Tillyard, 1918)
7	*Cueta* (Navás, 1911)
8	*Delfimeus* (Navás, 1912b)
9	*Dendroleon* (Brauer, 1866)
10	*Distoleon* (Banks, 1910)
11	*Echthromyrmex* (McLachlan, 1867)
12	*Epacanthaclisis* (Okamoto, 1910)
13	*Gatzara* (Navás, 1915b)
14	*Geyria* (Esben-Peterson, 1920)
15	*Hagenomyia* (Banks, 1911a)
16	*Indoclystus* (Banks, 1941)
17	*Indoleon* (Banks, 1913)
18	*Indopalpares* (Rambur, 1842)
19	*Indophanes* (Banks, 1940)
20	*Iranoleon* (Holzel, 1968)
21	*Layahima* (Navás, 1912a)
22	*Macronemurus* (Costa, 1855)
23	*Myrmecaelurus* (Costa, 1855)
24	*Myrmeleon* (Linnaeus, 1767)
25	*Negrokus* (Navás, 1930a)
26	*Neuroleon* (Navás, 1909a)
27	*Nohoveus* (Navás 1918)
28	*Nuglerus* (Navás, 1912c)
29	*Palparellus* (Navás, 1912c)
30	*Palpares* (Rambur, 1842)
31	*Solter* (Navas, 1912a)
32	*Stenares* (Hagen, 1866)
33	*Stiphroneura* (Gerstaecker, 1885)
34	*Syngenes* (Kolbe, 1897)
35	*Tomatares* (Hagen, 1866)

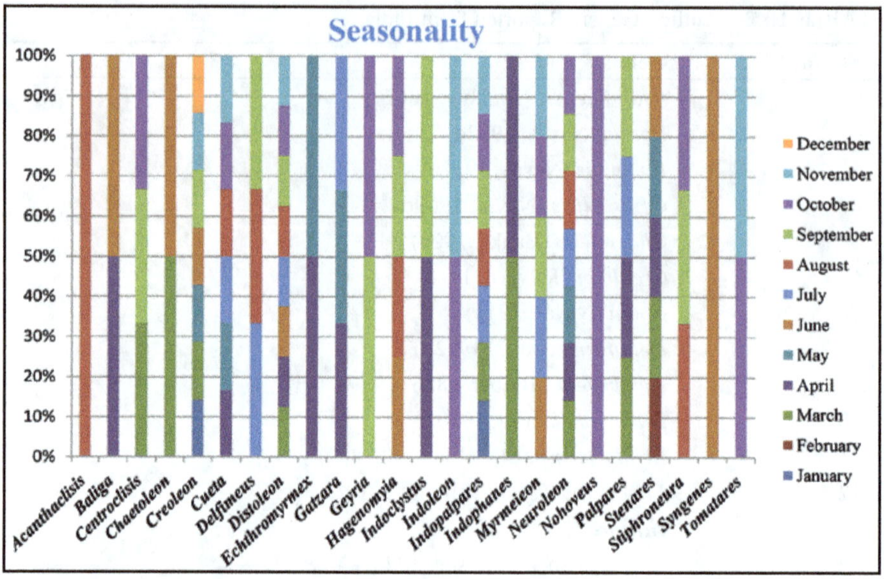

FIGURE 10.1 Seasonality of antlion genera in India.

Parauchenoglanis zebratus (Rambur, 1842), *M. contractus* = *Layahima contracta* (Walker, 1860), *M. montanus* = *Myrmeleon (Myrmeleon) trivialis* (Gerstaecker, 1885), *D. cubitalis* = *Banyutus cubitalis* (Navás, 1914a), *N. roscidus* = *Distoleon nefandus* (Walker, 1853)). Some antlion like *Myrmecaelurus afghanus* (Kimmins, 1950) has a taxonomic change which is now treated as *Iranoleon afghanus* (Kimmins, 1950). Another species such as *Myrmecaelurus implexus* (Walker, 1853) now treated as *Nohoveus implexus* (Walker, 1853), *Baliga sagax* (Walker, 1853) as *Hagenomyia sagax* (Walker, 1853), and *Indoleon audax* (Walker, 1853) as *Indophans audax* (Walker, 1853). *Chaetoleon, Nohoveus, Indophanes, Hagenomyia*, and *Iranoleon* are the newly recorded genera from India after the taxonomic changes. The number of species in each genus and its tribe, and subfamily is represented in Table 10.3. After those taxonomic changes, India has a total of 124 species belonging to 35 genera and 9 tribes of antlions.

TABLE 10.3 Number of Antlion Species in Each Genus.

Subfamily	Tribe	Genera	Species
Ascalaphinae	Palparini	*Palpares*	6
		Stenares	5
		Palparellus	1
		Tomatares	2
		Indopalpares	1
Dendroleontinae	Dendroleontini	*Dendroleon*	1
		Epacanthaclisis	1
		Gatzara	3
		Indoclystus	1
		Layahima	1
		Nuglerus	1
	Dimarini	*Echthromyrmex*	1
Nemoleontinae	Nemoleontini	*Indoleon*	2
		Macronemurus	2
		Indophanes	1
		Neuroleon	9
		Banyutus	2
		Creoleon	6
		Geyria	1
		Delfimeus	1
		Distoleon	19
		Negrokus	1
Myrmeleontinae	Myrmeleontini	*Baliga*	5
		Myrmeleon	19
		Hagenomyia	3
	Acanthaclisini	*Acanthaclisis*	1
		Centroclisis	5
		Stiphroneura	1
		Syngenes	2
	Myrmecaelurini	*Myrmecaelurus*	3
		Solter	2
		Iranoleon	1
		Nohoveus	1
	Nesoleontini	*Cueta*	12
	Brachynemurini	*Chaetoleon*	1
Total	**9**	**35**	124

10.4.1 BIOECOLOGY OF ANTLIONS

Ecology studies of antlion in India are mainly restricted to its pit-building larvae. Only some of the genera of the antlions such as *Myrmeleon* sp., *Baliga* sp., and *Hagenomyia* sp. show the pit-building habit (Stange, 2002). In India, the ecology work is restricted to the genus especially *Myrmeleon*. Recently, Misiriya et al. (2021) and Anila and Kakkassery (2022) worked on this genus to understand the larval behavior. Ecology of adult antlions needs to be studied in detail.

10.4.2 ENDEMISM AND CONSERVATION MEASURES

Antlion genus like *Indoclystus* is endemic to India while the remaining genera are reported from other countries. Out of 124 species of antlions, 51 species are endemic to India based on Oswald, 2022 database. It is represented in Table 10.4.

Many of these species need to be rediscovered from the country and should be thoroughly redescribed because a large number of species was not collected after their discovery. We have to take suitable conservation measures to protect the rare endemic antlions of India since 41.12% of the total antlion species from India are endemic.

The larval stages of major antlion species are voracious predators, so they play a vital role in decreasing the insect pest population. Due to the lack of knowledge of biology and ecology of antlions, its application as biological control agents is least studied. Intensive surveys on antlions will yield discoveries and new distributional reports to the Science. This will also establish the fact that the earlier described species have either been threatened or extinct. Hence further explorations including studies on biology, ecology, conservation, and taxonomical aspects are in high demand for this group.

ACKNOWLEDGMENTS

The authors are grateful to the Principal, Christ College (Autonomous), Irinjalakuda, Kerala for providing us with the facilities for undertaking this work. The first author offers sincere gratitude to CSIR, Government

TABLE 10.4 Endemic Species of Antlion from India.

S. No.	Genus	Species
1	*Baliga*	*Baliga jamduarensis* (Ghosh, 1984)
2		*Baliga montana* (Navás, 1930b)
3		*Baliga monticola* (Navás, 1937)
4	*Centroclisis*	*Centroclisis bandrensis* (Navás, 1934b)
5		*Centroclisis lutea* (Navás, 1912a)
6	*Creoleon*	*Creoleon fulvinervis* (Navás, 1932)
7		*Creoleon littoreus* (Navás, 1914a)
8	*Cueta*	*Cueta abdominalis* (Navás, 1930b)
9		*Cueta angulensis* (Ghosh, 1984)
10		*Cueta bolangirensis* (Ghosh, 1984)
11		*Cueta infensa* (Walker, 1853)
12		*Cueta levis* (Navás, 1931a)
13		*Cueta maindroni* (Navás, 1923)
14		*Cueta perpunctata* (Banks, 1931)
15		*Cueta punctulata* (Rambur, 1842)
16		*Cueta salai* (Navás, 1929)
17	*Distoleon*	*Distoleon bisoiensis* (Ghosh, 1984)w
18		*Distoleon cerdo* (Gerstaecker, 1894)
19		*Distoleon lebasinus* (Navás, 1931a)
20		*Distoleon levis* (Navás, 1914b)
21		*Distoleon marcida* (Banks, 1939)
22		*Distoleon pallidus* (Navás, 1934a)
23		*Distoleon pallipenne* (Banks, 1939)
24		*Distoleon plebejus* (Navás, 1914b)
25		*Distoleon pugnax* (Walker, 1853)
26		*Distoleon rhegmalis* (Navás, 1931a)
27		*Distoleon rhodocerus* (Navás, 1929)
28		*Distoleon umbratus* (Navás, 1930b)
29	*Gatzara*	*Gatzara benaci* (Navás, 1935)
30		*Gatzara insolita* (Walker, 1860)
31	*Indoclystus*	*Indoclystus singularis* (Westwood, 1847)
32	*Indoleon*	*Indoleon infestus* (Walker, 1853)
33	*Macronemurus*	*Macronemurus punjabensis* (Ghosh, 1982)
34		*Macronemurus trivittatus* (Banks, 1911b)
35	*Myrmeleon*	*Myrmeleon (Myrmeleon) fulvescens* (Navás, 1934b)

TABLE 10.4 *(Continued)*

S. No.	Genus	Species
36		*Myrmeleon (Myrmeleon) inanis* (Gerstaecker, 1894)
37		*Myrmeleon (Myrmeleon) indicus* (Navás, 1921)
38		*Myrmeleon (Myrmeleon) marginicollis* (Gerstaecker, 1894)
39		*Myrmeleon (Myrmeleon) mediatus* (Navás, 1931b)
40		*Myrmeleon (Myrmeleon) oberthuri* (Navás, 1923)
41		*Myrmeleon (Myrmeleon) ursinus* (Fabricius, 1798)
42		*Myrmeleon (Myrmeleon) periculosus* (Walker, 1853)
43	*Neuroleon*	*Neuroleon (Neuroleon) pallidus* (Banks, 1939)
44		*Neuroleon (Neuroleon) reticulatus* (Navás, 1930b)
45		*Neuroleon (Neuroleon) apicalis* (Navás, 1915a)
46		*Neuroleon (Neuroleon) guptaii* (Ghosh, 1984)
47		*Neuroleon (Neuroleon) unpunctatus* (Ghosh, 1981)
48	*Palpares*	*Palpares tigroides* (Walker, 1860)
49	*Stenares*	*Stenares acutus* (Ghosh, 1990)
50		*Stenares frazeri* (Banks, 1931)
51	*Syngenes*	*Syngenes palpalis* (Banks, 1931)

of India, for financial support in the form of the CSIR Junior Research Fellowship (08/376(0010)/2019-EMR-I).

KEYWORDS

- **neuroptera**
- **myrmeleontidae**
- **antlion**
- **India**

REFERENCES

Anila, K.; Kakkassery, F. K. Physio-Chemical Properties of Pit Soil of antlion larvae (Genus *Myrmeleon* sp.) with Special Reference to its Population Density. *Int. J. Entomol. Res.* **2022,** 7 (1), 81–85.

Banks, N. Myrmeleonidae from Australia. *Ann. Entomol. Soc. Am.* **1910,** *3*, 40–44. [BotN ref#45] https://doi.org/10.1093/aesa/3.1.40

Banks, N. Notes on African Myrmeleonidae. *Ann. Entomol. Soc. Am.* **1911a,** *4*, 1–31. [BotN ref#52] https://doi.org/10.1093/aesa/4.1.1

Banks, N. Notes on Indian Neuropteroid Insects. *Proc. Entomol. Soc. Washington* **1911b,** *13*, 99–106. [BotN ref#51]

Banks, N. Synopses and Descriptions of Exotic Neuroptera. *Trans. Am. Entomol. Soc.* **1913,** *39*, 201–242. [BotN ref#59]

Banks, N. New Neuropteroid Insects. *Bull. Mus. Comp. Zool.* **1920,** *64*, 297–362. [BotN ref #73]

Banks, N. Some Oriental Neuropteroid Insects. *Psyche* **1931,** *38*, 56–70. [BotN ref #86]

Banks, N. New Genera and Species of Neuropteroid Insects. *Bull. Mus. Comp. Zool.* **1939,** *85*, 439–504. [BotN ref#103]

Banks, N. Report on Certain Groups of Neuropteroid Insects from Szechwan, China. *Proc. U. S. Natl. Mus.* **1940,** *88*, 173–220. [BotN ref#106] https://doi.org/10.5479/si.00963801.88-3079.173

Banks, N. Some New and Interesting Neuroptera in the American Museum of Natural History. *Am. Mus. Novit.* **1941,** *1143*, 1–5. [BotN ref #110]

Bhagat, R. C. Biodiversity and Systematic Checklist of Neuroptera (Insecta: Neuropterida) of Jammu & Kashmir State (India*). Indian J. Fundam. Appl. Life Sci.* **2013,** *3*, 14-16. [BotN ref #16527]

Brauer, F. In *Reise der Osterreichischen Fregatta Novara um die Erde, in den jahren 1857, 1858, 1859, unter den befehlen des Commodore B. von Wüllerstorf-Urbair, Zoologischer Theil*; von Scherzer, H. K., Ed.; *Bd. 2, No. 4 (Neuroptera). K. Gerold, Wien*, 1866; p 104. [BotN ref #1682]

Chandra, K.; Majumder, A.; Raha, A.; Halder, S. New Records of Antlions (Neuroptera: Myrmeleontidae) from Chhattisgarh, India. *Entomol. Appl. Sci. Lett. J.* **2014,** *1* (4), 70–74. [BotN ref #16168]

Chandra, K.; Sharma, R. M. In *Checklist of Indian Neuropterids (Insecta: Megaloptera; Raphidioptera; Neuroptera);* Zoological Survey of India: Central Zone Regional Centre Jabalpur, Madhya Pradesh, 2009; p 22. [BotN ref#13800]

Chandra, K.; Thilak, J. Insecta: Neuroptera. In *Fauna of Bandhavgarh Tiger Reserve (Madhya Pradesh)*; Director, Zoological Survey of India, Ed.; Conservation Area Series; Zoological Survey of India: Kolkata, 2009; vol 40; pp 71–73. [BotN ref #14496]

Chandra, K.; Thilak, J.; Sidhu, A. K. Neuroptera. In *Fauna of Madhya Pradesh (including Chhattisgarh)*; Meeting: State Fauna Series, Zoological Survey of India, Part 3; Director, Zoological Survey of India; Ed.; Zoological Survey of India: Kolkata, 2011; vol 15, pp 151–165. [BotN ref #16164]

Costa, A. In *Ricerche entomologiche sopra I monti Partenii*; Stamperie e Calcografia, Napoli, 1855; p 29. [BotN ref#7873]

Esben-Petersen, P. Revision of Some of the Type-Specimens of Myrmeleonidae, Described by Navas and Placed in the Vienna Museum. *Annales de la Société Entomologique de Belgique* **1920,** *60*, 190–196. [BotN ref#146]

Fabricius, J. C. In *Supplementum Entomologiae systematicae*; impensis Chr. G. Proft.: Hafniae, 1798; p 572. [BotN ref#2312] https://doi.org/10.5962/bhl.title.125869

Gerstaecker, A. Zwei fernere Decaden Australischer Neuroptera Megaloptera. In *Mitt[h]eilungen aus dem Naturwissenschaftlichen Verein für Neu-Vorpommern und Rugen*, 1885; vol *16*, pp 84–116. [BotN ref#2557]

Gerstaecker, A. Über neue und weniger gekannte Neuropteren aus der familie Megaloptera Burm. In *Mitt[h]eilungen aus dem Naturwissenschaftlichen Verein für Neu-Vorpommern und Rugen*, 1894; vol *25*, pp 93–173. [BotN ref#2559]

Ghosh, S. K. Fauna of Rajasthan, India Neuroptera. *Rec. Zool. Surv. India* **1977,** *72*, 309–313. [BotN ref#2573]

Ghosh, S. K. New Record of the Species *Indophanes barbarus* (Walker) [Myrmeleontidae: Neuroptera] from India. *Bull. Zool. Surv. India* **1980,** *3*, 119. [BotN ref #2578]

Ghosh, S. K. On a Small Collection of Neuroptera from Andaman and Nicobar Islands. *Rec. Zool. Surv. India* **1980,** *77*, 247–254. [BotN ref#2579]

Ghosh, S. K. New Records of Neuroptera Planipennia from the Laccadive Islands, India, Coupled with a Short Note on the Taxonomic Variation of *Centroclisis eustalacta* (Gerst.). *Bull. Zool. Surv. India* **1981,** *3*, 269–270. [BotN ref#2581]

Ghosh, S. K. On New and Little-Known Species of Planipennia (Order Neuroptera) from India. *Bull. Zool. Surv. India* **1981,** *4*, 131–139. [BotN ref#2583]

Ghosh, S. K. A New Species of *Macronemurus costa* (Myrmeleonidae: Planipennia, Neuroptera) from India. In *Entomon: Internationale Zeitschrift für die gesamte Insektenkunde*, 1982; vol *7*, pp 111–113. [BotN ref#2584]

Ghosh, S. K. Contribution to the Taxonomical Studies of Neuroptera (Suborder Planipennia) from Eastern India. 1. Family Myrmeleontidae. In *Records of the Zoological Survey of India*; Miscellaneous Publications, Occasional Paper, 1984; vol *52*, pp 1–63. [BotN ref#2587]

Ghosh, S. K. Insecta: Neuroptera: Planipennia: Myrmeleontidae: A Check-List. In *Fauna of Orissa*; Part 1, State Fauna Series, No. 1; Lamba, B. S., Ed.; Zoological Survey of India: Calcutta, 1987; pp 183–192. [BotN ref#8904]

Ghosh, S. K. Description of an Unknown Female Along With Two New Locality Records of the Family Myrmeleontidae (Neuroptera) from Lakshadweep Islands, India. *Rec. Zool. Surv. India* **1990,** *87*, 259–261. [BotN ref#9487]

Ghosh, S. K. On New and Little-Known Species of *Stenares* Hagen (Myrmeleontidae: Neuroptera) from India. *Rec. Zool. Surv. India* **1990,** *86*, 453–459. [BotN ref#9485]

Ghosh, S. K. Interesting Species of Neuroptera from Himachal Pradesh. *Rec. Zool. Surv. India* **1991,** *89*, 89–94. [BotN ref#9490]

Ghosh, S. K. Neuroptera. In *Fauna of Nanda Devi Biosphere Reserve: A World Heritage Site* Fauna of Conservation Areas; Alfred, J. R. B., Ed.; Zoological Survey of India: Calcutta, 1997; vol *9*, p 166. [BotN ref#10046]

Ghosh, S. K. Faunal Diversity in India: Neuroptera. In *Faunal Diversity in India: A Commemorative Volume in the 50th Year of India's Independence*; Alfred, J. R. B., Das, A. K., Sanyal, A. K., Ed.; Zoological Survey of India: Calcutta, 1998; p 497. [BotN ref#10047]

Ghosh, S. K. Insecta: Neuroptera. In *Fauna of Meghalaya, Part 7 (Insecta)*; Director, Zoological Survey of India; Ed.; Zoological Survey of India, 2000; pp 81–115. [BotN ref#13378]

Ghosh, S. K. Neuroptera Fauna of North-East India. In *Records of the Zoological Survey of India, Occasional Paper*, 2000; vol *184*, pp i–xiii + 1–179. [BotN ref#11664]

Ghosh, S. K.; Sen, S. Check-list of Indian Planipennia (Order Neuroptera). *Rec. Zool. Surv. India* **1977,** *73* (1–4), 277. [BotN ref#2592]

Hagen, H. A. Hemerobidarum Synopsis synonymica. *Stettiner Entomologische Zeitung* **1866,** *27*, 369–462. [BotN ref#460]

Halder, S.; Mukherjee, T.; Gupta, D.; Chandra, K. Insecta : Neuroptera. In *Faunal Diversity of Indian* Himalaya; Director, Zoological Survey of India, Ed., 2018; pp 591–596.

Hölzel, H. Zur Kenntnis der Myrmeleoniden des Iran (Planipennia, Myrmeleonidae). *Stuttgarter Beiträge zur Naturkunde* **1968,** *181*, 1–32. [BotN ref#2960]

Kaur, S.; Pandher, M. S.; Chandra, K. Description of the Male of *Distoleon sambalpurensis* Ghosh, 1984 (Neuroptera: Myrmeleontidae) from India. *Zootaxa* **2019,** *4661* (3), 587–593. https://doi.org/10.11646/zootaxa.4661.3.11

Kimmins, D. E. The 3rd Danish Expedition to Central Asia. Zoological Results 4. Odonata, Ephemeroptera and Neuroptera (Insecta) from Afghanistan. *Videnskabelige Meddelelser fra Dansk Naturhistorisk Forening, Kobenhaven* **1950,** *112*, 235–241. [BotN ref#3278]

Kolbe, H. J. Neuropteren. Die Netzflügler. In *Die Thierwelt [Deutsch-] Ost-Afrikas und der Nachbargebiete*; *Band IV (Wirbellose Thiere)*; Möbius, K. A., Ed.; Dietrich Reimer: Berlin, 1897; pp 1–42. [BotN ref#3434]

Linnaeus, C. Systema natura per regna tria naturae secundum classes, ordines, genera, species, cum characteribus, differentiis, synonymis, locis; Tom. I, Pars II; 12th Ed.; Laurentii Salvii: Holmiae, 1767; p 795. [BotN ref#3733] https://doi.org/10.5962/bhl.title.157601

Machado, R. J.; Gillung, J. P.; Winterton, S. L.; Garzón-Orduña, I. J.; Lemmon, A. R.; Lemmon, E. M.; Oswald, J. D. Owlflies are Derived Antlions: Anchored Phylogenomics Supports a New Phylogeny and Classification of Myrmeleontidae (Neuroptera). *Syst. Entomol.* **2019,** *44* (2), 418–450. [BotN ref#18370] https://doi.org/10.3897/dez.64.11704

McLachlan, R. New Genera and Species, &c., of Neuropterous Insects; and a Revision of Mr. F. Walker's British Museum Catalogue of Neuroptera, part ii. (1853), as Far as the End of the Genus *Myrmeleon. J. Linnean Soc. London Zool.* **1867,** *9*, 230–281. https://doi.org/10.1111/j.1096-3642.1867.tb00195.x [BotN ref #342]

Misiriya, A. M.; T. B. Suryanarayanan; Bijoy C. Feeding Activity of Antlion of Myrmeleon sp. in Different Soil Samples from Thrissur District, Kerala; *Bugs R All #220*; *Zoo's Print.*, 2021; vol *36* (12), pp 29–34.

Navás, L. Neurópteros nuevos de la fauna ibérica. In *Actas y Memorias del Primer Congreso de Naturalistas Españoles (held in Zaragoza, October 1908)*; Zaragoza, **1909a;** vol *1909*, pp 143–158. [BotN ref#495]

Navás, L. Notas neuropterológicas. XI. Mirmeleónido nuevo de Madagascar, Series 1. *Butlletí de la Institució Catalana d'Història Natural* **1909b,** *9*, 71–72. [BotN ref#498]

Navás, L. Notes sur quelques Névroptères. I. In *Insecta*; Rennes, 1911**;** vol *1*, pp 239–246. [BotN ref#536]

Navás, L. Notas sobre Mirmeleónidos (Ins. Neur.). *Brotéria (Zoológica)* **1912a,** *10* (29–75), 85–97. [BotN ref#555]

Navás, L. Notes sur quelques Névroptères. IV. *Névroptères d'Algérie recueillis par M. le Baron P. Surcouf. Insecta, Rennes*, 1912b; vol 2, pp 217–228. [BotN ref#558]

Navás, L. Myrméléonides nouveaux ou peu connus (Ins. Névr.). *Annales de la Societe Scientifique de Bruxelles* **1912c,** *36*, 203–248. [BotN ref#551]

Navás, L. Myrméléonides nouveau de l'extréme Orient (Neuroptera). *Revue Russe d'Entomologie* **1912d,** *12*, 110–114. [BotN ref#552]

Navás, L. Myrméléonides (Ins. Névr.) nouveaux ou critiques [I]. *Annales de la Société Scientifique de Bruxelles* **1914a,** *38*, 229–254. [BotN ref#601]

Navás, L. Neuroptera asiatica, III Series. *Revue Russe d'Entomologie [=Русское Энтомологическое Обозрение; =Russkoe Entomologicheskoe Obozrenie]* **1914b,** *14*, 6–13. [BotN ref#603]

Navás, L. Neurópteros nuevos o poco conocidos, Cuarta [IV] Serie. *Memorias de la Real Academia de Ciencias y Artes de Barcelona* **1915a,** *11* (3), 373–398. [BotN ref#638]

Navás, L. Neurópteros nuevos o poco conocidos, Quinta [V] Serie. *Memorias de la Real Academia de Ciencias y Artes de Barcelona* **1915b,** *11* (3), 455–480. [BotN ref#639]

Navás, L. Quelques Névroptéres de l'Algerie; *Insecta, Rennes*, 1918; vol *8*, pp 167–176. [BotN ref#709]

Navás, L. Comunicaciones entomológicas. 4. Insectos exóticos nuevos, críticos o poco conocidos. *Revista de la [Real] Academia de Ciencias Exactas Fisico-Quimicas y Naturales de Zaragoza* **1921,** *6* (1), 61–81. [BotN ref#725]

Navás, L. Insecta Orientalia, I Series. In *Memorie dell'Accademia Pontifica dei Nuovi Lincei, Rome*, Series 2, 1923; vol *6*, pp 29–41. [BotN ref#763]

Navás, L. Comunicaciones entomológicas 10. Insectos de la India. Serie 1. *Revista de la Academia Ciencias Exactas Fisico-Quimicas y Naturales de Zaragosa* **1929,** *12*, 177–197. [BotN ref#846]

Navás, L. Insecta Nova, Series XV. In *Memorie dell'Accademia Pontifica dei Nuovi Lincei, Rome*, Series 2, 1930a; vol *14*, pp 409–418. [BotN ref#879]

Navás, L. Comunicaciones Entomológicas, 12. Insectos de la India 2.a serie. *Revista de la Academia Ciencias Exactas Fisico-Quimicas y Naturales de Zaragosa*, Series 1, 1930b; vol *13*, pp 29–48. [BotN ref#873]

Navás, L. Comunicaciones Entomológicas, 13. Insectos de la India. 3. a serie. *Revista de la [Real] Academia de Ciencias Exactas Fisico-Quimicas y Naturales de Zaragoza* 1931a, *14* (1), 74–92. [BotN ref#891]

Navás, L. Décadas de insectos nuevos, Decada 1. *Revista de la Real Academia de Ciencias Exactas Fisicas y Naturales de Madrid* **1931b,** *26*, 60–69. [BotN ref#892]

Navás, L. Comunicaciones entomológicas, 14. Insectos de la India. 4.a serie. *Revista de la [Real] Academia de Ciencias Exactas Fisico-Quimicas y Naturales de Zaragoza* **1932,** *15* (1), 11–41. [BotN ref#908]

Navás, L. Comunicaciones entomológicas, 16. Insectos de la India. Quinta [V] serie. *Revista de la [Real] Academia de Ciencias Exactas Fisico-Quimicas y Naturales de Zaragoza* **1934a,** *17* (1), 29–48. [BotN ref#946]

Navás, L. Insecta Orientalia, XIII Series. *Memorie dell'Accademia Pontifica dei Nuovi Lincei, Rome* **1934b,** *1* (3), 217–228. [BotN ref#951]

Navás, L. Neurópteros exóticos, 2. a serie. *Memorias de la Real Academia de Ciencias y Artes de Barcelona* **1935,** *25* (3), 37–59. [BotN ref#968]

Navás, L. Myrméléonidés nouveaux. In *Comptes Rendus XIIe Congres International de Zoologie*; Meeting: 1935, Lisbon, Portugal, 1937; vol *2*, pp 1472–1479. [BotN ref#986]

Okamoto, H. Die Myrmeleoniden Japans. *Wiener Entomologische Zeitung* **1910,** *29*, 275–300. [BotN ref#4978]

Oswald, J. D. Neuropterida Species of the World. Lacewing Digital Library, Research Publication No. 1, 2022. http://lacewing.tamu.edu/SpeciesCatalog/Main. (Accessed 21 Feb, 2022.

Rambur, M. P. In *Histoire naturelle des Insectes: Nevropteres*; Fain et Thunot: Paris, 1842; p 534. [BotN ref#5314]

Sharma, R. M.; Chandra, K. Insecta: Neuroptera. In *Fauna of Maharashtra*; Director, Zoological Survey of India, Ed.; State Fauna Series; Zoological Survey of India: Kolkata, 2012; vol *20* (2), pp 485–488. [BotN ref#16165]

Sharma, R. M.; Chandra, K. Insecta: Neuroptera. In *Fauna of Karnataka*; Director, Zoological Survey of India, Ed.; State Fauna Series;Zoological Survey of India: Kolkata, 2013; vol 21, pp 163–165. [BotN ref#16166]

Singh, L. R. K.; Ahmed, I.; Chandra, K.; Gupta, D. Insecta: Neuroptera and Megaloptera. In *Faunal Diversity of Biogeographic Zones of India: Western Ghats*, Published by the Director, Zool. Surv.: India, Kolkata, 2020; pp 501–508.

Stange, L. A. Family Myrmeleontidae. in Penny, N. D. (ed.). A Guide to the Lacewings (Neuroptera) of Costa Rica. *Proc. Calif. Acad. Sci.* **2002,** *53* (4), 275–289 (text), 433–446 (figures).

Stange, L. A. A Systematic Catalog, Bibliography and Classification of the World Antlions (Insecta: Neuroptera: Myrmeleontidae). In *Memoirs of the American Entomological Institute*, 2004; vol *74*, pp 1–565. [BotN ref#11168]

Tillyard, R. J. Mesozoic Insects of Queensland. No. 3. Odonata and Protodonata. *Proc. Linnean Soc. New South Wales* **1918,** *43*, 417–436. [BotN ref#318]

Walker, F. In *Catalogue of the Specimens of Neuropterous Insects in the Collection of the British Museum,* Part II (Sialidae- Nemopterides); Newman: London, 1853; p 284. [BotN ref#6194]

Walker, F. Characters of Undescribed Neuroptera in the Collection of W.W. Saunders. *Trans. Entomol. Soc. London* **1860,** *10*, 176–199. [BotN ref#6195] https://doi.org/10.1111/j.1365-2311.1860.tb01844.x

Westwood, J. O. In *The Cabinet of Oriental Entomology: Being a Selection of the Rarer and More Beautiful Species of Insects, Natives of India and the Adjacent Islands*. The greater Portion of which are now, for the First Time, Described and Figured; Smith: London, 1847; p 88 . [BotN ref#6259] https://doi.org/10.5962/bhl.title.34273

CHAPTER 11

Industrial Entomology: Progress and Economic Perspectives

RAMAN JASROTIA, MENAKSHI DHAR, SEEMA LANGER, and NEHA JAMWAL

Department of Zoology, University of Jammu, Jammu, Jammu & Kashmir, India

ABSTRACT

Large number of insects are known to colonize the earth; however, the percentage of harmful insects is less than 1%. Insects are known to play a pivotal role in the field of pharmaceuticals, nutraceuticals, food, and therapies. They play an extensive role in remodeling the lives of humans. Insects such as honeybees, silkworms, lac insects, and scale insects are commercially beneficial and so are used on large scale. Industrial entomology has a great role in the extraction of proteins, vitamins, minerals, and chitin from insects so that they can be used in the food sector after proper processing. Apiculture, a small-scale industry, is embraced by a large number of people across the globe. Bees are used as micromanipulators that gather nectar from the flowers to produce honey. Industrial entomology is the research-based management of natural events that includes the gamut areas of research and technology. Sericulture, an agro-industry, plays a tremendous role in the rural economy. The new perspectives offer the utilization of the products obtained from the insects in the sustenance of life, biomedical, and bioengineering fields. These rural cottage and agro-based industries generate employment for the people of such areas. Lac-industry, a forest-based industry, has a significant role in the

Insect Diversity and Ecosystem Services, Volume 1: Importance, Threats, Conservation and Economic Perspectives. Younis Ahmad Hajam, Sajad Hussain Parey, and Rouf Ahmad Bhat (Eds)

economy. There are profuse applications of the lac and it has become the paramount item of export. Insects are promising assets for the production of novel antibiotics, as antibiotic-resistant human pathogens are currently on the rise. The mass rearing and the merchandising of useful insects are expanding gradually.

11.1 INTRODUCTION

Insects are the dominant form of life on the Earth and these are the most diverse and numerous among the invertebrate classes. They play a critical role in sustaining the functions of an ecosystem and furnish indispensable ecological services to mankind. Insects are of great economic importance owing to their high diversity, role in the ecosystem, and their impact on agriculture and human health. Insects are found well-nigh in every conceivable habitat. They have appeared on Earth approximately 390 million years back and since then these insects have diversified into various species. Insects are the dominant forms of life on Earth. According to new research findings, it has been estimated that there are about 5.5 million insect species on Earth, among which 1 million species have been identified and that represent 20% of the total species. Insects comprise 90% of the total animal species in the world (Chapman, 2011). They render valuable ministrations to humanity along with the ecosystem to a large extent. The most diversified group of organisms is that of insects and the number of insects is more than any other animal group. Insects epitomize 80% of the species recorded across the globe. Approximately, there are 1.4 billion insects for every human on this planet according to the survey conducted by National Geographic and all of them have a crucial place in an ecosystem. Insects provide a lot of ecological services and so they are known as the level pullers of the world. They form an important source of food for various species of animals. Trout, bass, and breams feed on mayflies and stoneflies. Insects have remained uncredited for their function in the food web. These creatures play an important role as decomposers, help in recycling the animal and plant wastes, and thus keep the ecosystem balanced and healthy. Museum workers utilize this behavior of carpet beetles to immaculate the skeletons kept in the museums (Malti, 2017). Insects form a part of the nutritious diet of various insectivorous birds such as bluebirds, nighthawks, swallows, warblers, and woodpeckers. Besides birds, numerous reptiles and amphibians are also insectivorous.

Human civilizations have been perpetuated by the multifarious services rendered through the useful insects; they help in the natural regulation of insect pests and produce certain beneficial products. Insect pollinators play an important role in the reproduction of various crops and flowering plants. Many insects help in the eradication of unwanted weeds by feeding on them. Certain insects prove beneficial for maintaining the quality of the soil. Insects such as dung beetles make tunnels in the soil and this way they improve the aeration in the soil. The use of insects and their products on an industrial scale for the welfare of mankind is known as industrial entomology. The scale insect, *Dactylopius* found on the Mexican cacti produces carmine that is used to dye foods, cosmetics, textiles, and drugs. Shellac, obtained from the lac insect, *Lacciferlacca* (Homoptera) is important among the naturally occurring resins. Paramount insects of industrial importance are silkworm, (order-Lepidoptera), honey bee (Hymenoptera), and lac insect (Hemiptera). Brain surgery is being revolutionized by tumor paint, a kind of venom obtained from scorpions. Beeswax and lacwax are used on commercial scale. Chemicals extracted from ants, maggots, and bees are of great use in the medical field. Cantharidin obtained from the blister beetles has been used to treat warts and it also has antitumor and aphrodisiac properties. Resistance developed toward the antibiotics is being resolved by the use of cockroaches. Crickets and insect larvae are used as fishing bait. Insects such as locusts, grasshoppers, beetles, grubs, caterpillars, ants, and wasps are the rich source of vitamins, minerals, and proteins. About 80% of the trees and bushes present on the Earth's surface are pollinated by insects. More than 150 species of plants take advantage of the insects for the dispersal of seeds over long distances. The economic significance of the insects is indefinable; they play an important role in biological pest control. In the United States, the biological control of pest species is valued at 5.4 billion dollars per year. Dung beetles and termites help in the decomposition of animal and plant matter in the agricultural fields and thus release nutrients that are utilized by the plants. Insects are of crucial significance by the virtue of their diverseness, abundance, and their role in the ecosystem (Scudder, 2017). Insects that feed on corpses have shown increased importance in the field of forensic entomology. Such benefits provided by the insects have been acclaimed as a kind of industrial application of the insects for the use by humans.

11.2 INDUSTRIAL APPLICATION

The utility of insects in the industries have been broadly categorized into four classes—use for insect pest management; use for the production of useful products; use for the production of honey, silk, and lac; and use of insects for their specific properties (Fig. 11.1).

11.2.1 USE IN INSECT PEST MANAGEMENT

The main objective of insect pest management is to support the balance of an ecosystem and keep it healthy by turning down the application of pesticides in crops. Worldwide, the use of pesticides has shown a great increase to 3.5 billion kg per year since the last 20 years, thus adding up to 45 billion dollars to the world market; this cost can be reduced by utilizing the individual insects against various pests. The technique of using natural enemies for pest management is in use since early times. During the past few years, sundry industries have been set up that mass produce umpteen natural enemies for insect pest management. By and large, spider mites and parasitoid wasps are produced in large numbers in laboratories for the management of insect pests.

A Brazilian company, named Bug AgentesBiologicos or Bug, holds a leading position in the biological pest control industry, and by this way, they are proceeding toward reducing the use of chemicals in the agricultural sector. This company was set up in Sao Paulo, in 2000, by Diogo Rodrigues Carvalho and Heraldo Negri de Oliveira. They developed a technology for the mass production of a tiny wasp, which when sprayed over the fields prey upon caterpillars and other insect pests. Various insects employed for biological pest control are represented in Table 11.1.

11.2.2 USE FOR THE PRODUCTION OF USEFUL PRODUCTS

11.2.2.1 USE IN THE FOOD INDUSTRY

Insects are a part of human food as their nutritional value has been widely accepted all over the world. Raheem et al. (2018) reported that the indigenous people of Asia, South America, Africa, and Europe consumed various insect species for hundreds of years. Various species of insects that

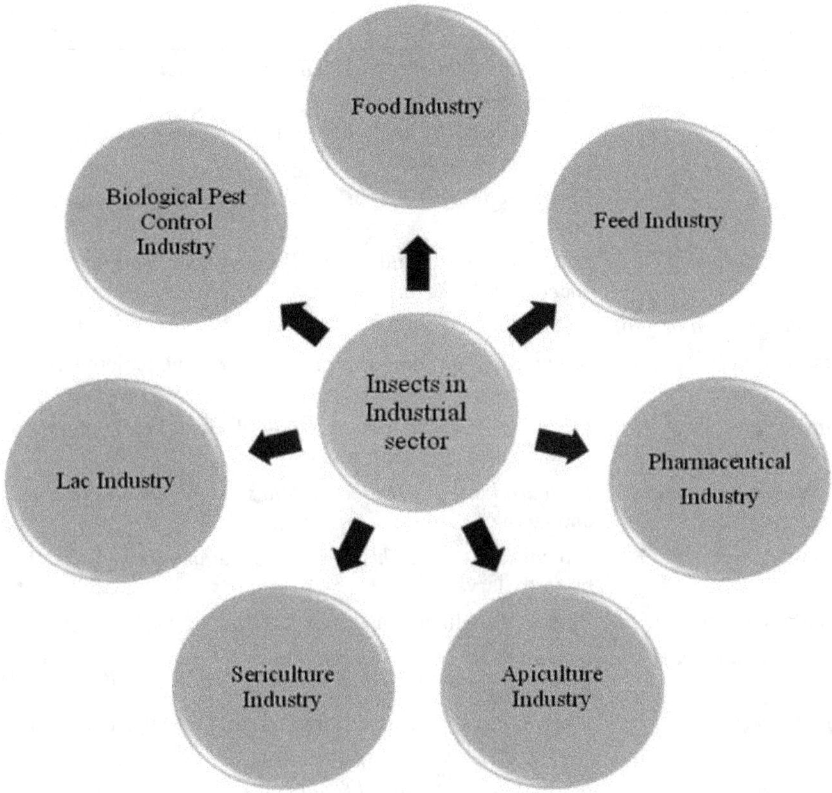

FIGURE 11.1 Role of insects in various industries.

are habitually consumed comprise of beetles, bees, ants, crickets, grasshoppers, locusts, and caterpillars. According to Yen (2009), roundabout 2000 species of insects are eaten up in at least 113 countries across the globe. Protein derived from the insect species has more valuation in the market than other sources of protein (Dobermann et al., 2017). In some countries, insects are readily available in high-class restaurants, cafeterias of the schools, and open markets, thus establishing the profit-making business. The establishment of the edible insect industry requires the support of governmental and non-governmental organizations as they can aware the masses about the nutritional value of the novel food products, developed by the use of insects. All over the world, in 2018, the edible insect

TABLE 11.1 Insects Used for Biological Pest Control.

Insects	Preys on
Coccinellasps.	*Aphis spp., Trialeurodesabutilonea, Leptinotarsa decemlineata,* and mites
Mantis religiosa	Larva of butterfly and moths, midges, Orthopteran, and Coleopteran insects
Spiders	*Cimex lectularius*, aphids, roaches, Orthopteran, and dipteran insects
Calleida decora	Larva of butterfly and moths, *Formica spp., Leptinotarsa decemlineata, Agrotisipsilon,* and *Ariolimax*
Aphidoletesaphidimyza	Aphids
Atanycolus spp.	*Manduca sexta, Manduca quinquemaculata,* lepidopteran larva, and sap-sucking insects
Nabis spp.	*Leptinotarsa decemlineata*, *Trichoplusiani*, Lepidopteran larvae, mites, and sap-sucking insects
Chrysoperlacarnea	*Dialeurodes spp., Graphocephala spp., Pseudococcus spp.*, and larva of moths
Orius	*Tetranynchus spp.*, *Thrips spp*., eggs of insects, moth, and butterfly larvae
Chauliognathuslugubris	Eggs of grasshopper and sap-sucking insects
Trichopodapennipes	*Lymantria dispar, Popillia japonica , Agrotisipsilon,* and *Anasa tristis*
Allograptaobliqua	*Dactylopius spp.*, leaf sucking insects, butterfly, and moth larvae
Cryptolaemus montrouzieri	Some species of genus *Pseudococcus*
Phytoseiulus persimilis	*Tetranychus spp.*
Geocoris	*Nephotettixspp.*, insect bugs, and spider mites
Meteorusautographae	Larval forms of beetles and flies, eggs of moths, pupae of various insects, and various adult insects

market was 406 million USD and it is expected to increase to 1.8 billion USD by 2023 (Ahuja and Mamtani, 2020). To ameliorate the technique of insect processing, various strategies for their production and storage are burgeoning. Various technologies used in insect processing industries are freeze-drying, oven drying, dry fractionation, freezing, marination, and fermentation. To enhance the shelf-life of food products that are derived from insects, numerous methods are utilized in the industries. Bees, silk-worms, and cochineals are considered as fully domesticated species. The edible insect industry is advancing and the want for new products is also increasing. Mass production begins with the post-harvesting of raw insects and terminates with the formation of foodstuffs and other by-products. The by-products produced are released and these are then re-used (Raheem et al., 2019). Recently, three-dimensional food printing techniques have been employed to yield insect foodstuffs. The consumption of insects began about 700 years ago (Ramdos-Elorduy, 2009). There are about 2300 species of insects belonging to 18 orders that are known to be edible. The use of insects in the food industry is to be increased as the production of animal protein is facing a huge crisis with an increasing population. Besides the production of the silk from the silkworm, their pupae are also utilized as food product in several countries of the world. Farming and processing of insects need to be systematized to certify the quality of the products obtained from the insects. Bodo tribe in Assam consumes a large number of insects such as termites, grasshoppers, beetles, and caterpillars every day by just drying them up in the sun. Insect meat is more economical and environment friendly than the meat of any other animal, when cultivated on an industrial scale. According to Food and Agriculture organization (FAO), at present, insects form the diet of about two billion people. A silk federation in Madhya Pradesh, India has been mass-producing canned silkworm pupae for human consumption.

The use of edible insects in the food industry is increasing rapidly as they are excellent sources of nutrients. They have high protein content than the other conventional food sources. According to Srivastava (2009), the protein content (in gm) of giant water beetle, red ant, silkworm pupae, mealworms, fly larvae, cricket, caterpillar, termite, and weevil are 19.8, 13.8, 9.6,20.27, 15.58, 21.32, 6.7, 14.2, and 6.7, respectively. Various types of carbohydrates such as triose, glycogen, fructose, and ketoheptose are present in edible insects. Total carbohydrate content varies from 1% to 10%. Total carbohydrate content (in gm) of giant water beetle, red ant, silkworm pupae, cricket is 2.1, 2.9, 2.3, and 5.1, respectively. The carbohydrate content of the lychee giant stink bug are 0.15%. Chitin, the most important component of the cell wall of insects, when consumed helps in boosting immunity and also helps in the treatment of high blood pressure. Chitin content varies between 15% and 18% in the body of an edible insect. Insects are considered as an important source of lipids, proteins, minerals, and vitamins. Larvae and pupae stages of an insect have higher fat content than their adult form. Edible insects have a higher amount of unsaturated fatty acids and palmitic acid. The insects of order-Lepidoptera have higher linolenic acid and that of Coleoptera have higher oil acid content (Yin et al., 2016). The protein content of insects such as redants, predaceous diving beetle, and the grasshopper is higher than pork, beef, chicken, fish, and eggs.

Approximately 255 species of edible insects are so far on the record from different states of India. The highest consumption of insects, i.e., about 158 species are consumed as a part of food in Arunachal Pradesh, followed by Manipur, Nagaland, and Assam where almost 16–40 species are considered as edible insects. However, in Kerala, Madhya Pradesh, Karnataka, Odisha, and Tamil Nadu only six insect species are categorized as edible insect species. About 34% of the insects consumed belong to the order-Coleoptera. 24% of the species consumed belong to the order-Orthoptera, 17% to order-Hemiptera, 10% to order-Hymenoptera, 8% to order-Odonata, 4% to order-Lepidoptera, 2% to order-Isoptera, and only 1% to order-Ephemeroptera (Chakravorty, 2014). Tribes of Arunachal Pradesh prefer to take the Coleopteran species. The termites that belong to the order-Isoptera are consumed throughout the country. All over the world, approximately 468 species of Coleoptera, 351 species of Hymenoptera, 267 species of Orthoptera, 253 species of Lepidoptera, 102 species of Hemiptera, 78 species of Homoptera, 61 species of Isoptera, 34 species of Diptera, 29 species of Odonata, 19 species of Ephemeroptera, 10 species of Trichoptera, 5 species of Neuroptera, 3 species of Anoplura, and 1 species of Thysanura are consumed as food (Ramos Elorduy, 2005).

11.2.2.2 USE IN THE FEED INDUSTRY

The pupal stage of the silkworm is the valuable source of protein for fish and poultry. The utilization of the plentitude of spent silkworm pupae as a feed can help to decrease the environmental problems (Wang et al., 2010). Insect meal is processed in enormous amounts in the feed industry as it is a remarkable part of the animal diet. Generally, black soldier flies, common housefly larvae, and mealworms are used in the production of animal feed. According to FAO, insect meal could substitute between 25% and 100% of the fishmeal in the animal diet without any detrimental effect. Recent studies indicated no change in the aroma and the texture of the fish fed in the larvae of *Hermetiaillucens*. Egg production in the domesticated birds was observed to increase by 2.4% when the fish meal was replaced by the dried mealworm (Wang et al., 1996). Commercial insect rearing industries and small-scale household industries exist worldwide for the production of insect feed. Maggot meal is reported to be an important substitute for fishmeal for the chickens that are raised for meat production (Hwangbo et al., 2009). The production of maggots assists in reducing the accumulation of manure. The growing aquaculture and poultry sector prophesies the development and the progress of the insect feed industry. Few greenhouse gases (GHGs) are emitted by insect feed production which helps to propel the growth of the insect feed market. According to the research findings of Mleck et al. (2014), edible insects consist of carotene, thiamine, riboflavin, pyridoxine, ascorbic acid, calciferol, tocopherol, phylloquinone, and menaquinone. Rumpold and Schluter (2013) found out that the insects belonging to the order-Orthoptera and Coleoptera have a plenteous amount of Vitamin B9. It was observed that the meat of the chicken that was fed on grasshopper containing diet has low cholesterol and high protein content (Makkar, 2014). At present, insects are reared in small-scale industries and for the production of insect meals, cost-efficient and sophisticated mass insect rearing industries are required to be set up. Insects are the optimistic source of protein for animal feeds and can meet the escalating global demand likely. The black soldier fly, house fly, and yellow mealworm have gained significant attention as they can feed on different organic wastes and convert them into high-quality nutrients. Insect-based protein can be used as a substitute for fishmeal or soybean meal and can be used as a feed for pigs, poultry, and fish. Larvae of black soldier fly and housefly are nowadays reared at large-scale, as they can be used as feed ingredients. Mealworms are raised on an industrial scale and fed to the broiler chickens in poultry farms. The enthusiasm for insect farming is elevating gradually as people are becoming aware of the commercial and environmental benefits of insects.

11.2.2.3 USE IN THE PHARMACEUTICAL INDUSTRY

The presence of physiologically vital substances in edible insects increases their application in the field of the medical industry. Chen et al. (2009) reported the immune-stimulatory effects in termites; anti-cancer properties in caterpillar fungus; and aphrodisiac properties in male *Antheraea pernyi*. The medical use of insects is called entomotherapy and it has been originated from traditional medicine. Approximately 5000 microbe-free maggots cultured in the laboratory are supplied to the hospitals in the United States every week, as they aid in cleaning the open wounds. Larval forms of lepidopteran insect, Jatropha leaf miner, are mashed into paste form and is applied topically to the affected areas of the skin. This treatment is also used to lower the body temperature, gastrointestinal disorders, and to induce lactation. Cantharidin, obtained from the bodies of blister beetles, is used for the treatment of mild skin disease, molluscum contagiosum, caused by the poxvirus (Bairagi, 2019). Resilin, an elastomeric protein that is present in many insects, is used in the medical field for the repairment of arteries. The gut of the larvae of nettle caterpillar is used for the treatment of rashes. The cooked larva of the *Eucalyptus* carpenter worm is used to get relief from asthma and excessive bleeding during childbirth. Thus, insects have a potential source in the pharmaceutical industry and can be used for the treatment of various diseases. Some of the important insects having pharmaceutical utility are enlisted in Table 11.2.

11.2.3 USE FOR THE PRODUCTION OF HONEY, SILK, AND LAC

11.2.3.1 APICULTURE

Bee-keeping is a skill and a charismatic science. Worldwide, the apiculture industry is expected to reach 11 billion dollars by 2026. It is the agro-industrial small-scale management appertained to the apiary by the application of the various scientific and commercial techniques. The growing use of honey as a substitute for artificial sweeteners is thrusting the success of the apiculture industry. A huge expansion is exhibited by the Indian apiculture market during the year 2015–2020. Apart from the production of honey, various other by-products such as beeswax, bee venom, propolis, honeycomb, and royal jelly are also related to apiculture. These products find their application in the food, pharmaceutical, and cosmetic industry (Fig. 11.2). Bee-keeping, an agro-based industry, is exemplary for development as the rural and tribal population earns money by developing such projects. Apiculture provides employment to more than 2,50,000 individuals and is spread over 40,000 villages of India (Sivaram, 2019). India provides the habitat to *Apis cerana, Apis dorsata, Apis mellifera*, and

TABLE 11.2 Use of Common Insects in the Pharmaceutical Industry.

Common name	Treatment of diseases
Stingless bee	Cures cough and acne
Ghost moth	Used as a fortifier
Cicada	Treat migraine and ear infection
May Beetle	Anemia, rheumatic disorders
Mud wasps	Provide lime to the fetus
Palm beetle	Used to treat earache
Pod borer weevil	Panacea stomach pains
Housefly	Cures baldness and eye cysts
Red velvet mite	Treats malaria, urogenital disorders, paralysis; bears aphrodisiac property
Ground beetle	Treats suture wounds
Giant skipper	Treatment of rheumatic disorders and has aphrodisiac property
Cattle tick	Chickenpox
Peanut beetle	Treats asthma, tuberculosis, sexual impotence, and arthritis
Polyrachislamellidens	Used to get relief from inflammatory disorders
Locust	Asthma, chronic cough, and post-childbirth anemia
Apis dorsata	Used as appetizer and cures cold, cough, diarrhea, and vertigo
Ant Lion	Cures boils and warts
Treehoppers	Relieves from body ache and jaundice
Rice field cricket	Used for treatment of headache, indigestion, and malaria
Chinese mantis	Used to treat warts

Apis florae. *Apis millifera* is remarkably apt for commercial beekeeping as it is the introduced species to India and can resist various viral diseases.

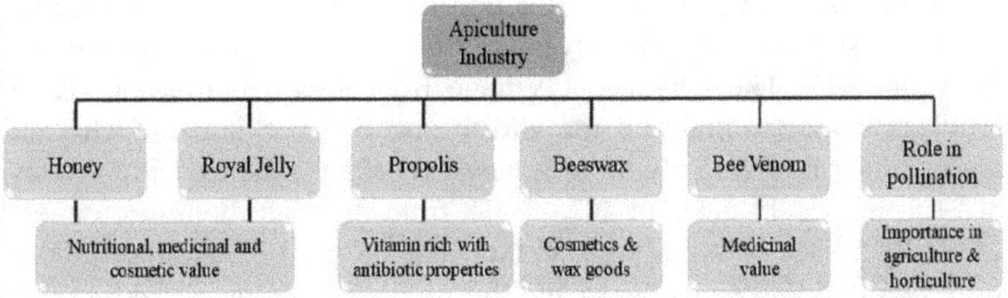

FIGURE 11.2 Importance of apiculture.

At present, there are approximately 1.5 million colonies of honey bees in India, and 55,000 tons of honey is produced every year (Mitta et al., 2017). According to data provided by FAO, 2.13 million tonnes of honey was produced across the world during the year 2013. India stands at the seventh position in terms of honey production. Setting up the apiculture industry requires efficacious collaboration among beekeepers, exporters, traders, governmental agencies, and private agencies. In the year 2020, the apiculture market attained a value of INR 18,836.2 million in India. Recent statistics generated by FAO predicted that Asia and Africa are the major producers and consumers of beeswax. A wide range of waxes such as beeswax, cannuba wax, emulsifying wax, and paraffin wax are used in various industries in the manufacture of creams, lotions, and lipsticks. Beeswax finds its use in the candle, food, pharmaceutical, and cosmetic industry. In the pharmaceutical industry, beeswax is used to make medicines, balms, soft gels, creams, and in the cosmetic industry; it is used for making skin moisturizers, creams, skin-care products, and soaps. With the expansion of such industries, the demand for beeswax is escalating continuously. Worldwide, beeswax production was reported to be 42,307 tonnes in the year 2017. India and Ethiopia are the leading producers of beeswax all over the globe. The leading exporter and importer of beeswax are China and Germany, respectively. Some of the institutions set up by the Indian Government in collaboration with state government of various states of India are:

a. Agricultural Products Export Development Authority
b. Central Bee Research and Training Institute
c. The Export Inspection Council
d. Tribal Cooperative Marketing Development Federation of India Limited.

In the last ten years, the production of honey has shown an increase by 200%, supported by initiatives such as Honey Mission, introduced in the year 2017. The Khadi and Village Industries Commission (KVIC) provides basic training and tools to the people working in such small-scale industries. The volume of honey export has reached upto the level of 51.5 thousand tonnes in 2017–2018 from the 29.6 thousand tonnes in 2014–2015. Bee colonies are also increasing. About 22 lakh bee colonies were reported in the year 2015 and this number raised to 34 lakh in 2018. Of this, 5 lakh bee colonies were formed by *Apis cerana,* and the rest were formed by *Apis millifera*. During this period, honey production also increased from 80 to 105 thousand tonnes. In 2019, India exported 61,333 metric tonnes of honey, thus raising India's rank in the production of honey.

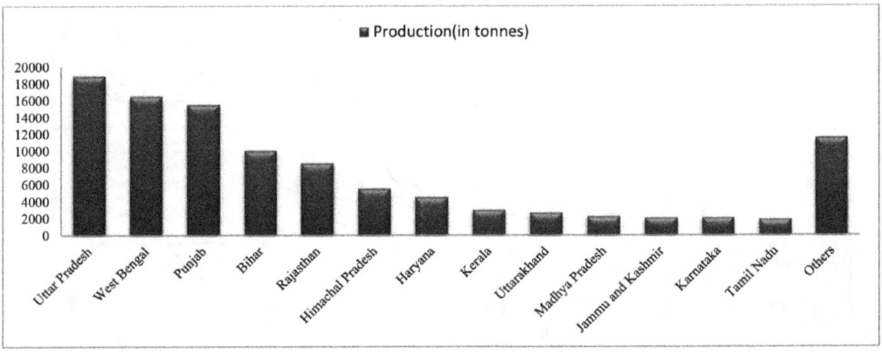

FIGURE 11.3 Production of honey (in tonnes) reported from different states of India in 2017–2018 (modified by National Bee Board, MoA&FW, GoI).

According to FAO Corporate Statistical Database, the data generated for the year 2017–2018 depicts that China is the leading producer of the honey in the world and had a record production of about 551.4 thousand tonnes, followed by Turkey and Argentina that produce approximately 114.5 and 76.4 thousand tonnes of honey, respectively. Maximum honey production in India was recorded to be 64.9 thousand tonnes in the year 2017–2018.

People are also making their way into the bee venom industries as it helps them to earn huge profits. These industries act as the source of living for the venom collectors. The bee venom industry is becoming popular among the masses as the equipment for the extraction of venom from the hives is convenient and accessible these days. Bee venom is produced in several countries such as America, Eastern Europe, and Southeast Asia. The bee venom is expensive and has a value of about 80 dollars per gram. It is a complex mixture of several compounds: enzymes, polypeptides, phosphatides,carbohydrates, pheromones, water, and minerals. Some of the well-known vendors of the global bee venom extract merchandise are: Bee Whisper, Frenz, Abecco pure, and ApiHealth NZ Ltd. These vendors have made collaboration with various government and non-governmental organizations to procure a firm footing in the market.

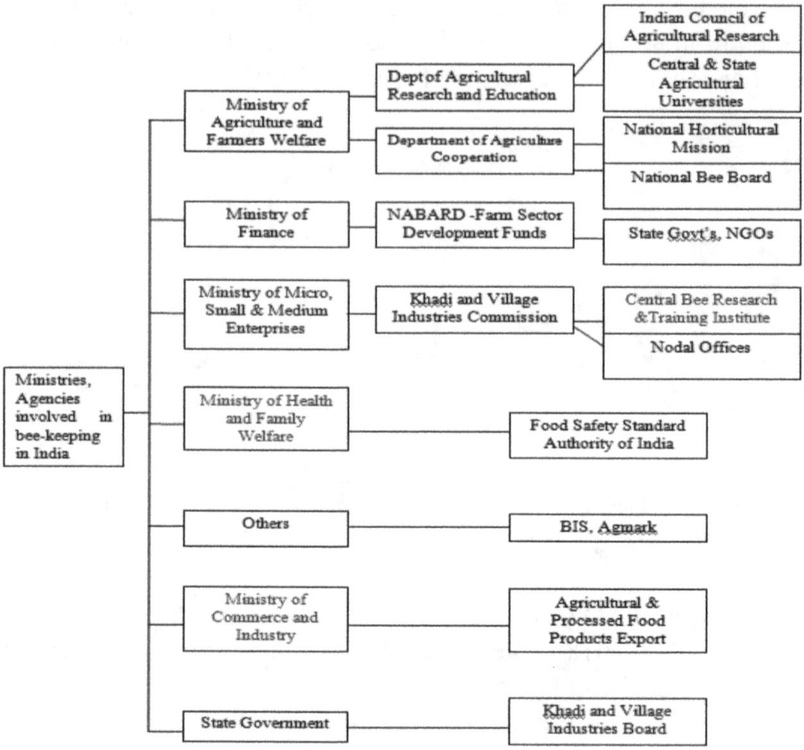

FIGURE 11.4 Various organizations involved in the apiculture industry.

Royal jelly is a gelatinous substance secreted by honey bees. It has commercial value in the healthcare industry due to its anti-bacterial, anti-allergic, anti-tumor, anti-inflammatory, and healing properties. Royal jelly is rich in amino acids, proteins, vitamins, and other nutrients and so its use in the food industry is augmenting gradually. Royal Jelly has a medical significance too and it is used to treat post-menopausal symptoms. Durham's Bee Fram, Puritan's Pride, Organic Bee Farms, NOW Health Group, and Jiangshanm Bee Enterprises are some of the well-known participants in the royal jelly market across the world. The market value of royal jelly is much more than honey, propolis, venom, and wax. Production of Royal Jelly in small-scale industries has become of living for beekeepers all across the world (Ramadan and Al-Ghamdi, 2012; Ghramh et al., 2020). Annually, China produces and exports 4000 tonnes of royal jelly which accounts for 90% of the royal jelly production worldwide.

Remarkable efforts are being made by scientists to improve the tools and techniques for ameliorating the production of royal jelly in the industries.

11.2.3.2 SERICULTURE

The relation of insects with human beings cannot be neglected. Insects are the climacteric integrants of an ecosystem. Silk was recognized as a valued product since the year 2640 B.C. in China. Sericulture is an agro-based cottage forest industry that has a vital role in improving economy and living standards of people of any country. The silk industry is an amalgamation of agronomics, insect rearing, and pure textile activities. This is the silk-producing technique and requires little investment. The sericulture industry is labor-intensive and can provide employment to the rural population. Countries such as China, India, Japan, Brazil, the Republic of Korea, Thailand, Iran, and Vietnam are the major producers of silk in the world. Cocoons and raw silk are produced by some other countries such as Kenya, Nigeria, Bangladesh, Colombia, Nepal, Turkey, Zimbabwe, Malaysia, Japan, Romania, Uganda, and Zambia, who also produce in small amounts. In India, silk industry is flourishing and it generates job opportunities for approximately 7.56 million people. Recent findings have suggested that sericulture may be developed as an extremely beneficial agro-industry. India stands at second position in terms of silk production all over the world. According to the data generated for the year 2015–2016; approximately 28,294 tonnes of raw silk was produced in India of which mulberry silk was 20,478 million tonnes, tasar silk was 2819 million tonnes, eri silk was 5060 million tonnes, and muga silk was 166 million tonnes. About 90% of the total silk produced in India is mulberry silk (Geetha and Indira, 2011) According to Kasi (2000), production of silk is an agriculture-based activity. It includes cultivation of mulberry, leaf-harvesting, silkworm rearing, cocoon production, and reeling cocoons. Plantation of mulberry plants and rearing of silkworms are the agricultural activities and production of cocoon and reeling of silk are the industrial activities that are done either in cottage industries or in large manufacturing industries. Since time immemorial, the silkworm *Bombyx mori* has been appraised intensively owing to its pertinence with the production of silk.

According to International Sericultural Commission, maximum silk produced in China during the year 2015, 2016, 2017, 2018, and 2019

TABLE 11.3 Sericultural Statistics in India (Modified by Central Silk Board, Bengaluru).

Particulars	Unit	2013–14	2014–15	2015–16	2016–17	2017–18	2018–19
Mulberry Cocoon Production	Million tones	1,42,487	1,59,259	1,51,787	1,55,688	1,61,684	1,85,560
Mulberry silk production	Million tonnes	19,476	21,390	20,478	21,273	22,066	25,344
Tasar silk production	Million tonnes	2619	2434	2819	3268	2988	2981
Eri silk production	Million tonnes	4237	4726	5060	5637	6661	6910
Muga silk production	Million tonnes	148	158	166	170	192	233
Total non-mulberry silk production	Million tonnes	7004	7318	8045	9075	9840	10,124
Total Raw silk production	Million tonnes	26,480	28,708	28,523	30,348	31,906	35,468
Employment	Lakh person	78.50	80.30	82.50	85.10	86.04	91.78

was 1,70,000.00, 1,58,400.00, 1,42,000.00, 1,20,000.00, and 1,20,000.00 million tonnes, respectively. However, the silk produced in India during these years was recorded to be 28,523.00, 30,348.00, 31,906.00, 35,261.00, and 35,820.00 metric tonnes, respectively.

11.2.3.3 LAC CULTURE

The practice of rearing lac insects for the production of lac is lac culture. A natural resin, lac is secreted by lac insects, *Lacciferlacca* and *Kerria lacca*. India ranks first in the production of the lac and 75% of the lac produced is exported (Shah et al., 2015). Lac industry has socio-economic importance and it provides employment to 3–4 million people. Over 20,000 tons of lac is produced every year. Lac is used at the industrial level for the mass production of various commodities. Lac dye is used for coloring wool and silk. Lac wax is in the food and medical industry. Bleached shellac finds its use in various industries such as paint, textile, and aluminum industries. Dewaxed bleached shellac is beneficial for coating in cosmetic industries; bangles are made from lac in various small-scale industries due to its adhesive and molding properties. Lac insects flourish on 400 species of plants and have great economic, medicinal, and social significance. Lac cultivation provides employment to millions of lac growers. India alone contributes to 70% of the lac produced globally. Myriad industrial applications of the lac have increased its demand across the globe.

11.3 USE OF INSECTS FOR THEIR SPECIFIC PROPERTIES

Insects are great benefactors to the economy of any country. The insects offer significant help in resolving the emerging challenge of organic waste disposal. They convert the waste into reusable material and reintroduce them into the local food value chain. Insects have been recognized as the efficacious resource of various industries specifically for the mass production of cellulose, chitinase enzymes, medicine, microbial insecticides, and various other substances. Over the last 30 years, insect biotechnology has shown tremendous growth. Various biotechnological techniques are used on the insects or insect cells for the production of various products to be used in the agriculture and pharmaceutical industries. Insects are exploited in various ways for boosting the bio-economy. Secretions of the larva of the green bottle fly, in vitro, reveals to improve the closure of any kind of wound. Chitinase enzyme obtained from the *Manduca sexta* and *Bombyx mori* (Order-Lepidoptera) is used for biological pest control. Antimicrobial peptides generated by insects are used in the production of new drugs in the health care industry. Drosomycin, an anti-fungal peptide, is obtained from the *Drosophila melanogaster*. Secretions of *Formica aquilonia* are known to exhibit anti-inflammatory and antioxidant properties. The venom of several insects has important components such as perillyl alcohol and perillaldehyde that can be used for the treatment of cancer (Stratton et al., 2010). Transgenic silkworms have great industrial applications as they are utilized in the production of silk protein with modified properties. Efforts are on the progress to manufacture the drugs by making the use of transgenic silkworms (Takeda, 2013). The use of whole insects, their organs, or cells in the fields of medicine, agriculture, and industry is called yellow biotechnology or insect biotechnology. Yellow-colored hemolymph of the insects has produced lots of chemicals and proteins and thus is used in medical, pharmaceutical, and agricultural industries.

11.4 CONCLUSION

The growing value of insects in the industries would help in the conservation of insect biodiversity. Insects seem to harbor a great potential and so can be used in various sectors. Insects are the absolute solution to the food and nutrition challenges. To understand the future prospectives of insects, much work is required to be done. There is an intense demand to advance the sericulture and apiculture industry for the well-being of people associated with these industries, which will also help in augmenting the economy of the country. The benefits of insects to the human race can in no way be trivialized.

KEYWORDS

- **antibiotics**
- **apiculture**
- **lac industry**
- **nutraceuticals**
- **pharmaceuticals**
- **sericulture**

REFERENCES

Ahuja, K.; Mamtani, K. Edible Insects Market Size By Product (Beetles, Caterpillars, Grasshoppers, Bees, Wasps, Ants, Scale Insects & Tree Bugs), By Application (Flour, Protein Bars, Snacks), Industry Analysis Report, Regional Outlook, Application Potential, Price Trends, Competitive Market Share & Forecast, 2020–2026. *Global Market Insights* 2020, 1–170.

Bairagi, S. H. Insects with Potential Medicinal Significance: A Review. *J. Sci. Tech. Res.* **2019,** *16,* 12024–12027.

Chakravorty, J. Diversity of Edible Insects and Practices of Entomophagy in India: An Overview. *J. Biodivers. Biopros. Dev.* **2014,** *1,* 1–6.

Chapman, A. D. *Numbers of Living Species in Australia and the World*, 2nd ed).; Report for the Australian Biological Resources. 2011.

Chen, X.; Feng, Y.; Chen, Z. Y. Common Edible Insects and Their Utilization in China. *Entomol. Res.* **2009,** 39, 299–303.

Dobermann, D.; Swift, J. A.; Field, L. M. Opportunities and Hurdles of Edible Insects for Food and Feed. *Nutr. Bulletin.* **2017,** *42,* 293–308.

Geetha, G. S.; Indira, R. Silkworm Rearing by Rural Women in Karnataka: A Path to Empowerment. *Indian J. Gend. Stud.* **2011,** *18,* 89–102.

Ghramh, H. A.; Khan, K. A.; Ahmed, Z.; Ansari, M. J. Quality Evaluation of Saudi Honey Harvested from the Asir Province by Using High-performance Liquid Chromatography (HPLC). *Saudi J. Biol. Sci.* **2020,** *227,* 2097–2105.

Hwangbo, J.; Hong, E. C.; Jang, A.; Kang, H. K.; Oh, J. S.; Kim, B. W.; Park, B. S. Utilization of House Fly-Maggots, a Feed Supplement in the Production of Broiler Chickens. *J. Environ. Biol.* **2009,** *30,* 609–614.

Kasi, E. Development and Change due to Sericulture—A Study in Chittoor District. Unpublished M.Phil Dissertation. University of Hyderabad, Hyderabad, Andhra Pradesh, India, 2000.

Makkar, H. P. S.; Tran, G.; Heuzé, V.; Ankers, P. State-of-the-Art on Use of Insects as Animal Feed. *Anim. Feed Sci. Technol.* **2014,** *197,* 1–33.

Malti. Importance of Insects for Humans and the Environment. *Glob. J. Res. Anal.* **2017,** *6,* 530–531.

Mitta, K. T.; Ramaiah, A.; Mishra, G.; Srinivisan, M. R. Bee Keeping in India. In *Indust Entomol*, 2017.

Mlcek, J.; Rop, O.; Borkovcova, M.; Bednarova, M. A Comprehensive Look at the Possibilities of Edible Insects as Food in Europe—A Review. *Pol. J. Food Nutr. Sci.* **2014,** *64,* 147–157.

Raheem, D.; Carrascosa, C.; Oluwole, O. B.; Nieuwland, M.; Saraiva, A.; Millán, R.; Raposo, A. Traditional Consumption of and Rearing Edible Insects in Africa, Asia and Europe. *Crit. Rev. Food. Sci. Nutr.* **2019,** *59,* 2169–2188.

Ramadan, M.; Al-Ghamdi, A. Bioactive Compounds and Health-Promoting Properties of Royal Jelly: A Review. *J. Funct. Foods* **2012,** *4,* 39–52.

Ramos Elorduy, J. Insects: A Hopeful Food Source. In *Ecological Implications of Minilivestock; Role of Rodents, Frogs, Snails, and Insects for Sustainable Development*; Paoletti, M. G., Ed.; Science Publishers: New Hampshire, 2005; pp 263–291.

Ramos-Elorduy, J. Anthropo-Entomophagy: Cultures, Evolution and Sustainability. *Entomol. Res.* **2009,** *39,* 271–288.

Rumpold, B. A.; Schluter, O. K. Nutritional Composition and Safety Aspects of Edible Insects. *Mol. Nutr. Food Res.* **2013,** *57,* 802–823.

Scudder, G. G. E. The Importance of Insects: Science and Society Chapter 2. In *Insect Biodiversity*, 2017.

Shah, T. H.; Thomas, M.; Bhandari, R. Lac Production, Constraints and Management—A Review. *Inter. J. Curr. Res.* **2015,** *7,* 13652–13659.

Sivaram, V. Status, Prospects and Strategies for Development of Organic Beekeeping in the South Asian Countries. Division of Apiculture and Biodiversity Department of Botany Bangalore University: Bangalore ñ 560056, India. 2019**.**

Srivastava, S. K.; Babu, N.; Pandey, H. Traditional Insect Bioprospecting as Human Food and Medicine. *Indian J. Tradit. Knowl.* **2009,** *8,* 485–494.

Stratton, S. P.; Alberts, D. S.; Einspahr, J. G.; Sagerman, P. M.; Warneke, J. A.; Curiel-Lewandrowski, C.; Myrdal, P. B.; Karlage, K. L.; Nickoloff, B. J.; Brooks, C.; Saboda, K.; Yozwiak, M. L.; Krutzsch, M. F.; Hu, C.; Bowden, G. T.; Bartels, P. H. A Phase 2a Study of Topical Perillyl Alcohol Cream for Chemoprevention of Skin Cancer. *Cancer Prev. Res. (Phila).* **2010,** *3,* 160–169.

Takeda, S. New Field of Insect Science: Research on the Use of Insect Properties. *Entomol. Sci.* **2013,** *16,* 125–135.

Wang, J.; Wu, F. A.; Liang, Y.; Wang, M. Process Optimization for the Enrichment of ?-linolenic Acid from Silkworm Pupal Oil Using Response Surface Methodology. *Afr. J. Biotechnol.* **2010,** *9,* 2956–2964.

Wang, Y.; Chen, Y.; Li, X.; Xia, J.; Du, Q.; Zhi, C. Study on the Rearing Larvae of *Tenebrio moliter* Linne and the Effects of its Processing and Utilizing. *Acta Agric. Univ. Henansis* **1996,** *30,* 288–292.

Yen, A. L. Edible Insects:Traditional Knowledge or Westernphobia?. *Entomol. Res.* **2009,** *39,* 289–229.

Yin, W.; Liu, J.; Liu, H.; Lv, B. Nutritional Value, Food Ingredients, Chemical and Species Composition of Edible Insects in China. *Intech. Open.* **2017,** 27–53

CHAPTER 12

Forensic Entomology: Progress and Implications

RAHUL DAS[1], MUSKAN SINGHAL[1], PRASHANT SINGH[2], ROHIT VERMA[3], MAHIPAL SINGH SANKHLA[1], MANSI SAXENA[1], EKTA B. JADHAV[4], KAPIL PARIHAR[5], KUMUD KANT AWASTHI[6], and CHANDRA SHEKHAR YADAV[6]

[1]*Department of Forensic Science, Vivekananda Global University, Jaipur, Rajasthan, India*

[2]*School of Forensic Sciences, National Forensic Sciences University, Gujarat, India*

[3]*Dr. APJ Abdul Kalam Institute of Forensic Science & Criminology, Bundelkhand University, Jhansi, India*

[4]*Department of Forensic Chemistry and Toxicology, Government Institute of Forensic Science, Aurangabad, Maharashtra, India*

[5]*State Forensic Science Laboratory, Jaipur, Rajasthan, India*

[6]*Department of Life Science, Vivekananda Global University, Jaipur, Rajasthan, India*

ABSTRACT

Insects and arthropods are found to be of utmost importance during the investigation of criminal cases. Forensic entomology involves the study of various insect species along with their characteristics for their application in legal context. Medico-legal entomology plays a major role in forensic aspect, as compared to stored-product entomology and urban entomology.

Insect Diversity and Ecosystem Services, Volume 1: Importance, Threats, Conservation and Economic Perspectives. Younis Ahmad Hajam, Sajad Hussain Parey, and Rouf Ahmad Bhat (Eds)

It is associated with the study and examination of insects encountered at different types of crime scenes. The estimation of time since death (TSD) of an individual is aided by medico-legal entomology. The insects which are found in the crime scene, either from the corpse or its vicinity, can also help in the detection of drugs and other toxic substances from the body of the deceased. Beetles and flies are among the two most crucial species of insects that are utilized in forensic entomology. These insects are used to determine the post-mortem interval (PMI) of a body, through the study of their development and succession patterns. Some other implications of forensic entomology include the determination of site, identification of abuse, archae-entomology, and molecular analysis of entomological evidences. There are certain factors which affect the examination of entomological evidences. The major factor affecting the examination is weather condition of the place where the corpse is present. However, the study of the various aspects of insects has a major contribution in the investigation of a case.

12.1 INTRODUCTION

Forensic entomology makes use of insects and arthropods to assist in cases under the criminal and civil justice system. This is because these insects exhibit an important effect upon the decomposition of corpses (Singh et al., 2016). Forensic entomology is an interdisciplinary field which is considered as a branch of both arthropod science and forensic science (Keh, 1985; Catts and Goff, 1992; Amendt et al., 2004). It is the study of the succession pattern of the stages of development of different species of arthropods found on the decomposed cadavers and it is used most commonly during the investigations of death. It is a crucial parameter to scientifically estimate the time since death (TSD). Entomology has been used for solving crimes since centuries. Research in the field of forensic entomology in the recent decades has resulted in its advancement (Anderson et al., 2001; Benecke, 2001; Williams and Villet, 2006). Forensic entomologists are able to establish the geographical location of the place where death occurred if the body was killed in one location and then moved to another depending upon the insect species present (Catts and Haskell, 1990). Entomological evidences usually link a suspect to a specific location or a suspect and victim. Forensic entomologists are able to identify sites of perimortem or post-mortem trauma in cases of advanced decomposition or extreme post-mortem change. Insects are also utilized as an alternative for DNA and toxicology samples (Goff et al., 1989; Wells et al., 2001). The determination of time of death is valuable in criminal investigations as it can be used to identify the victim and individual(s) responsible for the victim's death (Gennard, 2012). Forensic pathologists assess the time of death by witnessing the state of the body after death. The characteristics under concerns are algor mortis, livor mortis, and rigor mortis. However, these methods are only applicable during the first 48–72 h after death (Campobasso et al., 2001). Estimation of time of death after 48–72 h using techniques associated with forensic medicine become less accurate (Young et al., 2013). Succession data of the insects can be used to estimate the time range for which the body was exposed and the development data can be utilized to calculate the time period for which that the body was being colonized by insects (Amendt, et al., 2007). The various stages involved in insect succession can be recognized while the corpse decomposes and a range of different invertebrate fauna is linked with each stage (Byrd and Tomberlin, 2019). Blowflies are the insects which have been studied the most with regard to forensics. They are the insects which are most commonly associated with dead bodies. Blowflies are used to provide the most accurate information with respect to post-mortem interval (PMI), which means the time that has elapsed since death of an individual (Gullan and Cranston, 2014).

12.2 FORENSIC ENTOMOLOGY

12.2.1 STORED-PRODUCT ENTOMOLOGY

Stored entomology is the branch of entomology that deals with the study of insects associated with raw and processed food. These insects cause qualitative and quantitative damage to the food that further result in product alteration, customer complaint, health issue, and difficulty in cost assessment and management (Hagstrum and Subramanyam, 2009). Raw and processed food should be kept away from the reach of rodents, insects, birds, microorganisms, and mites. Therefore, harvested grain should be stored in grain reserves that inhibit the arthropods to make colonies. It involves arthropods contamination in various products, for example, flies in sauces and spiders in the different areas (Catts and Goff, 1992). Some of the packed food develops beetles (Coleoptera) on them. Beetles feed on

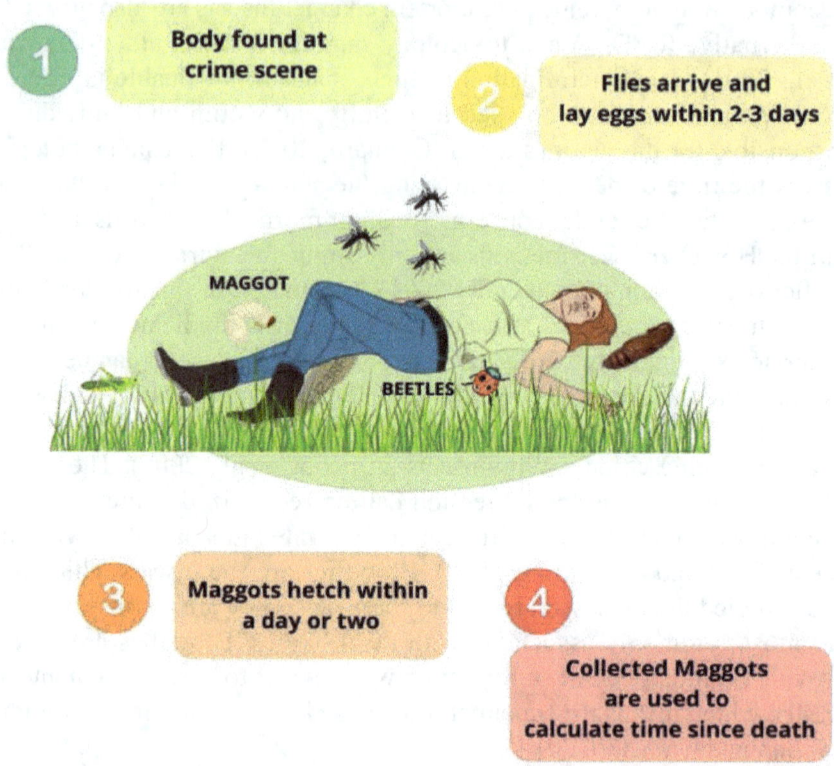

FIGURE 12.1 Illustration of crime scene timeline.

the growing fungi and mold on these substrates and contaminate the food that causes high skin irritation to the person coming in the contact with the contaminated substrate. The signature habitat of these beetles includes dry, shelter areas. Large larva of Coleoptera develops on stored grain like wheat, rice, and maize. Many species of cylas develops and grows into tubers, stems of herbs, and plants like sweet potato. Cylas lays eggs on the soil that are infested in the tubers by the natural phenomenon in the form of larvae and then the pupate goes into a tunnel. Further development takes place during harvesting and storing. Stored food which has high starch content, like bamboo, can develop Bostrichidae as this starts developing within 3–5 days after the storing. Larva of Bostrichidae develops on drugs, maize, cacao, dry food, and flour (Harwood and James, 1979). Therefore, mainly the species belonging to the order Coleoptera, Hymenoptera,

Lepidoptera, dipteral, and Hemiptera, are associated with the stored food (Hagstrum and Subramanyam, 2016).

12.2.2 URBAN ENTOMOLOGY

Urban entomology is the branch of entomology that deals with the associated study of insects like a beetle, cockroaches, ants, and bugs within the urban environment (Gondhalekar, 2019). An urban environment is a complex form of habitat developed by humans from agricultural and natural sites. There are mainly two types of habitat—one that is made intentionally like park, centuries, food store, street trees, and some are natural, like standing water, garbage, landfill, indoor house, etc. Every habitat supports a different group of arthropods depending on its local conditions like climate, temperature, humidity, availability of resources, etc. (Harwood and James, 1979). Every group of insects has its particular temperature and humidity conditions for their survival. Insects like cockroaches and house dust mites grow in indoor temperature and relative humidity. Therefore, they are easily observed in the indoor household. These indoor insects produce various allergens that can result in serious health issues like asthma (Robinson, 2020). Habitat containing large trees like mature oaks, maples, and beech trees has high chances to develop Azalea, bugs, and Japanese beetles (Held, 2019). Order Psocoptera can cause high loss as it makes large colonies in warm and humid conditions (Olsen, 1981). Therefore, pest status in the human environment is sometimes a matter of concern as humans are affected directly or indirectly, i.e., by stinging or contaminating the food. Garbage and waste dumping sites are the houses of maggots and houseflies as these arthropods are attracted to the odor of decomposition (Robinson, 2020).

12.2.3 MEDICO-LEGAL ENTOMOLOGY

It is the most important aspect of both forensic and medico-legal entomology. It deals with the systematic analyses of insects at different crime scenes like rape, murder, suicide, etc. (Smith, 1986). The most important aspect of this is to know the TSD. TSD is the time between death and corpse discovery. Arthropod occurrence at crime scene gives the idea of time, type of scene, causes, and manner of death (Harwood and James,

1979). Insects are the creatures that develop first on the decomposing corpse, especially blowflies. Blowflies identify the corpse and deposit on the carrion within a few hours of the death (Catts and Goff, 1992). The life cycle of the arthropod changes differently around the corpse, it has its fixed succession changes that help to estimate PMI.

Study suggests that the most important post-mortem invaders are Calliphoridae, Muscidae, and Sarcophagidae (Kulshrestha et al., 1987). In the later stage of decomposition, beetles or Coleoptera present an important role as they are developed in later decomposition stage in unique succession pattern (Kulshrestha and Satpathy, 2001).

12.2.4 ENTOMOTOXICOLOGY

It is the branch of forensic entomology that deals with the detection of drugs and other toxicants in the body with the help of insects. It also deals with the effects of drugs on arthropods (Introna et al., 2001). Accidental and incidental cases, i.e., drug-related deaths, are solved by principle of entomotoxicology (Goff, Lord et al., 1994). Toxicological analysis mainly deals with the traditional sources like blood, urine, and organ to estimate the time of death. But in the corpse, during the later decomposition stages, there is no availability of these sources (de Carvalho, 2009). Analyses of toxicants attached to the larvae is done by radio-immune analysis (RIA), gas chromatography, thin-layer chromatography, high pressure liquid, chromatography, and gas–mass analysis (Introna et al., 2001). Diptera larva mainly grows in the presence of drugs or toxins. If the body shows the growth of Diptera larva, it indicates the intake of alcohol or drug by the person when he was alive, i.e., incidental death (Goff et al., 1994). The growth of Calliphoridae larvas in the skeletal remain/body indicates the presence of barbiturates (Beyer et al., 1980).

12.2.5 ARCHAE-ENTOMOLOGY

Archaeological studies of crime scenes are helpful in locating and defining the type of scene and in recovering the maximum that is supporting and connecting crime scenes. Moreover, these studies are also used to identify the chain of custody and PMI (Dirkmaat and Adovasio, 1997; Killam, 2004). Re-construction of scenario according to perimortem and post-mortem events requires the historical value of that scene that can be verified by archaeology. Cold evidence (evidence that is more than 20-year-old) are recovered and reconstructed by the archaeological study as this field of science has developed the methodologies regarding location, collection, interpretation, and analysis of the evidence (Dirkmaat and Adovasio, 1997).

12.3 INSECTS OF FORENSIC SIGNIFICANCE

The proper identification of the insect and arthropod species of forensic importance is the most crucial element in the field of forensic entomology. It is species identification that allows the proper developmental data and distribution ranges to be applied to an investigation. If the species determination is incorrect, or otherwise in error, the estimated PMI is invalid. A few insect species exist that have growth and development rates so similar that one can be applied to the other with little or no error in the resulting PMI determination. However, some species that appear similar to the naked eye have drastically different growth rates, behaviors, and habitat preferences. It is not possible to make comparisons between insect appearance and developmental physiology. Therefore, a qualified forensic entomologist should always make the species determinations. It is important to note that insects may play various roles in the process of a death investigation. The ones included in this chapter are species whose growth rate and developmental history, behavior, and geographic distribution are frequently utilized by forensic entomologists. There are situations when any insect can be of potential forensic importance. For example, a moth larva recovered from a plant seed attached to a blanket used in a rape or sexual assault may provide evidence linking the crime to a particular location. Insect evidence also has other valuable uses beyond aiding in the estimation of a PMI. Sometimes entomological evidence is found as broken and fragmented insect parts in the clothing or personal belongings of the suspect or victim. These fragmentary remains are sometimes utilized successfully to create a link between the victim, suspect, and scene (Byrd and Tomberlin, 2019). Whereas the adult insect body is composed of different structures with distinctive and diagnostic characters, the majority of the larvae, especially in Diptera of forensic interest, are, from a morphological point of view, very simple organisms. The general body of an adult insect is composed of a head with sensorial and feeding organs, a thorax with its locomotors appendixes, and the abdomen containing digestive apparatus and reproductive organs (Vanin et al., 2017). Most often these are fly maggots, primarily blowflies and flesh flies. Those showing the longest period of development are assumed to manifest the PMI provided the corpse was exposed and conditions were suitable for insect activity at the beginning of the period (Catts et al., 1990).

12.3.1 FORENSIC SIGNIFICANCE OF BEETLES

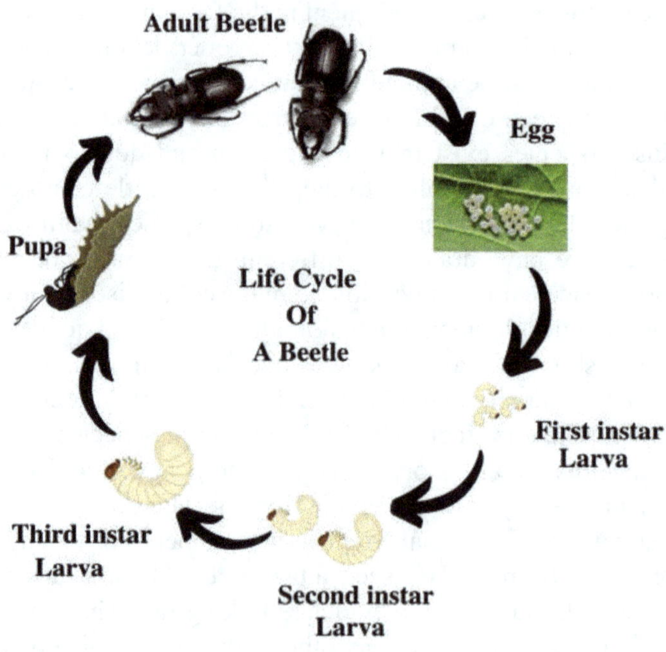

FIGURE 12.2 Illustration of life cycle of a beetle.

Carrion beetles fall under the suborders Adephaga and Polyphaga. They roughly correspond to the carrion guilds of predators and necrophages. The marked phylogenetic trends are not easily identifiable in case of beetles as seen in flies. However, Staphylinoidea and Scarabaeoidea are represented well in carrion communities. There are about two dozen beetle families that have been found in carrion (Louw et al., 1993; Braack, 1986; Tantawi et al., 1996; Ekanem and Dike, 2010). Insects make use of dead bodies in the form of food source for the growth of their larvae. This makes them forensically significant in the investigation process (Goff and acarology, 2009). Various xenobiotics like drugs and other toxic substances which are present in the tissue of the cadaver get transferred to the metabolic system of larvae while the insects feed on the cadaveric tissue. These xenobiotics are further transferred to other arthropods through the food chain, when the arthropods predate on the larvae. Insects can also serve as an appropriate toxicological sample since it is present in numerous count on the cadaver, and the puparial case also remains unchanged and unaffected for a prolonged period of time (Bourel et al., 2001; Introna et al., 2001). Figures indicate that a large number of deaths occur due to substance abuse, and in several cases, death bodies are recovered after many weeks to months (Goff et al., 1989; Carvalho et al., 2001). While actively feeding on cadaveric tissue, xenobiotics such as drugs and other toxic substances that exist in the tissue gets transferred to the metabolic system of larvae. Further, these xenobiotics are transferred through food chain to other arthropods that predate on the larvae. The larvae of insects are present in numerous quantities on the cadaver and they serve as a suitable toxicological sample. The case of the pupa also remains unchanged and unaffected for a long period of time (Bourel et al., 2001; Introna et al., 2001). The insect undergoes metamorphosis, which is a change from one life stage to the other stage of the insect. The initial step involves the laying of eggs in large amount on the corpse. The maggot then gets emerged from the eggs and feed on the corpse. They finally undergo three instar stages of larvae. The mature maggots move away from the source of food in order to pupate in a suitable site, usually in the soil during the third instar. The adult insect finally emerges out of the pupa and the empty cases of puparia are left behind. These puparia cases are often found on the clothing of the deceased, under room carpets, or under the soil (Byrd and Tomberlin, 2019).

12.3.2 FORENSIC SIGNIFICANCE OF FLIES

There are more than two million species of insects in the earth's ecosystem and they constitute a dominant life form. According to some researchers, the era that we are living in has been described as the era of insect. The

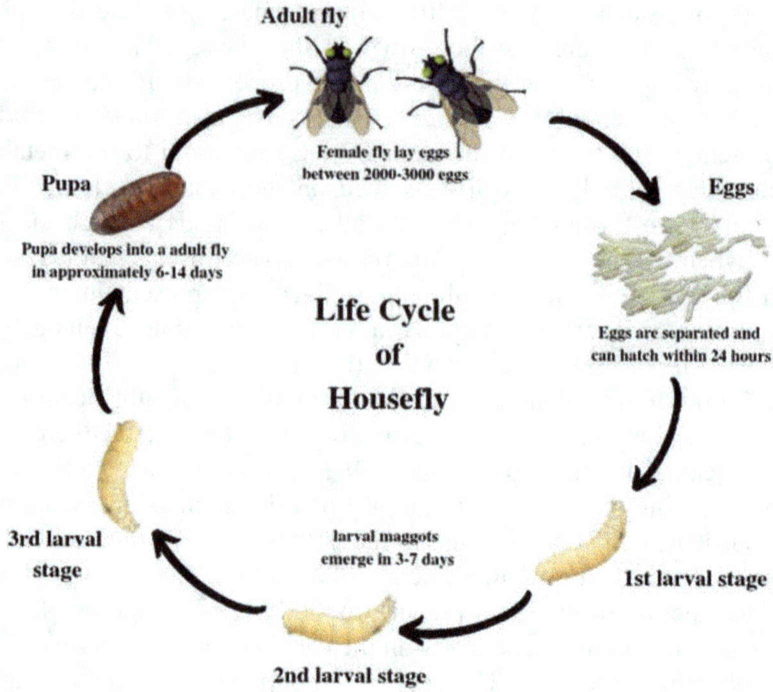

FIGURE 12.3 Illustration of life cycle of housefly.

insects carry out many tasks starting from the substance cycle to the process of pollination. They are denoted as the labors of the ecosystem. There are some other species that can be categorized as rare with regard to the total number of species. They include the species of forensic significance, agricultural insect pests, and parasite vectors. These species of insects have been researched more because of their interaction with humans in various ways. They are of great significance in the determination of the time of death and forensic toxicology. The transport conditions of the corpse can also be determined (Tüzün et al., 2010). House flies are most commonly considered when the term fly is mentioned. However, there are a number of articles about Diptera species such as gasterophilus, drosophila, and gnats. Sinantropic Dipteria species exist in nutritional elements, habitat, agricultural lands of human beings, and wastes. Thus, a close relation has been found between these species and human beings (Harwood and James, 1979; Mullen and Durden, 2009).

Flies exhibit holometabolical development and it has been found that there exist significant differences in between the holometabolical metarmorphosis of adult and newly formed living creatures. Holometabolical metamorphosis consists of egg, larva, pupa, and adult phases, respectively. The larva of the fly is termed as grub. The shape of larva is cylindrical and is pointed toward the edge. Larva is frequently saphorak. Saphorak larva feeds on yeast, bacteria, and organic substances. It lives in a semi-liquid environment. Flies which have forensic and medical significance belong in this group. The pupa and adult phases follow after the larva phase. The characteristic feature of its adult is the transformation of the second pair of wings to a structure called Halteres organ. It differentiates them from other winged insects (Greenburg, 1973).

12.3.3 FORENSIC SIGNIFICANCE OF OTHER INSECTS

There has been an increase in the number of deaths due to drugs in USA and rest of the world. The dead bodies are often discovered after a gap period and the corpse could be highly de-composed or in skeletonized form. There will not be enough tissues for toxicological analysis under such circumstances. It is, however, possible to detect various toxins and controlled substances by the analysis of the insects, their larval, or puparial skin present near the corpse (Sankhla et al., 2017). Entomologists have started the detection of drugs in insects with a view that it would be useful in forensic investigations (Beyer et al., 1980; Introna et al., 2001). The knowledge of local insect assemblages and their population dynamics is essential for an effective legal investigation using forensic entomology. Entomotoxicology involves the analysis of toxic substances in arthropods, mainly flies and beetles that feed on carrion. Investigators can determine the presence of toxins in a body at the time of death with the aid of arthropods present on a carrion or at a scene of crime (Introna et al., 2001). The study of the effects of drugs on the rate of development of carrion-feeding insects and the use of these insects as an alternative sample in the absence of other tissues is dealt in forensic entomotoxicology (Byrd and Tomberlin, 2019). The past years has encountered a renaissance of interest in forensic investigations by entomologists. Catts and Goff (1992) identified three categories of forensic entomology: urban, stored-product, and medicolegal entomology. Urban forensic entomology comprises litigations and civil law actions concerning arthropods in dwellings or as house and garden pests. Law suits dealing with the misuse of pesticides are included in urban entomology. Stored-product forensic entomology generally deals with arthropod influx or contamination of a wide range of commercial products. Like urban entomology, this category also involves litigation. Medicolegal forensic entomology deals with arthropod involvement in events involving crimes, usually crimes such as murder, suicide, and rape which are violent. It also includes other harms such as physical abuse and contraband trafficking (Erzinclioglu, 1987). A more accurate name for this type of forensic entomology is medico-criminal forensic entomology (Smith, 1986).

12.4 INSECT DEVELOPMENT AND SUCCESSION ON CARRION

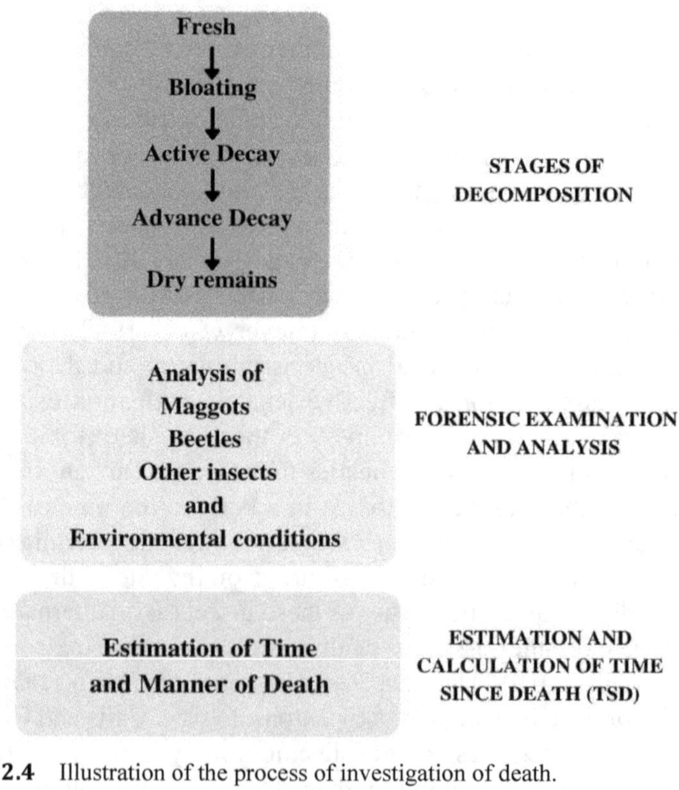

FIGURE 12.4 Illustration of the process of investigation of death.

12.5 ESTIMATION OF PMI

Most important aspect of forensic entomology is the calculation of PMI. PMI is the time needed by the larvae to achieve its maximum growth (Adams and Hall, 2003). Growth of the insects depends upon the external environmental factors. Colonization of species differs species to species and environment to environment (Pohjoismäki et al., 2010). For the estimation of PMI, evidence related to forensic entomology are photographed, identified, relabeled, measured, and preserved in 70% of ethanol (García-Rojo et al., 2013). Most common species included in the process of decomposition of corpse are Calliphoridae (blowflies), Sacrophagidae (flesh flies), and Muscidae (house Flies); these flies arrive within minutes after decomposition (Joseph et al., 2011).

Estimation of PMI is done by estimating the age of fly and species (Catts and Goff, 1992). Accurate species identification is a challenge for many entomologists. When an unidentified mass of insects is discovered or a corpse is discovered at an early stage (i.e., during egg or pupa formation), it is difficult to estimate the species. Therefore, species are incubated in the lab at the particular condition of the crime scene and allowed to grow till the differential factors are noticed (Byrd et al., 2010; Joseph et al., 2011).

PMI can be estimated in two ways.

i. Development of insect
ii. Decomposition

12.5.1 DEVELOPMENT OF INSECT

As the body decomposes, every stage specie needs particular time for its growth and development. Therefore, estimating the stage that is found on the corpse can be useful. But it is assumed that flies started their life cycle soon after the death. And it is believed/assumed that decomposition of flies is independent to weather and humidity. Therefore, in the case of secondary crime scene, there are chances of false result (Kumar et al., 1990; Catts and Goff, 1992).

12.6 INDICATION OF ABUSE

Do we often think about what happened at the crime scene? What happened to the person before his death? The insect presents at the scene of the crime tells much about the scenario.

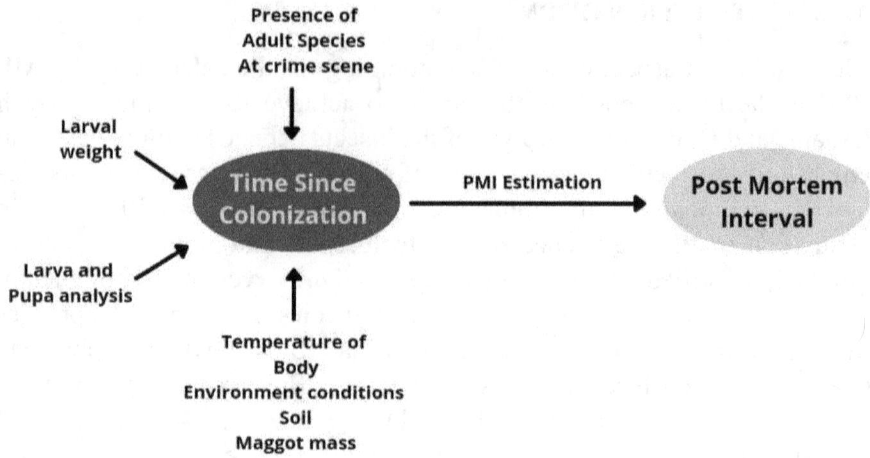

FIGURE 12.5 Illustration of data needed for the estimation of PMI.

Abuse and rape can be identified by the presence of a particular type of fly at a particular temperature. Sexual assault before the death results in the flow of blood in the genito-anal region and proves the act of sexual assault before the death. Moreover, some insects develop by the process of coetaneous myiasis, i.e., the cases in which victims are still alive, but maggots-infested. Their type of cases mainly occurs with young children and seniors. Forensic entomologists can identify these cases by observing circumstances (Kenawy et al., 2019).

In cases of neglected children, growth of the maggots is observed in the anal-genital region, i.e., under the diaper (Anderson, 1999; Kenawy et al., 2019).

In cases of neglected elderly person, presence of maggots is there in the wounds. These wounds will reflect the neglect of care (Benecke, 2001).

In cases of neglected animals, arthropods begins to colonize in animal cadaver with family Calliphoridae (mainly Blowflies) and then family Sascophagidae (flesh flies and then dipterans families) (McGarry et al., 2018). Condition of myiasis, i.e., infestation by dipteran larva on the animal is quite common. Myiasis is the result of veterinary problems. In the cases of injury, there will be larva colonization of adult fly. Adult flies will lay eggs in the tissue (Anderson and Huitson, 2004).

12.6.1 DETERMINATION OF SITE

Variations in the species of insects help in determining the site of crime. However, there are only few insects that accurately confirm the site or strongly support the given described habitat (Anderson, 1999; Matuszewski et al., 2013). Largest class of insects is Metazoan class and it can survive in almost every habitat (Price, 1997). Blowflies (Calliphoridae) are found almost at every crime scene, and works as a "Biological clock in the determination of corpse." Larva of *Lucilia sericeta* grows faster at 25 °C (Amendt et al., 2007). Sacrophigidae larva grows in the habitat of constant temperature. Presence of *Cochlimyia macellaria* and *Phaenicia clevia* indicates the winter season (habitat temperature below 0°C). Moreover, *Calliphora vicina* is all year species, i.e., it can be found in every year and in every season. Besides these *C. vomitoria, Musina Stabulas*, and *Dermestes* are the spring and autumn species. And *Phormia regima* and *Sacrophaga* are the summer species (Schroeder et al., 2003; Frost et al., 2009). Whereas, larva of *Calliphora vicina* lay eggs only on the corpse. Blattoder (cockroaches) and Diptera (house flies) belong to the Anthropophilic arthopod and are associated with indoor crime scene. Arthropods like fireflies, biting midges, and mosquitoes, which are mainly associated with dumping crime scene. Common insects that are found or indicate the human quarters are stable fly, lesser fly, i.e., *Muscina stabulan, Famnia Canicularis*, respectively (Frost et al., 2009). Whereas, arthopods like *Licila serceta, Dermestes frischi, Dermetes laniaria*, and *Necobia rufipes* are mainly found at outdoor crime scene (Matuszewski et al., 2013).

12.6.2 COLLECTION, PRESERVATION, AND MORPHOLOGICAL SPECIES IDENTIFICATION

Standards and guidelines for the correct collection, preparation, storage, and interpretation of the entomological evidence have been published by the European Association for Forensic Entomology (EAFE) (Amendt et al., 2007) and by other Forensic Associations for their members. In general, a correct sampling requires the collection of all the developmental stages present, not only on the body but also at the crime scene (Vanin et al., 2017).

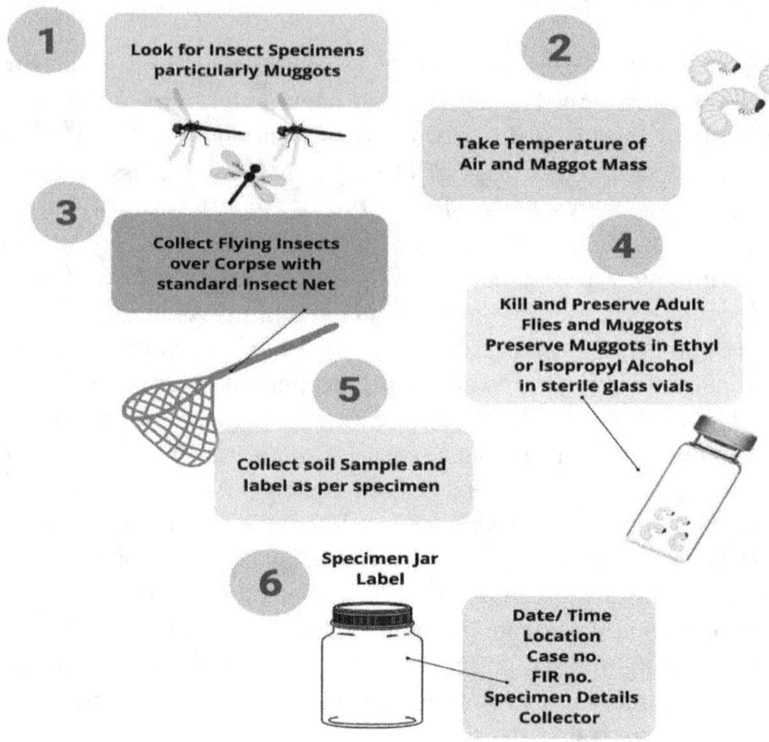

FIGURE 12.6 Illustration of the process of collection, preservation, and storage of insects.

12.6.3 COLLECTION OF ENTOMOLOGICAL EVIDENCE

Invasion of distinct bodily areas can occur in a random order. Natural inlets are the first parts of a body to be occupied (eyes, nose, mouth, anal, or genital area). A wound site might be indicated by any other mass appearance of larvae in the initial phases of decomposition. Several orifices and body areas of a single carcass may be infested at random with distinct species, which should all be investigated and sampled. If the corpse was packed or contained (e.g., in a carpet or bag), the enclosing items and the victim's clothing should be inspected for insects. Since young insects often scatter away from a body after eating (e.g., for pupation), the region around the body must be thoroughly investigated. In the field, based on the soil type, the leafy detritus, and soil beneath and around the body must be checked for flies to a range of 2–10 m (Anderson, 1999). Wandering caterpillars or maggots may also be found beneath and/or in carpets, mats, or rugs, as well as in furniture such as couches, cabinets, chests, and potentially beneath floorboards or baseboards. Each and every crime scene must be thoroughly examined for nutrition sources other than the body (e.g., animal cadavers, natural wastes), since they may provide the breeding ground for corrupting bug evidence. Collection from the body might take place both during the healing process and during the autopsy (Amendt et al. 2011).

12.6.4 PRESERVATION

Multiple samples must be obtained at least 5–10 cm deep. To inhibit further growth of most species, storage at extremely cold temperatures (e.g., in a fridge at 4°C) is advised once they are studied (Amendt et al., 2007). Insects of every type, size, and shape (e.g., eggs, small and large larvae, pupae, and adults) should be collected, using different vials for each type of insect and sampling site. The sampling should not be restricted to just the largest larvae or putative oldest stages: small larvae of one species could be older than large larvae of another species (Amendt et al., 2011). Empty puparia should also be collected since they plainly show the completion of at least the first "wave" of colonisation. However, according Amendt et al. (2007), size of the sample will vary based on the amount of larvae detected, but as a general reference, it should vary from all of the larvae where fewer than 100 are accessible to 1–10% of the larvae if thousands are accessible. Adult or immature specimens that have previously died should be kept in 70–95% ethanol. If samples can be collected alive, they should be stored in vials kept at low temperatures (e.g., in a fridge at 4°C). The vials must allow air in but prevent insects from escaping. The secretions released by the insects can be absorbed using fine sand like tissue paper. Within 24 h, all surviving specimens should be handed to an expert for raising (Amendt et al., 2011). Specimens that are still alive must be exterminated as quickly as feasible. It is advised to decapitate fly larvae in incredibly hot (80 C) water and beetle larvae in freezing temperatures, i.e., in a freezer at −20°C (Adams and Halls, 2003; Midgley and Villet, 2009). It is advised that deceased specimens be preserved in 70–95% ethanol. Formalin/formaldehyde must be prevented for concerns about health and safety as well as because storing samples in them increases the probability of degraded DNA (Amendt et al., 2011). A more thorough examination of the clothing can produce a more comprehensive sample (Vanin et al., 2017). Insect specimens can also be preserved in hot water (>80°C for 30 s) or submerged in various preservation chemicals (e.g., KAA, XAA, Carnoy's solution 7, Ethanol, and so on) (Amendt et al. 2011).

12.6.5 MORPHOLOGICAL IDENTIFICATION

As said before, species recognition necessitates the use of a professional. It is worth noting that only a multidisciplinary team (medical examiner, forensic entomologist, forensic botanist, forensic anthropological) can offer the finest recreation of the incident that occurred during the peri-mortem phase. While the mature insect body is formed of several structures with distinct and distinguishing characteristics, the bulk of the larvae, particularly in Coleoptera of forensic significance, are morphologically relatively basic animals. A mature insect's body is made up of a skull with sensory and eating organs, a thoracic with propulsive annexures, and a midsection with digestive and reproductive systems. Larvae are invariably flightless and typically limbless in their immature phases. The cephalic area of most forensically interesting larvae is not sclerotized. Hatchlings are difficult to distinguish visually at the genus and species due to their basic structure. In these instances, molecular characterization is vital and critical (Vanin et al., 2017). For such objective, many genes (cytb, COI, COII, EF-1, period, tRNA, 12S, 18S, 28S) have now been employed. COI has been the most extensively examined of them. Regardless of the fact that the DNA extraction procedure requires certain adaptations, the scientific approach has been employed successfully on ancient materials (Raoult et al., 2008). Considering evolution, phenology, distribution, and ecology are particular to single organism or diversity of species, recognizing the life forms in both a forensic and archaeological context is critical. The initial phase in any scientific or forensic recreation, known as the "6Ws" (Who, When, Where, What, How, Why), is based on accurate identification of species (Vanin et al., 2017). Insect maturation is species-specific, population-specific, and temperature-dependent (Richards et al., 2009). Insects are poikilothermic creatures, which means that their core temperature is mostly determined by the temperature of their surroundings. Temperature and sunlight influence insect growth and behavior, including movement, flying, mating, and nutrition seeking. Moreover, other factors like as social contact and nutrient availability might have an impact on the species' overall genetics and behaviors (Vanin et al., 2017).

12.7 APPLICATION OF MOLECULAR METHODS TO FORENSIC ENTOMOLOGY

The most prevalent use of forensic entomology involves the investigations related to death of an individual. The major elements present in these types of investigations are TSD or PMI, manner and cause of death, association of suspects with the death scene, and movement of the corpse. All these elements connect to arthropod occurrence and activities (Catts and Goff, 1992).

12.7.1 PMI ESTIMATION

The assessment of mPMI is determined by two factors: the period of development and the phases of colonisation associated to food and moisture content preferences. The following two techniques can be used to assess mPMI. The first technique is mostly appropriate for deaths that occur in temperate areas, days/weeks even before body is discovered, and takes into consideration the rate of growth of the initial coloniser species. The nature of the fauna gathered from the body is taken into account in the second technique. The first method focuses on fly species and developmental phases from the Calliphoridae, Sarcophagidae, and Muscidae families. The period since the implantation of the eggs is estimated using the phase of development, which is connected with the temperature during which the body was subjected prior to its finding (Vanin et al., 2017). Accumulated degrees days (ADDs) method can be utilized for determining the correlation, developmental stage, time, temperature, etc. The temperature history and species identification must be reconstructed in a proper manner for correct estimation of the time of colonization (=mPMI). Thermal reconstruction of the crime scene can be done using temperature records from the meteorological station closet to the scene or by using a data logger (Vanin et al., 2017). The composition of the entomological community found on the body can be useful in the estimation of the mPMI of cadavers found after a long time after death. A decomposing body can be compared in terms of physical and chemical point of view to an altering environment. Each decay stage in a changing environment is visited by different species of arthropods signifying the waves of colonization. These species can be used for mPMI estimation as they colonize the body in a predictable order (Smith and Cornell University Press, 1986). Certain species can be used as an indicator of the season of death due to their strict phenology. This will be mostly significant in cases where cadavers are found several months after death, and in archaeological context (Singh et al., 2016).

12.7.2 MOLECULAR ANALYSIS

For many genera, individual species cannot be distinguished based only on larval appearance. Adult determination may be achievable, but it may take time and may be unusual owing to rearing issues or dead larval evidence (Amendt et al., 2011). As a response, scientific approaches based on species specific nucleotide sequences of some genes are being used as predictive tools for the authentication of forensically relevant species (Zehner et al., 2004). The genotype for subunit I of mitochondrial transcribed cytochrome oxidase (CO I) is well regarded for this function, and reference sequences for a wide range of species are available. In addition to identifying complete animals, several molecular approaches may be used to analyze insect pieces or empty puparia (Mazzanti et al., 2010). A three-step approach is often used to analyze the sequence of nucleotides: (1) isolation of DNA, (2) multiplication of the gene of interest (for most instances CO I), and (3) determination of the target nucleotide sequence of the produced PCR product. Segments of 350–650 bp are usually sufficient for testing. The fidelity of the DNA extraction is an essential metric that is heavily influenced by the insect specimen's storing circumstances. Using ethanol (70–100%) preserved specimens or entirely dry matter, such as mounted adults, good results must be achieved. Since microbes that develop in humid environments can damage DNA, storage conditions are critical. Despite months of storage, the DNA could be recovered using traditional procedures such as phenol-chloroform separation or widely viable extracting kits, as indicated. However, for killing and preservation, ethyl acetate and formaldehyde should be disregarded since these compounds might harm DNA. Following Pcr analysis of the target gene, commonly available kits may be used to readily analyze the nucleotide sequence, which is then accompanied by capillary electrophoresis and data processing. If the nucleotide pattern of the item of relevance examined resembles a reference sequence, the species in issue is typically identified. However, it cannot be ruled out because related species may have very similar nucleotide sequences, complicating unambiguous species identification. As a result, extensive knowledge of all important species is required (Amendt et al., 2011). CO I sequencing data from Lucilia caesar/L. Illustris and Chrysomya megacephala/C. Saffranea are two examples (Nelson et al., 2008). Extensive phylogenetic analyses may be useful in certain circumstances (Amendt et al., 2011).

12.7.3 HUMAN DNA ANALYSIS FROM ENTOMOLOGICAL EVIDENCE

Examination of human DNA is also carried out from the digestive tract of insects in certain cases (Campobasso et al., 2005). Under circumstances where the dietary supply of the maggots being gathered at the crime site is in doubt, human DNA analysis from entomological evidences is performed (Amendt et al., 2011). It is also performed in cases wherein larvae are found but there is absence of corpse in the scene. The inclusion of human DNA in the stomach larvae will specify that there was presence of decomposing body in that site which has been subsequently relocated. Forensic STR typing can be used to identify the person's identity by comparing it to known DNA profiles (Zehner et al., 2004). Non-feeding phases of insects can also be subjected to this form of DNA study (Carvalho et al., 2005). In addition to DNA sequence analysis, gene expression studies also show characters of a potential future tool which can aid in the determination of the age of developing individuals. Since the genomic pattern is reliant on the age of the growing insect, the monitoring of any variation in particular gene expressions, which are critical in specific developmental stages, can provide a time scale for age calculation. This is achieved by finding out the amount of mRNA of specific genes through the extraction of RNA, followed by reverse transcription into cDNA and the quantification of cDNA. For example, the utilization of real time PCR system. Insect larva do not change size during growth, and the pupal phase accounts for around half of their whole lifespan. Puparia can be dehydrated to see the physical changes linked with the growth of the larvae inside, which is connected to age. Following vast and groundbreaking Drosophila developmental investigations, necrophagous flies has become focus of this study (Amendt et al., 2011). Tarone et al. (2007) studied the transcriptional activity of genes associated in L. Sericata eggs and predicted their generation based on changing expression rates. For a few gene products, age-dependent genetic makeup was also found in pupae (Zehner et al., 2004). However, it should be noted that the quality of the results derived from this type of analysis depend upon the reference database used for the comparison of sequences and associated taxonomy (Chimeno et al., 2019). The formation of a high-quality reference DNA barcode database is an important requirement to achieve accurate results (Spelda et al., 2011). There have been observed variances in the types of insects engaged with the rotting body in various locations and ecosystems. A careful inspection can show species variety, since species associated with one kind of environment found on a body are frequently different from those found when the corpse is relocated after burial (Benecke et al., 2004). Finding the proper species is a crucial initial aspect in estimating the age of larvae (Amendt et al., 2011). Physical examination is commonly employed for identification of species, which needs specialized knowledge and takes a bit of time (Benecke et al., 2004). To avoid this issue, polymerized chain reaction amplification of relevant portions of the larval genes and comparing with reference records can be used to confirm species characterization (Chen et al., 2004).

12.8 CONCLUSION

The use of insects for forensic purposes can serve as an important tool in the investigation process of various crimes. Forensic entomology makes use of insects and their characteristics to help the criminal justice system. It has been in picture since a long time. However, recent developments in the field have led to its proper utilization in the recent years. Among the three major types of forensic entomology, viz., stored-product entomology, which is the study of insects related to raw and processed food, urban entomology, which refers to the study of insects within the urban environment and medico-legal entomology, which is the study of insects in order to determine the TSD; medico-legal entomology plays the most crucial role in forensic science. The drugs and other toxic substances which might be present in the body of the deceased during the time of death can also be detected with the help of insects. The larvae are selected for the analysis of toxicants. It is achieved through high-pressure liquid chromatography, gas chromatography, RIA, thin-layer chromatography, and gas–mass analysis. Archaeological study of a crime scene also helps in finding out the cold evidences which are present since ages in the scene of crime. The cold evidences are recovered and reconstructed in order to aid in the investigation. The insects which are more commonly encountered in the crime scenes are beetles and flies. The various stages of the life cycle of insects are studied to find out the time interval since the death of a person. The succession of insects on carrions and their developmental stages also help in the estimation of PMI. Forensic entomology can also be utilized to find out whether any sort of physical abuse has been committed upon an individual or not. Variation among the species of insects helps in determination of site of crime. Regardless of the drawbacks, forensic entomology is an important and crucial discipline in terms of criminal justice system.

KEYWORDS

- **time since death**
- **post-mortem interval**
- **insect succession**
- **endotoxicology**
- **entomology**

REFERENCES

Adams, Z. J.; Hall, M. J. Methods Used for the Killing and Preservation of Blowfly Larvae, and Their Effect on Post-Mortem Larval Length. *Forensic Sci. Int.* **2003**, *138,* 50–61.

Amendt, J.; Campobasso, C. P.; Gaudry, E.; Reiter, C.; LeBlanc, H. N.; Hall, M. J. Best Practice in Forensic Entomology—Standards and Guidelines. *Int. J. Legal Med.* **2007**, *121,* 90–104.

Amendt, J.; Krettek, R.; Zehner, R. Forensic Entomology. *Naturwissenschaften.* **2004,** *91,* 51–65.

Amendt, J.; Richards, C. S.; Campobasso, C. P.; Zehner, R.; Hall, M. J. Forensic Entomology: Applications and Limitations. *Forensic Sci. Med. Pathol.* **2011,** *7,* 379–392.

Anderson, G. S.; Byrd, J. H.; Castner, J. L. *Insect Succession on Carrion and Its Relationship to Determining Time of Death*; CRC Press, 2001; pp 143–176.

Anderson, G. S. Forensic Entomology: The Use of Insects in Death Investigations. Unpublished manuscript, 1999.

Anderson, G. S.; Huitson, N. R. Myiasis in Pet Animals in British Columbia: The Potential of Forensic Entomology for Determining Duration of Possible Neglect. *Can. Vet. J.* **2004,** *45,* 993.

Benecke, M. A Brief History of Forensic Entomology. *Forensic Sci. Int.* **2001,** *120,* 2–14.

Benecke, M.; Josephi, E.; Zweihoff, R. Neglect of the Elderly: Forensic Entomology Cases and Considerations. *Forensic Sci. Int.* **2004,** *146,* 195–199.

Beyer, J. C.; Enos, W. F.; Stajić, M. Drug Identification Through Analysis of Maggots. *J. Forensic Sci.* **1980,** *25,* 411–412.

Bourel, B.; Tournel, G.; Hedouin, V.; Deveaux, M.; Goff, M. L.; Gosset, D. Morphine Extraction in Necrophagous Insects Remains for Determining Ante-mortem Opiate Intoxication. *Forensic Sci. Inter*. **2001,** *120,* 127–131.

Braack, L. E. O. Arthropods Associated with Carcasscs in the Northern Kruger National Park. *S. Afr. J. Wildl. Res.* **1986,** *16,* 91–98.

Byrd, J. H.; Lord, W. D.; Wallace, J. R.; Tomberlin, J. K. *Collection of Entomological Evidence During Legal Investigations*; CRC Press, 2010**;** pp 127–176.

Byrd, J. H.; Tomberlin, J. K. *Forensic Entomology: The Utility of Arthropods in Legal Investigations*, CRC Press, 2019.

Campobasso, C. P.; Di Vella, G.; Introna, F. Factors Affecting Decomposition and Diptera Colonization. *Forensic Sci. Inter.* **2001,** *120,* 18–27.

Campobasso, C. P.; Linville, J. G.; Wells, J. D.; Introna, F. Forensic Genetic Analysis of Insect Gut Contents. *Am. J. Forensic Med. Pathol*. **2005**, *26,* 161–165.

Carvalho, F.; Dadour, I. R.; Groth, D. M.; Harvey, M. L. Isolation and Detection of Ingested DNA from the Immature Stages of Calliphora Dubia (Diptera: Calliphoridae). *Forensic Sci. Med. Pathol*. **2005,** *1,* 261–265.

Carvalho, L. M.; Linhares, A. X.; Trigo, J. R. Determination of Drug Levels and the Effect of Diazepam on the Growth of Necrophagous Flies of Forensic Importance in Southeastern Brazil. *Forensic Sci. Inter*. **2001,** *120,* 140–144.

Catts, E. P. Analyzing Entomological Data. In *Entomology and Death: A Procedural Guide*, **1990**; pp 124–137.

Catts, E. P.; Goff, M. L. Forensic Entomology in Criminal Investigations. *Annu. Rev. Entomol*. **1992,** *37,* 253–272.

Catts, E. P.; Haskell, N. H. Entomology & Death: A Procedural Guide. *Forensic Entomol. Assocs*. **1990**.

Chen, W. Y.; Hung, T. H.; Shiao, S. F. Molecular Identification of Forensically Important Blow Fly Species (Diptera: Calliphoridae) in Taiwan. *J. Med. Entomol*. **2004,** *41,* 47–57.

Chimeno, C.; Morinière, J.; Podhorna, J.; Hardulak, L.; Hausmann, A.; Reckel, F.; Haszprunar, G. DNA Barcoding in Forensic Entomology–Establishing a DNA Reference Library of Potentially Forensic Relevant Arthropod Species. *J. Forensic Sci*. **2019,** *64,* 593–601.

de Carvalho, L. M. L. *Toxicology and Forensic Entomology*; Springer, 2009; pp 163–178.

Dirkmaat, D. C.; Adovasio, J. M. The Role of Archaeology in the Recovery and Interpretation of Human Remains from an Outdoor Forensic Setting. In *Forensic Taphonomy: The Postmortem Fate of Human Remains*; 1997, pp 39–64.

Ekanem, M. S.; Dike, M. C. Arthropod Succession on Pig Carcasses in Southeastern Nigeria. *Papeis Avulsos de Zoologia* **2010,** *50,* 561–570.

Erzinclioglu, Y. Z. *A Manual of Forensic Entomology*; Smith, K. G. V., Ed.; British Museum (Natural History) and Cornell University Press, 1987.

Frost, C. L.; Braig, H. R.; Amendt, J.; Perotti, M. A. Indoor Arthropods of Forensic Importance: Insects Associated with Indoor Decomposition and Mites as Indoor Markers. In *Current Concepts Forensic Entomol*; 2019, pp 93–108.

García-Rojo, A. M.; Martínez-Sánchez, A.; López, R.; de la Vega, J. G.; Rica, M.; González, M.; Disney, R. H. L. A Mathematical Model Applied for Assisting the Estimation of PMI in a Case of Forensic Importance. First Record of Conicera Similis (Diptera: Phoridae) in a Corpse. *Forensic Sci. Inter*. **2013,** *231,* 11–18.

Gennard, D. *Forensic Entomology: An Introduction*; *John Wiley & Sons*, 2012.

Goff, M. L. Early Post-mortem Changes and Stages of Decomposition in Exposed Cadavers. *Exp. Appl. Acarol.* **2009,** *49,* 21–36.

Goff, M. L.; Lord, W. D. Entomotoxicology. A New Area for Forensic Investigation. *Am. J. Forensic Med. Pathol.* **1994,** *15,* 51–57.

Goff, M. L.; Omori, A. I.; Goodbrod, J. R. Effect of Cocaine in Tissues on the Development Rate of Boettcherisca Peregrina (Diptera: Sarcophagidae). *J. Med. Entomol.* **1989,** *26,* 91–93.

Gondhalekar, A. D. 2018 Highlights of Urban Entomology. *J. Med. Entomol.* **2019,** *56,* 1188–1193.

Greenburg, B. Flies and Disease. *Biol. Dis. Trans.* **1973.**

Gullan, P. J.; Cranston, P. S. *The Insects: An Outline of Entomology*; John Wiley & Sons, 2014.

Hagstrum, D.; Subramanyam, B. *Stored-product Insect Resource*; Elsevier, 2016.

Hagstrum, D. W.; Subramanyam, B. A Review of Stored-product Entomology Information Sources. *Am. Entomol.* **2009,** *55,* 174–183.

Harwood, R. F.; James, M. T. *Entomology in Human and Animal Health*; Macmillan Publishing Co. Inc. 1979.

Held, D. *Urban Landscape Entomology*; Academic Press, 2019.

Introna, F.; Campobasso, C. P.; Goff, M. L. Entomotoxicology. *Forensic Sci. Inter*. **2001,** *120,* 42–47.

Joseph, I.; Mathew, D. G.; Sathyan, P.; Vargheese, G. The Use of Insects in Forensic Investigations: An Overview on the Scope of Forensic Entomology. *J. Forensic Dental Sci*. **2011,** *3,* 89.

Keh, B. Scope and Applications of Forensic Entomology. *Annu. Rev. Entomol.* **1985,** *30,* 137–154.

Kenawy, M. A.; Abdel-Hamid, Y. M. Medico-Legal Forensic Entomology "Use of Insects in Death Investigations"—A Review. *Egypt. Acad. J. Biol. Sci., A Entomol.* **2019,** *11,* 59–74.

Killam, E. W. *The Detection of Human Remains*; Charles C Thomas Publisher, 2004.

Kulshrestha, P.; Chandra, H. Time Since Death. An Entomological Study on Corpses. *Am. J. Forensic Med. Pathol.* **1987,** *8,* 233–238.

Kulshrestha, P.; Satpathy, D. K. Use of Beetles in Forensic Entomology. *Forensic Sci. Inter*. **2001,** *120,* 15–17.

Louw, S. V. D. M., Van der Linde, T. C. Insects Frequenting Decomposing Corpses in Central South Africa. *Afr. Entomol*. **1993,** *1,* 265–269.

Matuszewski, S.; Szafałowicz, M.; Jarmusz, M. Insects Colonising Carcasses in Open and Forest Habitats of Central Europe: Search for Indicators of Corpse Relocation. *Forensic Sci. Inter*. **2013,** *231,* 234–239.

Mazzanti, M.; Alessandrini, F.; Tagliabracci, A.; Wells, J. D.; Campobasso, C. P. DNA Degradation and Genetic Analysis of Empty Puparia: Genetic Identification Limits in Forensic Entomology. *Forensic Sci. Inter*. **2010,** *195,* 99–102.

McGarry, J.; Ratsep, E.; Ressel, L.; Leeming, G.; Ricci, E.; Verin, R.; Yeates, J. Introducing Forensic Entomology in Cases of Suspected Animal Neglect. *Veter. Record*. **2018,** *182,* 139–139.

Midgley, J. M.; Villet, M. H. Effect of the Killing Method on Post-mortem Change in Length of Larvae of Thanatophilus Micans (Fabricius 1794) (Coleoptera: Silphidae) Stored in 70% Ethanol. *Inter. J. Legal Med*. **2009,** *123,* 103–108.

Mullen, G. R.; Durden, L. A. *Medical and Veterinary Entomology*; Academic press, 2009.

Nelson, L. A.; Wallman, J. F.; Dowton, M. Identification of Forensically Important Chrysomya (Diptera: Calliphoridae) Species Using the Second Ribosomal Internal Transcribed Spacer (ITS2). *Forensic Sci. Inter*. **2009,** *177,* 238–247.

Olsen, A. R. List of Stored-product Insects Found in Imported Foods Entering United States at Southern California Ports. *Bullet. ESA* **1981,** *27,* 18–20.

Pohjoismäki, J. L.; Karhunen, P. J.; Goebeler, S.; Saukko, P.; Sääksjärvi, I. E. Indoors Forensic Entomology: Colonization of Human Remains in Closed Environments by Specific Species of Sarcosaprophagous Flies. *Forensic Sci. Inter*. **2010,** *199* (1–3), 38–42.

Raoult, D.; Reed, D. L.; Dittmar, K.; Kirchman, J. J.; Rolain, J. M.; Guillen, S.; Light, J. E. Molecular Identification of Lice from Pre-Columbian Mummies. *J. Infect. Dis*. **2008,** *197,* 535–543.

Richards, C. S.; Crous, K. L.; Villet, M. H. Models of Development for Blowfly Sister Species Chrysomya Chloropyga and Chrysomya Putoria. *Med. Vet. Entomol*. **2009,** *23,* 56–61.

Robinson, W. H. Urban Entomology: Insect and Mite Pests in the Human Environment. *Garland Sci*. **2020**.

Samnol, A.; Kumar, R.; Sankhla, M. S. Medico-Legal Importance of Entomological Methods for Determination of Time Since Death of Decomposed Corpse. *Seybold Rep*. 153*3,* 9211.

Sankhla, M. S., Sharma, K.; Kumar, R. Future Trends in Forensic Entamotoxicology. **2017,** *6*.

Schroeder, H.; Klotzbach, H.; Püschel, K. Insects; Colonization of Human Corpses in Warm and Cold Season. *Legal Med*. **2003,** *5,* 372–374.

Singh, R.; Sharma, S.; Sharma, A. Determination of Post-burial Interval Using Entomology: A Review. *J. Forensic Leg. Med*. **2016,** *42,* 37–40.

Smith, K. G. V. *A Manual of Forensic Entomology*; British Museum (Natural History), London, and Cornell University Press, 1986.

Spelda, J.; Reip, H. S.; Oliveira-Biener, U.; Melzer, R. R. Barcoding Fauna Bavarica: Myriapoda—A Contribution to DNA Sequence-based Identifications of Centipedes and Millipedes (Chilopoda, Diplopoda). *Zookeys* **2011,** *156,* 123–139.

Tantawi, T. I.; El-Kady, E. M.; Greenberg, B.; El-Ghaffar, H. A. Arthropod Succession on Exposed Rabbit Carrion in Alexandria, Egypt. *J. Med. Entomol.* **1986,** *33,* 566–580.

Tarone, A. M.; Jennings, K. C.; Foran, D. R. Aging Blow Fly Eggs Using Gene Expression: A Feasibility Study. *J. Forensic Sci.* **2007,** *52,* 1350–1354.

Tüzün, A.; Dabiri, F.; Yüksel, S. Preliminary Study and Identification of Insects' Species of Forensic Importance in Urmia, Iran. *Afr. J. Biotechnol.* **2010,** *9,* 3649–3658.

Vanin, S.; Huchet, J. B. Forensic Entomology and Funerary Archaeoentomology. Taphonomy of Human Remains: Forensic Analysis of the Dead and the Depositional Environment; 2017, pp 167–186.

Wells, J. D.; Introna, F.; Di Vella, G.; Campobasso, C. P.; Hayes, J.; Sperling, F. A. Human and Insect Mitochondrial DNA Analysis from Maggots. *J. Forensic Sci.* **2001,** *46,* 685–687.

Williams, K. A.; Villet, M. H. A History of Southern African Research Relevant to Forensic Entomology. *S. Afr. J. Sci.* **2006,** *102,* 59–65.

Young, S. T.; Wells, J. D.; Hobbs, G. R.; Bishop, C. P. Estimating Postmortem Interval Using RNA Degradation and Morphological Changes in Tooth Pulp. *Forensic Sci. Inter.* **2013**.

Zehner, R.; Amendt, J.; Krettek, R. STR typing of Human DNA from Fly Larvae Fed on Decomposing Bodies. *J. Forensic Sci.* **2004,** *49,* 1–4.

Zehner, R.; Amendt, J.; Schütt, S.; Sauer, J.; Krettek, R.; Povolný, D. Genetic Identification of Forensically Important Flesh Flies (Diptera: Sarcophagidae). *Int. J. Leg. Med.* **2004,** *118,* 245–247.

Index

A

Above mean sea level (AMSL), 191
Acanthoplusspiseri Brancsik, 108
Adult hoverflies, 175
AgenceFédérale pour la Sécurité de la ChaîneAlimentaire (AFSCA), 124
Agrochemical pollutants, 59
Aldehyde oxidase, 121
Americanum, 118
Antlions
  India
    bioecology of, 208
    diversity and distribution, 202, 204
    endemism and conservation measures, 208
    flight period and seasonality, 204
    taxonomic studies, 204
Apis cerana, 228
Apis millifera, 227
Aquatic insects, 40
  aquaculture
    role, 50–53
  aquatic ecosystems
    order coleoptera—beetles and weevils, 42–43
    order diptera—true flies, 41–42
    order ephemeroptera—mayflies, 43–44
    order hemiptera—water bugs, 42
    order odonata—dragonflies and damselflies, 43
    order plecoptera—stoneflies, 44
    order trichoptera—caddisflies, 44–45
  conservation
    laws and regulations, 53
  economic importance
    animals, food, 46
    biological environmental monitoring, 47–49
    decaying matter, 49
    food for fish, 45–46
    muse for, 47
    pests and weeds, 46–47
    predators, 49–50
Ascorbic acid, 112

B

Bee venom industry, 229
Beeswax, 228
Beetles, 244–245
Bioactive/biological components
  aldehyde oxidase, 121
  Americanum, 118
  ascorbic acid, 112
  Chlorophorusannularis, 118
  Chorthippuscurtipennis, 114
  Colophospermum mopane, 118
  Cytochrome p450, 121
  Doratiferaquadriguttata, 114
  enzymes, 118
  gluten degrading enzyme, 121
  Imbrasiabelina, 118
  Lumbricusterrestris, 118
  Lymantria monacha, 114
  Malacosoma americanum, 114
  Melanoplus borealis, 114
  NPK fertilizer, 118
  Orgyialeucostigma, 114
  Paropsisatomaria, 114
  Pinus sylvestris, 114
  protein structural, 113
  T. molitor, 118
  *Trichoderma hamatum*in, 118
  Trirhabda virgata, 114
  wood-feeding, 121
  yellow mealworm beetle, 113
  Zophobasmorio, 114
Biodiesel production, 91
  adult and larval stages, 98
  clear and sustainable fuel, 83

environmental degradation, 99
fat bodies, 84
insect utilization, 98
production, 91
Biological control
Megaselia scalaris, 9
population, suppression, 8
Spodoptera frugiperda, 8
Biological environmental monitoring, 47–49
Biopolymer's manipulation, 111
Black soldier fly larvae (BSFL), 84, 85
Brachytrypesspp, 86
Bumblebee
above mean sea level (AMSL), 191
conservation measures, 196
diversity and distribution, 193
habitat and foraging plants, 191–192
identification and mimicry, 194–195
Indian Himalayan Region (IHR), 190
nestingbehaviour, 195
species in India, 193–194
threats, 195–196

C

Caddisflies. *See* Trichoptera
Chlorophorusannularis, 118
Chorthippuscurtipennis, 114
Chrysinaaurigans, 112
Chrysinachrysargyrea, 112
Colophospermum mopane, 118
Corporate Statistical Database, 229
Cytochrome p450, 121

D

Dactylopius, 219
Deutsche Landwirtschaftsgesellschaft (DLG), 124
Distributional Atlas of European Trichoptera (DAET), 166
Doratiferaquadriguttata, 114
Drusus, 163

E

Ecosystem
engineering, 7–8
pollinators and flora-bearing flowers, 6–7
role in, 10
underground systems, 5–6
valuation, 4
Edible insect species, 99
coleoptera (beetles), 102
hymenoptera, 102
lepidoptera (caterpillars), 100–101
nutrient content consumed, 103–104
orthoptera, 101
Entomological threats, 66, 68
decline
agricultural practices, 69
biodiversity, loss of, 68
climate change, 71
co-extinction, 72
fragmentation and habitat loss, 69
invasive species, 70–71
management, 70
overexploitation, 72
pesticides, 71
pollution, 70
urbanization, 69–70
insects, 62
conservation, 72–73
external feeders, 63
and humans, 61
internal feeders, 63–64
nuisance pests, 65–66
predators and parasitoids, 66
vectors, 64–65
modernization, 62
and stored grains, 62
Episyrphus balteatus, 174, 176, 178, 181
Episyrphus viridaureus, 181
EPT (Ephemeroptera, Plecoptera, and Trichoptera), 151
ETS (Ephemeroptera, Trichoptera, Simuliidae), 151
*Eupeodes lunigeri*s, 176

F

Federal Office of Food Safety and Veterinary Services (FFSVO), 124
Fine Particulate Organic Matter (FPOM), 150
Fine particulate organic matter (FPOM), 151

Food safety measurement, 121
AgenceFédérale pour la Sécurité de la ChaîneAlimentaire (AFSCA), 124
authorized consumption, 124
Deutsche Landwirtschaftsgesellschaft (DLG), 124
Federal Office of Food Safety and Veterinary Services (FFSVO), 124
International Platform on Insects for Food and Feed (IPIFF), 124
Forensic entomology, 237
archaeological studies, 242–243
carrion, 248
DNA and toxicology samples, 239
entomological, 238
entomotoxicology, 242
insects, 243
beetles, 244–245
entomotoxicology, 247
flies, 245–247
law suits, 248
medico-legal entomology, 241–242
molecular methods
adult determination, 256
DNA extraction, 256
human DNA analysis, 257–258
PCR analysis, 257
PMI estimation, 255–256
post-mortem interval (PMI), 238
collection, 251
determination of site, 251
development of insect, 249
indication of abuse, 250
morphological identification, 254–255
morphological species identification, 251
preservation, 251, 253–254
stored-product entomology, 239–241
time since death (TSD), 238
urban entomology, 241

H

Hermetiaillucens, 114
Holocentropus, 151
Honey, silk, and lac
Apiculture, 226
Apis cerana, 228
Apis millifera, 227
bee venom industry, 229
beeswax, 228
Corporate Statistical Database, 229
Khadi and Village Industries Commission (KVIC), 228
lac culture, 233
Royal jelly, 230
sericulture, 231, 233

I

Imbrasiabelina, 118
Indian Himalayan Region (IHR), 189, 190
Industrial entomology
Dactylopius, 219
defined, 219
economic significance, 219
honey, silk, and lac
Apiculture, 226
Apis cerana, 228
Apis millifera, 227
bee venom industry, 229
beeswax, 228
Corporate Statistical Database, 229
Khadi and Village Industries Commission (KVIC), 228
lac culture, 233
Royal jelly, 230
sericulture, 231, 233
industrial application
feed industry, 224–225
food industry, 221, 222, 223, 224
pest management, 220
pharmaceutical industry, 225–226
insects, uses, 233
drosomycin, 234
secretions, 234
transgenic silkworms, 234
Lacciferlacca (Homoptera), 219
Insect biodiversity conservation
biological control
Megaselia scalaris, 9
population, suppression, 8
Spodoptera frugiperda, 8

ecosystem
engineering, 7–8
pollinators and flora-bearing flowers, 6–7
role in, 10
underground systems, 5–6
valuation, 4
food provision
humankind, 9
insect conservation
habitats and landscapes, 17, 20
monitoring, 16–17
Lepidopteran insects, 3
sensors with, 18–19
threats to insects
biological control, 15
co-extinction, 14
environmental pollution, 10–11
invasive alien organisms, pervasion, 12–13
natural habitat, 11–12
overexploitation, 14
worldwide climate change, 13–14
Insect-based value-added products
animal feed, 104
M domestica, 105
Trinervitermes spp., 105
bioactive/biological components
aldehyde oxidase, 121
Americanum, 118
ascorbic acid, 112
Chlorophorusannularis, 118
Chorthippuscurtipennis, 114
Colophospermum mopane, 118
Cytochrome p450, 121
Doratiferaquadriguttata, 114
enzymes, 118
gluten degrading enzyme, 121
Hermetiaillucens, 114
Imbrasiabelina, 118
Lumbricusterrestris, 118
Lymantria monacha, 114
Malacosoma americanum, 114
Melanoplus borealis, 114
NPK fertilizer, 118
Orgyialeucostigma, 114
Paropsisatomaria, 114
Pinus sylvestris, 114
protein structural, 113
T. molitor, 118
*Trichoderma hamatum*in, 118
Trirhabda virgata, 114
wood-feeding, 121
yellow mealworm beetle, 113
Zophobasmorio, 114
biodiesel production, 91
adult and larval stages, 98
clear and sustainable fuel, 83
environmental degradation, 99
fat bodies, 84
insect utilization, 98
production, 91
biological compounds isolated, 115–117
edible insect species, 99
coleoptera (beetles), 102
hymenoptera, 102
lepidoptera (caterpillars), 100–101
nutrient content consumed, 103–104
orthoptera, 101
entomophagy, 81, 99
implementation, 82
food safety measurement, 121
AgenceFédérale pour la Sécurité de la ChaîneAlimentaire (AFSCA), 124
authorized consumption, 124
Deutsche Landwirtschaftsgesellschaft (DLG), 124
Federal Office of Food Safety and Veterinary Services (FFSVO), 124
International Platform on Insects for Food and Feed (IPIFF), 124
and hazardous effects, 121
industrialization, 105
Acanthoplusspiseri Brancsik, 108
adaptation strategies, 109
advancement, 108
artificial diets, 109
BFS, 109
biopolymer's manipulation, 111
and biotechnological approaches, 111
Chrysinaaurigans, 112
Chrysinachrysargyrea, 112
grinding and milling methods, 108
larval rearing, 108

nanotechnology, 111
selection, 111
semi liquids, 109
solid diets, 109
termites, 108
insect species
frass, 119–120
types of, 106–107
insect-rearing process, 100
legislations, 121
lipid content
black soldier fly larvae (BSFL), 84, 85
Brachytrypesspp, 86
Calliphoravicina, 86
Chrysomyamegacephala, 84
extraction method, 86–91
Hermetiaillucens, 84
Musca domestica, 84, 86
Sarcophagacrassipalis, 85
manure, 92–97
metabolic rates, 99
sustainability exhibit techniques, 82
International Platform on Insects for Food and Feed (IPIFF), 124
International Union for Conservation of Nature (IUCN), 159, 165
Ischiodon scutellaris, 181

K

Khadi and Village Industries Commission (KVIC), 228

L

Lacciferlacca (Homoptera), 219
Larvae of Helicopsychidae, 152
Lipid content
black soldier fly larvae (BSFL), 84, 85
Brachytrypesspp, 86
Calliphoravicina, 86
Chrysomyamegacephala, 84
extraction method, 86–91
Hermetiaillucens, 84
Musca domestica, 84, 86
Sarcophagacrassipalis, 85
Lumbricusterrestris, 118
Lymantria monacha, 114

M

Malacosoma americanum, 114
Melanoplus borealis, 114

O

Orgyialeucostigma, 114

P

Paropsisatomaria, 114
Pinus sylvestris, 114
Plectrocnemia, 151
Pollination, 189
bumblebee
above mean sea level (AMSL), 191
conservation measures, 196
diversity and distribution, 193
habitat and foraging plants, 191–192
identification and mimicry, 194–195
Indian Himalayan Region (IHR), 190
nesting behaviour, 195
species in India, 193–194
threats, 195–196
Post-mortem interval (PMI), 238
collection, 251
determination of site, 251
development of insect, 249
indication of abuse, 250
morphological identification, 254–255
morphological species identification, 251
preservation, 251, 253–254
Potamophylax nigricornis, 163
Potamophylax winneguthi, 163

R

Rhyacophila species, 165
River continuum concept (RCC), 151
Royal jelly, 230

S

Scaeva pyrastri, 181
Soft-sucking insect pests, 176
Stored-product entomology, 239–241
Syrphidae (Insecta: Diptera), 174
adult hoverflies, 175
Episyrphus balteatus, 174, 176
Eupeodes lunigeris, 176

materials and methods, 177
results and discussion, 177
Episyrphus balteatus, 178, 181
Episyrphus viridaureus, 181
Ischiodon scutellaris, 181
Scaeva pyrastri, 181
soft-sucking insect pests, 176
studies, 175
Toxomerus, 175

T

T. molitor, 118
Taxonomy in education
background
Aristotelian method, 30–31
western classification, 30
defined, 29
levels
alpha taxonomy, 31
beta taxonomy, 31
gamma taxonomy, 31
macro taxonomy, 31–32
micro taxonomy, 31
numerical taxonomy
biochemical taxonomy, 33
classical taxonomy, 32
cytotaxonomy, 33
evolutionary systematics, 32
neo-taxonomy, 32
phylogenetic systematics, 32
significance, 33
agriculture, 34–35
biodiversity conservation, 34
fisheries, 35
flora and fauna, 35
investigation and education, 34
in medicine, 35–36
parasitology, 35
and pest management, 34–35
and veterinary science, 35
Time since death (TSD), 238
*Trichoderma hamatum*in, 118
Trichoptera, 147
conservaiton
Distributional Atlas of European Trichoptera (DAET), 166
German national Red List, 166
International Union for Conservation of Nature (IUCN), 165
diversity, 148–149, 150
ecosystem services
Annulipalpia, 151
EPT (Ephemeroptera, Plecoptera, and Trichoptera), 151
ETS (Ephemeroptera, Trichoptera, Simuliidae), 151
Fine Particulate Organic Matter (FPOM), 150
Fine particulate organic matter (FPOM), 151
Holocentropus, 151
Larvae of Helicopsychidae, 152
Plectrocnemia, 151
river continuum concept (RCC), 151
Europe, diversity
Drusus, 163
mountain barrier, 165
Potamophylax nigricornis, 163
Potamophylax winneguthi, 163
Rhyacophila species, 165
historical resume, 148
importance of
European Water Framework Directive, 162
studies, 162–163
International Union for Conservation of Nature (IUCN), 159
main threats
deforestation, 169
hydropower plants, 168
investigations, 167
studies, 167
urban developments, 169
Water Framework Directive (WFD), 168
worldwide diversity and distribution
aquatic larvae, 160
species-rich families, 161–162
Trirhabda virgata, 114

U

Urban entomology, 241

W

Water Framework Directive (WFD), 168
Wood-feeding, 121
Worldwide diversity and distribution
  aquatic larvae, 160
  species-rich families, 161–162

Y

Yellow mealworm beetle, 113

Z

Zophobasmorio, 114

www.ingramcontent.com/pod-product-compliance
Lightning Source LLC
LaVergne TN
LVHW010620300926
849162LV00001B/14
* 9 7 8 1 7 7 4 9 1 5 8 0 6 *